Proceedings of the Steklov Institute of Mathematics

edited by

S. M. Nikol'skiĭ

Number 125 (1973)

Boundary Value Problems of Mathematical Physics. VIII

edited by

O. A. Ladyženskaja

Providence, Rhode Island

AMERICAN MATHEMATICAL SOCIETY

1975

Академия Наук
Союза Советских Социалистических Республик

ТРУДЫ
ордена Ленина
МАТЕМАТИЧЕСКОГО ИНСТИТУТА
имени В. А. СТЕКЛОВА
CXXV

Ответственный редактор
член-корреспондент АН СССР С. М. НИКОЛЬСКИЙ

КРАЕВЫЕ ЗАДАЧИ МАТЕМАТИЧЕСКОЙ ФИЗИКИ. 8

Сборник работ
под редакцией О. А. ЛАДЫЖЕНСКОЙ

Издательство "Наука"
Ленинградское отделение
Ленинград 1973

AMS (MOS) subject classifications (1970). Primary 35A05, 35A35, 43A85, 47A40, 35B40, 35B45, 10D15, 35D99, 76D05, 38G99, 35J10, 35J25, 35J55, 35J60, 35J70, 35K10, 35K55, 35M05, 65N05, 65N10, 35P10, 35P25, 35Q10, 35Q99; Secondary 81A45, 34B10, 34B15, 34B20, 34B30, 35D05, 35D10, 76D05, 35J25, 35K15, 35K20, 35L15, 35L20.

Library of Congress Cataloging in Publication Data (Revised) CIP
Main entry under title:
Boundary value problems of mathematical physics.

(Proceedings of the Steklov Institute of Mathematics, no.)
Translation of Kraevye zadachi matematicheskoĭ fiziki.
Includes bibliographies.
1. Boundary value problems. 2. Differential equations, Partial. I. Ladyzhenskaia Ol'ga Aleksandrovna, ed. II. Series: Akademiia nauk SSSR. Matematicheskiĭ institut. Proceedings, no. [etc.]
Title: Kraevye zadochi matematicheskoi fiziki. English.
QA1.A413 no. 83, etc. [QA379] 510'.8s [515'.353]
67-6187
ISBN 0-8218-3025-2

PROCEEDINGS OF THE STEKLOV INSTITUTE OF MATHEMATICS
IN THE ACADEMY OF SCIENCES OF THE USSR

(Труды математического института им. В. А. Стеклова, т. CXXV, 1973)

TABLE OF CONTENTS

Venkov, A. B. Expansion in automorphic eigenfunctions of the Laplace-Beltrami operator in classical symmetric spaces of rank one, and the Selberg trace formula. [Венков, А. Б. Разложение по автоморфным собственным функциям оператора Лапласа-Бельтрами в классических симметрических пространствах ранга один и формула следа Сельберга, 6–55] 1

Ivanov, A. V. On the solvability of the Dirichlet problem for some classes of second order elliptic systems. [Иванов, А. В. О разрешимости задачи Дирихле для некоторых классов эллиптических систем второго порядка, 56–87] 49

Ivanov, A. V. On the question of an admissible limiting growth of the right side of a quasilinear elliptic equation. [Иванов, А. В. К вопросу о допустимом предельном росте правой части квазилинейного эллиптического уравнения, 88–94] 80

Ivočkina, N. M. A priori estimates for solutions of the Dirichlet problem for multidimensional quasilinear equations of elliptic type. [Ивочкина, Н. М. Априорные оценки решений задачи дирихле для многомерных квазилинейных уравнений эллиптического типа, 95–105] 87

Korenev, N. K. On the strong convergence of solutions of difference equations to generalized solutions of second-order linear equations of three classical types. [Коренев, Н. К. О сильной сходимости решений разностных уравнений к обобщенным решениям линейных уравнений 2-го порядка трех классических типов, 106–126] 98

Krol', I. N. On solutions of the equation $D_{x_i}(|Du|^{p-2}D_{x_i}u) = 0$ with a singularity at a boundary point. [Кроль, И. Н. О решениях уравнения $D_{x_i}(|Du|^{p-2}D_{x_i}u) = 0$ с особенностью в граничной точке, 127–139] 118

Krol', I. N. On the behavior of the solutions of a quasilinear equation near null salient points of the boundary. [Кроль, И. Н. О поведении решений одного квазилинейного уравнения вблизи нулевых заострений границы, 140–146] 130

Oskolkov, A. P. On the asymptotic behavior of solutions of certain systems with a small parameter approximating the system of Navier-Stokes equations. [Осколков, А. П. Об асимптотическом поведении решений некоторых систем с малым параметром, аппроксимирующих систему уравнений Навье-Стокса, 147–163] 137

Oskolkov, A. P. On some convergent difference schemes for the Navier-Stokes equations. [Осколков, А. П. О некоторых сходящихся разностных схемах для уравнений Навье-Стокса, 164–172] 154

Rivkind, V. Ja. The grid method of solving problems of the dynamics of a viscous incompressible fluid. [Ривкинд, В. Я. Сеточный метод решения задач динамики вязкой несжимаемой жидкости, 173–186] 163

Skriganov, M. M. On the spectrum of the Schrödinger operator with a rapidly oscillating potential. [Скриганов, М. М. О спектре оператора Шредингера с быстро осциллирующим потенциалом, 187–195] 177

Solonnikov, V. A. and Ščadilov, V. E. On a boundary value problem for a stationary system of Navier-Stokes equations. [Солонников, В. А. и Щадилов, В. Е. Об одной краевой задаче для стационарной системы уравнений Навье-Стокса, 196–210] 186

Stupjalis, L. Boundary value problems for elliptico-hyperbolic equations. [Ступялис, Л. Краевые задачи для эллиптико-гиперболических уравнений, 211–230] 200

Preface

In the paper by A. B. Venkov spectral characteristics are considered of the Laplace-Beltrami operator on a fundamental region of discrete groups Γ acting on classical symmetric spaces of rank I: $S = SO_0(1,n)/SO(n)$, $SU(1,n)/S(U(1)\times U(n))$, $Sp(1,n)/Sp(1)\times Sp(n)$. The discrete groups satisfy the following conditions: 1) $\Gamma\ S$ is noncompact but has finite invariant volume; 2) the nonparabolic elements of Γ are essentially distinct from the parabolic elements. The main result consists in the proof of the eigenfunction expansion theorem. These themes were provided by the famous paper by A. Selberg on the trace formula. Venkov's results permit him to justify this formula completely for the cases of S and Γ he considers and also to calculate out fully all contributions in this formula for several examples. The corresponding arguments are presented in detail in the paper.

In the paper by M. M. Skriganov Schrödinger operators $\Delta u + q(x)u$ with rapidly oscillating potentials $q(x)$ unbounded at infinity are studied. For such operators sufficiency criteria are given assuring selfadjointness, semiboundedness, finiteness of the number of negative eigenvalues, and the absence of positive eigenvalues. A theorem on expansion in eigenfunctions of such operators is proved.

Five of the papers are devoted to quasilinear elliptic equations. In the paper by A. V. Ivanov *On the question of an admissible limiting growth of the right side of a quasilinear elliptic equation*, from equations of the form

(1) $$A_{ij}(x, u, u_x)u_{x_ix_j} = B(x, u, u_x)$$

the author selects a certain class (having nonempty intersection with the class of uniformly elliptic equations) for which the Dirichlet problem has a solution for right sides $B(x,u,p)$ increasing as $o(E_1 \ln|p|)$ when $|p|\to\infty$, where $E_1(x,u,p) = A_{ij}(x,u,p)p_ip_j$. In his second paper Ivanov studies elliptic systems of the form

$$A^s_{ij}(x, u^t, u^s_x)u^s_{x_ix_j} = B^s(x, u^t, u^s_x),\quad s = 1, \dots, N$$

and the form

$$A_{ij}(x, u^t, |u_x|)u^s_{x_ix_j} = B^s(x, u^t, u^t_x),\quad s = 1, \dots, N,$$

where $|u_x| = (\sum_{t,l}|u^t_{x_l}|^2)^{1/2}$, under the first boundary-value condition. To solve them, subject to a number of assumptions on the generators of their functions, a priori estimates of $\max_\Omega |u_x|$ are established that, together with known previous results, make it possible to investigate the question of solvability of the Dirichlet problem.

This paper generalizes the corresponding results on systems established by O. A. Ladyženskaja, N. N. Ural'ceva, N. M. Ivočkina and A. P. Oskolkov.

In her paper N. M. Ivočkina singles out a class of nonuniformly elliptic quasilinear equations in divergence form $da_i/dx_i = a$, for which, satisfying the first boundary-value condition, it is possible to give a priori estimates of $\max_\Omega |u_x|$ and the Hölder norms $|u_x|^{(\alpha)}_{\Omega'}$, $\overline{\Omega}' \supset \Omega$, that do not depend on $\partial a_i/\partial x_k$ for $k \neq i$.

In the paper by I. N. Krol' *On the behavior of the solutions of a quasilinear equation near null salient points of the boundary* it is proved that solutions of the Dirichlet problem for a quasilinear elliptic equation of the form

(2) $$(|u_x|^{p-2}u_{x_i})_{x_i} = 0,\quad p > 1,$$

having finite "energy" norm assume their boundary values at points where there are "null salients outside", with higher than power speed. To prove this a "barrier" function is constructed that is a nonnegative solution of (2) in the spherical cone $K(l)$ of angle l, equal to zero on $\partial K(l)$ and having finite energy norm near the vertex of $K(l)$. This function has the form $u(x) = |x|^{\lambda} f_{\lambda}(x|x|^{-1})$, $\lambda = \lambda(l)$. It is proved that $\lambda(l) = \pm L l^{-1} + O(1)$ as $l \to 0$, where L is the first zero of the solution of the Cauchy problem for a certain ordinary differential equation.

In his second paper Krol′ studies an increasing nonnegative solution of (2), equal to zero on $\partial K(l)$ and having the same form

$$u(x) = |x|^{\lambda} f_{\lambda}(\arccos(x_n |x|^{-1})).$$

The asymptotic behavior of $\lambda = \lambda(l)$ as $l \to \pi$ is determined.

Four papers are devoted to the Navier-Stokes equations. V. A. Solonnikov and V. E. Ščadilov study the Stokes linearized steady state problem under boundary conditions of the form

$$vn|_{S_1} = \alpha(s), \quad t - n(tn)|_{S_1} = b(s), \quad v|_{S_2} = a(s)$$

in the domain Ω, where $S_1 \cup S_2 = \partial\Omega$, n is the unit normal vector to $\partial\Omega$, t is the vector with components $t_i = \sum_{k=1}^{3} t_{ik}(s) n_k(s)$, and $t_{ik} = -\delta_{ik} p + \nu(\partial v_i/\partial x_k + \partial v_k/\partial x_i)$. This problem can be considered as being modeled on the motion of a fluid in the presence of a free boundary S_1. Unique solvability of this problem is established, and the existence of second-order derivatives of v is investigated.

In the paper by V. Ja. Rivkind new difference schemes for the Navier-Stokes equations are constructed and their convergence is studied. The explicit scheme turns out to be convergent for t-steps that are larger than in the corresponding papers of Temam and Ladyženskaja (namely for $\Delta t \sim (\Delta x)^2$). This is achieved at the expense of lopping off nonlinear terms, thanks to knowledge of an a priori estimate of $\max_{Q_T}|v|$ and $\max_{Q_T}|v_x|$ and certain other modifications of previous schemes.

In his paper *On the asymptotic behavior of solutions of certain systems with a small parameter approximating the system of Navier-Stokes equations* Oskolkov continues the investigation of systems he has proposed of elliptic and parabolic types passing as $\epsilon > 0$ into the steady state and nonsteady state Navier-Stokes system, respectively. Estimates are established for their solutions, depending on ϵ. In his second paper an explicit difference scheme is constructed on the basis of these approximations, permitting one in the limit to obtain a solution of an initial-boundary value problem for the Navier-Stokes equations.

The paper by N. K. Korenev is also devoted to the method of finite differences. For linear second-order equations of three classical types he proves strong convergence in the energy norm of solutions of difference schemes studied earlier by Ladyženskaja to generalized solutions of these equations having finite energy norm.

Finally, in the paper by L. Stupjalis unique solvability is established for an initial-boundary value problem involving a linear second-order equation that is elliptic in one part of the domain and hyperbolic in another. This is done under assumptions on the free terms that are broader than in previous work by Stupjalis and Ladyženskaja devoted to the same equation.

O. A. Ladyženskaja

Trudy Mat. Inst. Steklov.
125 (1973)

Proc. Steklov Inst. Math.
125 (1973)

EXPANSION IN AUTOMORPHIC EIGENFUNCTIONS OF THE LAPLACE-BELTRAMI OPERATOR IN CLASSICAL SYMMETRIC SPACES OF RANK ONE, AND THE SELBERG TRACE FORMULA

A. B. VENKOV

Abstract. We give a proof of the Selberg trace formula for three series of hyperbolic spaces $S: SO_0(1,n)/SO(n)$, $SU(1,n)/S(U(1)\times U(n))$, $\mathrm{Sp}(1,n)/(\mathrm{Sp}(1)\times \mathrm{Sp}(n))$. Ths discrete group Γ satisfies the following conditions, by assumption: a) $\Gamma\backslash S$ is noncompact and has finite invariant volume; b) a certain minimum principle is fulfilled for the set of nonparabolic elements of the group Γ. We perform a concrete computation of the additional contribution to the trace formula compared with the compact case for the space $SO_0(1,n)/SO(n)$ and for certain special groups Γ. As an intermediate result, this paper gives a proof of the theorem on expansion in eigenfunctions of the Laplace-Beltrami operator on the fundamental domain $\Gamma\backslash S$. In particular, we give a full characterization of the spectrum of this operator. In addition, we give a complete system of eigenfunctions for the continuous spectrum, which coincides with the system of analytically continued Eisenstein series.

Bibliography: 20 items.

Introduction

In 1956 in the fundamental paper [1], A. Selberg obtained many important number theoretic results using a certain formula, which as a result became known as the Selberg trace formula. This formula allows us to compute in group-theoretic terms the trace of the invariant operator k on the fundamental domain $\Gamma\backslash S$ of the discrete group Γ of transformations of the spaces $S=G/K$, where G is a locally compact group and K is a compact subgroup. The case when $\Gamma\backslash S$ is noncompact and has finite invariant volume is of special interest. In the same paper [1], Selberg presented without proof a form of the trace formula for $S=SL(2,R)/SO(2)$ and for any discrete subgroup $\Gamma\subset SL(2,R)$ with this property. The proof of the trace formula and its concrete computation is complicated by the fact that the operator k has a continuous spectrum. The problem of investigating the continuous spectrum turns out to be naturally connected with the analytic continuation of Eisenstein series. In his report [2], Selberg gave a brief description of a method which can be used for the analytic continuation of Eisenstein series. Since then, these ideas have been developed and generalized in Langlands' paper [3]. The Selberg-Langlands method is based on classical concepts of function theory.

In [4], L. D. Faddeev proposed another method for investigating the continuous spectrum using as an example the simple case of the upper half-plane. It is based on studying the resolvent of the selfadjoint extension of the Laplace-Beltrami operator to the fundamental domain $\Gamma\backslash S$ using the theory of perturbations of a continuous spectrum. This method gives complete information on the spectrum of this operator, and also on its system of

---

AMS (MOS) subject classifications (1970). Primary 43A85, 38G99, 35P10, 10D15.

eigenfunctions, and it allows us to prove the Selberg trace formula for the upper half-plane (see [5]). It is our hope that this method will be useful in much more general situations as well.

In the present article we give a proof of the Selberg trace formula for three series of irreducible rank 1 symmetric spaces: $SO_0(1,n)/SO(n)$, $SU(1,n)/S(U(1)\times U(n))$, $\mathrm{Sp}(1,n)/(\mathrm{Sp}(1)\times \mathrm{Sp}(n))$. The discrete group Γ is assumed to be such that its fundamental domain $\Gamma\backslash S$ is noncompact and has finite invariant volume (we note that, specifically in the case of rank 1 spaces, we know examples of discrete subgroups Γ of suitable real rank 1 groups, which are not Borel arithmetic groups). In addition, we assume that a certain minimum principle, analogous to the Petersson condition for $SL(2,R)/SO(2)$ (see [6]) is fulfilled for Γ; more precisely, for its set of nonparabolic elements. The arithmetic subgroups satisfy this principle.

We perform the concrete computation of the additional contribution to the trace formula compared with the compact case for the space $SO_0(1,n)/SO(n)$ and for a discrete group $\Gamma\subset SO_0(1,n)$ belonging to one of two types. In the first case, by assumption, the parabolic subgroups $\Gamma_\alpha\subset\Gamma$ do not contain any nontrivial elements of finite order; this is true, for example, for any congruence-subgroup of degree $\geqq 3$ of the group $SL(n+1,Z)\cap SO_0(1,n)$. In the second case we limit ourselves to $n=3$, $\Gamma=SL(2,Z+iZ)/\{\pm\left(\begin{smallmatrix}1&0\\0&1\end{smallmatrix}\right)\}$.

As an intermediate result, this paper gives a proof of the theorem on expansion in eigenfunctions of the selfadjoint extension of the Laplace-Beltrami operator to $\Gamma\backslash S$. In particular, we characterize the continuous spectrum of this operator, which turns out to be an m-fold Lebesgue spectrum, where m is the number of essential parabolic vertices of the domain $\Gamma\backslash S$, which is on the semi-axis $(-\infty,\ -1/4c_0^2]$, where the constant c_0 is given in §1. In addition, we give a complete system of eigenfunctions for the continuous spectrum, which coincides with the system of analytically continued Eisenstein series.

The results involving a real hyperbolic space were stated by the author in the note [7].

This paper consists of five sections. The first is devoted to computing the Iwasawa decomposition for our simple groups G, which is used for the coordinatization of the symmetric spaces, and also to other auxiliary propositions. In §§2 and 3 we give a description of the selfadjoint extension of the Laplace-Beltrami operator to $\Gamma\backslash S$, and we use the integral equation to investigate its resolvent. We then use these results to prove the expansion theorem. §4 is devoted to proving the Selberg trace formula. Finally, in §5 we give a concrete computation of the additional contribution to the trace formula (Examples 1 and 2).

Since all of these facts are proved similarly for all three of our hyperbolic spaces, we shall investigate in detail the most difficult case, namely

$$\mathrm{Sp}(1,n)/(\mathrm{Sp}(1)\cap \mathrm{Sp}(n)).$$

In conclusion, the author wishes to thank L. D. Faddeev, and also B. B. Venkov for suggesting the topic and helping with the research.

§1. Definitions and auxiliary propositions

1.1. *Definitions of the groups* $SO_0(1,n)$, $SU(1,n)$, $\mathrm{Sp}(1,n)$ *and some of their properties.*

We let I_n denote the unit matrix of order n. We introduce the following matrices (for the notation, see [8]):

$$I_{p,q}=\begin{pmatrix}-I_p & 0\\ 0 & I_q\end{pmatrix},\quad J_n=\begin{pmatrix}0 & I_n\\ -I_n & 0\end{pmatrix},\quad K_{p,q}=\begin{pmatrix}I_{p,q} & 0\\ 0 & I_{p,q}\end{pmatrix}.$$

If g is a matrix, then we let g^t and $\bar{g}$ denote the transpose matrix and the complex conjugate matrix to g, respectively. We consider the following three series of groups:

$$G_1^n=\{g\in GL(n+1,\ R)\,|\,g^tI_{1,n}g=I_{1,n},\ \det g=1\},$$
$$G_2^n=\{g\in GL(n+1,\ C)\,|\,g^tI_{1,n}\bar{g}=I_{1,n}\},$$
$$G_3^n=\{g\in GL(2(n+1),\ C)\,|\,g^tJ_{n+1}g=J_{n+1},\ g^tK_{1,n}\bar{g}=K_{1,n}\}.$$

The standard notation for them is as follows: $G_1^n=SO_0(1,n)$, $G_2^n=SU(1,n)$, $G_3^n=\mathrm{Sp}(1,n)$. We shall omit the index n in what follows. It is well known (see [8]) that G_1, G_2 and G_3 are real simple connected Lie groups.

The basic goal of this subsection is to compute explicitly the Iwasawa decomposition for the groups G_1, G_2 and G_3. The general definition of this decomposition for a connected real semisimple Lie group is contained, for example, in [9]. It is well known that the definition consists in the following. Let G be any of the groups G_1, G_2 or G_3. Let $K\subset G$ be a maximal compact subgroup. Let $\mathfrak{g}_0$ and $\mathfrak{k}_0$ denote the Lie algebras of the groups G and K, respectively, and let $\mathfrak{g}$ and $\mathfrak{k}$ denote their complexifications. We have $\mathfrak{g}_0=\mathfrak{k}_0+\mathfrak{p}_0$, where $\mathfrak{p}_0$ is the space of $\mathfrak{g}_0$ orthogonal to $\mathfrak{k}_0$ with respect to the Killing form. There exists an involution automorphism θ of $\mathfrak{g}$ such that $\theta\mathfrak{g}_0\subset\mathfrak{g}_0$, and such that for any $x\in\mathfrak{k}_0$ and $y\in\mathfrak{p}_0$ we have $\theta x=x$ and $\theta y=-y$. Further, let $\mathfrak{h}_{\mathfrak{p}_0}$ be a commutative subalgebra of $\mathfrak{g}_0$ contained in $\mathfrak{p}_0$ and having maximal possible dimension. Let $\mathfrak{h}_0$ be a maximal commutative subalgebra of $\mathfrak{g}_0$ containing $\mathfrak{h}_{\mathfrak{p}_0}$, and let $\mathfrak{h}$ be its complexification. Then $\mathfrak{h}$ is a Cartan subalgebra of $\mathfrak{g}$. We consider the set Ξ of all roots of $\mathfrak{g}$ relative to $\mathfrak{h}$, i.e. the set of all linear functions $\alpha(h)$ on $\mathfrak{h}$ such that $[h,x_\alpha]=\alpha(h)x_\alpha$ for any $h\in\mathfrak{h}$, where $x_\alpha\in\mathfrak{g}$, $x_\alpha\neq 0$. We have let $[\,,]$ denote multiplication in the Lie algebra $\mathfrak{g}$. It is easy to see that, if α is a root relative to $\mathfrak{h}$, then $\theta\alpha$ is also a root. We now consider the set Ξ_+ of all positive roots relative to some basis in $\mathfrak{h}$. We are interested in the subset $\Xi_+^0\subset\Xi_+$ given by the following condition: $\Xi_+^0=\{\alpha\in\Xi_+\,|\,\theta\alpha\neq\alpha\}$.

Further, let $\mathfrak{n}$ be the subspace of $\mathfrak{g}$ consisting of all root vectors x_α, $[h,x_\alpha]=\alpha(h)x_\alpha$, where $\alpha\in\Xi_+^0$ and runs through all of this set. It then turns out that $\mathfrak{n}$ is a nilpotent subalgebra of $\mathfrak{g}$, and, if we let $\mathfrak{n}_0$ denote the intersection $\mathfrak{n}_0=\mathfrak{n}\cap\mathfrak{g}_0$, then $\mathfrak{g}_0$ splits into a direct sum of three subalgebras: $\mathfrak{g}_0=\mathfrak{k}_0+\mathfrak{h}_{\mathfrak{p}_0}+\mathfrak{n}_0$. This is the Iwasawa decomposition for the algebra $\mathfrak{g}_0$. The corresponding decomposition of the group G as a product of three subgroups is obtained using the exponential mapping.

We compute the Iwasawa decomposition for the groups G_2 and G_3. This computation is carried out for G_1 in [**10**]. Further, in all the notation used in computing the decomposition we shall put a superscript 1, 2 or 3 so as to distinguish between the groups. For example, $\mathfrak{g}_0^2$ will denote the Lie algebra of the group G_2. It is well known (see [**8, 11**]) that $\mathfrak{g}_0^2$ consists of matrices of the form

$$\begin{pmatrix} z_1 & z_2 \\ \bar{z}_2^t & z_3 \end{pmatrix},$$

where z_1 and z_3 are skew-hermitian of orders 1 and n, respectively, $\operatorname{Tr} z_1 + \operatorname{Tr} z_3 = 0$, and z_2 is an arbitrary matrix. The subalgebra $\mathfrak{k}_0^2 \subset \mathfrak{g}_0^2$ consists of matrices of the form

$$\begin{pmatrix} z_1 & 0 \\ 0 & z_3 \end{pmatrix};$$

$\mathfrak{k}_0^2$ is isomorphic to the direct product $\mathfrak{su}(1) \times c \times \mathfrak{su}(n)$, where $\mathfrak{su}(k)$ denotes the Lie algebra of the group $SU(k)$ and c is the center of the algebra $\mathfrak{k}_0^2$. The maximal abelian subalgebra $\mathfrak{h}_{\mathfrak{p}_0}^2$ has the form

$$\mathfrak{h}_{\mathfrak{p}_0}^2 = \left\{ \begin{pmatrix} 0 & h & \\ h & 0 & \\ & & 0 \end{pmatrix} \right\},$$

where $h \in R$. The subalgebra $\mathfrak{h}_0^2$ consists of matrices of the form

$$H = \begin{pmatrix} u_1 & h & & & \\ h & u_1 & & 0 & \\ & & u_3 & & \\ & 0 & & \ddots & \\ & & & & u_{n+1} \end{pmatrix},$$

where $h \in R$, $u_j \in iR$ and $\operatorname{Tr} H = 0$. The Cartan subalgebra $\mathfrak{h}^2$ consists of matrices of type H, but with complex coefficients.

For a system of positive roots we choose a system of roots on $\mathfrak{h}^2$ which, in particular, preserve the sign of h. A simple verification shows that for the set Ξ_+^{02} we must choose the following system of roots:

$$\{e_1 - e_2,\ e_1 - e_3,\ \ldots,\ e_1 - e_{n+1},\ e_3 - e_2,\ e_4 - e_2,\ \ldots,\ e_{n+1} - e_2\},$$

where the $e_j = e_j(H)$ are the following linear functions:

$$e_1(H) = h - iu_1,$$
$$e_2(H) = -h - iu_1,$$
$$e_3(H) = iu_3,$$
$$\vdots$$
$$e_{n+1}(H) = iu_{n+1}.$$

Further, starting from the definition of the subalgebras $\mathfrak{n}$ and $\mathfrak{n}_0$, we find after elementary computations that $\mathfrak{n}_0^2$ consists of matrices of the form

$$\begin{pmatrix} ia & -ia & b_2 \dots b_n \\ ia & -ia & b_2 \dots b_n \\ \bar{b}_2 & -\bar{b}_2 & \\ \vdots & & 0 \\ \bar{b}_n & -\bar{b}_n & \end{pmatrix},$$

where $a \in R$ and $b_j \in C$, $j = 2, \cdots, n$.

Finally, applying the exponential mapping, we find that the corresponding subgroups have the form

$$K^2 = \left\{ \begin{pmatrix} 1 & 0 \\ 0 & k \end{pmatrix} \begin{pmatrix} e^{i\alpha} & 0 \\ 0 & e^{-\frac{i\alpha}{n}} I_n \end{pmatrix} \right\},$$

where $\alpha \in R$ and $k \in SU(n)$.

The commutative subgroup A_0^2 corresponding to the subalgebra $\mathfrak{h}_{\mathfrak{p}_0}$ has the form

$$A_0^2 = \left\{ \begin{pmatrix} \cosh t & \sinh t & 0 \\ \sinh t & \cosh t & \\ 0 & & I_{n-1} \end{pmatrix} \right\},$$

where $t \in R$.

Finally, the nilpotent subgroup U^2 corresponding to $\mathfrak{n}_0^2$ consists of the matrices

$$\begin{pmatrix} 1 + ia + \frac{|b|^2}{2} & -ia - \frac{|b|^2}{2} & b_2 \dots b_n \\ ia + \frac{|b|^2}{2} & 1 - ia - \frac{|b|^2}{2} & b_2 \dots b_n \\ \bar{b}_2 & -\bar{b}_2 & \\ \vdots & \vdots & \\ \bar{b}_n & -\bar{b}_n & I_{n-1} \end{pmatrix},$$

where $a \in R$, $b_j \in C$ $(j = 2, \cdots, n)$ and $|b|^2 = |b_2|^2 + \cdots + |b_n|^2$.

We have thereby obtained the desired Iwasawa decomposition $G_2 = U^2 A_0^2 K^2$. Similarly, in the case of the group G_3 the elements of the decomposition have the form: $\mathfrak{g}_0^3$ consists of matrices of the form

$$\begin{pmatrix} z_{11} & z_{12} & z_{13} & z_{14} \\ \bar{z}_{12}^t & z_{22} & z_{14}^t & z_{24} \\ -\bar{z}_{13} & \bar{z}_{14} & \bar{z}_{11} & -\bar{z}_{12} \\ \bar{z}_{14} & -\bar{z}_{24} & -z_{12}^t & \bar{z}_{22} \end{pmatrix},$$

where $z_{11}, z_{13} \in C$; z_{12} and z_{14} are rectangular $1 \times n$ matrices; z_{11} and z_{22} are skew-hermitian; and z_{13} and z_{24} are symmetric.

The subalgebra $\mathfrak{k}_0^3 \subset \mathfrak{g}_0^3$ consists of matrices of the form

$$\begin{pmatrix} z_{11} & 0 & z_{13} & 0 \\ 0 & z_{22} & 0 & z_{24} \\ -\bar{z}_{13} & 0 & \bar{z}_{11} & 0 \\ 0 & -\bar{z}_{24} & 0 & \bar{z}_{22} \end{pmatrix}.$$

The subalgebra $\mathfrak{k}_0^3$ is isomorphic to the direct product $(\mathfrak{sp}(1, C) \cap \mathfrak{u}(1)) \times (\mathfrak{sp}(n, C) \cap \mathfrak{u}(n))$, where $\mathfrak{sp}(k, C)$ is the Lie algebra of the group $\mathrm{Sp}(k, C)$. This isomorphism is given by the mapping

$$\left\{ \begin{pmatrix} z_{11} & z_{13} \\ -\bar{z}_{13} & \bar{z}_{11} \end{pmatrix}, \quad \begin{pmatrix} z_{22} & z_{24} \\ -\bar{z}_{24} & \bar{z}_{22} \end{pmatrix} \right\} \to \mathfrak{k}_0^3.$$

The maximal abelian subalgebra $\mathfrak{h}_{\mathfrak{p}_0}$ has the form

$$\begin{pmatrix} \begin{matrix} 0 & h \\ h & 0 \end{matrix} & & & \mathbf{0} \\ & \boxed{\begin{matrix} 0 \\ n\times n \end{matrix}} & & \\ & & \begin{matrix} 0 & -h \\ -h & 0 \end{matrix} & \\ \mathbf{0} & & & \boxed{\begin{matrix} 0 \\ n\times n \end{matrix}} \end{pmatrix},$$

where $h \in R$. The subalgebra $\mathfrak{h}_0^3$ consists of matrices of the form

$$H = \begin{pmatrix} u_1 & h & & & & & & & \mathbf{0} \\ h & u_1 & & & & & & & \\ & & u_3 & & & & & & \\ & & & \ddots & & & & & \\ & & & & u_{n+1} & & & & \\ & & & & & -u_1 & -h & & \\ & & & & & -h & -u_1 & & \\ & & & & & & & -u_3 & \\ & & & & & & & & \ddots \\ \mathbf{0} & & & & & & & & & -u_{n+1} \end{pmatrix},$$

where $h \in R$ and $u_j \in iR$, $j = 1, 3, 4, \cdots, n+1$.

The Cartan subalgebra $\mathfrak{h}^3$ consists of matrices of the same type, but with complex coefficients. The system of all roots relative to this subalgebra is given in [11]. For the system Ξ_+^{03} we must take the following system:

$$\{e_1 - e_2,\ e_1 - e_3,\ \ldots,\ e_1 - e_{n+1},\ e_3 - e_2,\ e_4 - e_2,\ \ldots,\ e_{n+1} - e_2,\ 2e_1,\ -2e_2\},$$

where $e_j = e_j(H)$, $H \in \mathfrak{h}$ and

$$e_1(H) = h - iu_1, \quad e_2(H) = -h - iu_1, \quad e_3(H) = -iu_3, \cdots, e_{n+1}(H) = -iu_{n+1}.$$

Here we again take the positive roots to be roots which, in particular, preserve the sign of h.

We further note that for $\theta = \theta(x)$, $x \in \mathfrak{g}^3$, we take the automorphism $\theta(x) = K_{1,n}xK_{1,n}$. The subalgebra $\mathfrak{n}_0^3$ consists of the matrices

$$\begin{pmatrix} ia & -ia & b_2 \dots b_n & d & d & f_2 \dots f_n \\ ia & -ia & b_2 \dots b_n & d & d & f_2 \dots f_n \\ \bar{b}_2 & -\bar{b}_2 & & f_2 & f_2 & \\ \vdots & \vdots & 0 & \vdots & \vdots & 0 \\ \bar{b}_n & -\bar{b}_n & & f_n & f_n & \\ -\bar{d} & \bar{d} & \bar{f}_2 \dots \bar{f}_n & -ia & -ia & -\bar{b}_2 \dots -\bar{b}_n \\ \bar{d} & -\bar{d} & -\bar{f}_2 \dots -\bar{f}_n & ia & ia & \bar{b}_2 \dots \bar{b}_n \\ \bar{f}_2 & -\bar{f}_2 & & -b_2 & -b_2 & \\ \vdots & \vdots & 0 & \vdots & \vdots & 0 \\ \bar{f}_n & -\bar{f}_n & & -b_n & -b_n & \end{pmatrix},$$

where $a \in R$ and the remaining parameters are complex.

Now, applying the exponential mapping, we compute the corresponding subgroups

$$K^3 = \left\{ \begin{pmatrix} \alpha & 0 \dots 0 & \beta & 0 \dots 0 \\ 0 & & 0 & \\ \vdots & A & \vdots & B \\ 0 & & 0 & \\ -\bar{\beta} & 0 \dots 0 & \bar{\alpha} & 0 \dots 0 \\ 0 & & 0 & \\ \vdots & -\bar{B} & \vdots & \bar{A} \\ 0 & & 0 & \end{pmatrix} \right\},$$

where $\alpha, \beta \in C$, $|\alpha|^2 + |\beta|^2 = 1$; A and B are matrices of order n such that $A^t\bar{A} + B^t\bar{B} = I_n$ and $A^tB = B^tA$.

The commutative one-dimensional subgroup A_0^3 corresponding to the subalgebra $\mathfrak{h}_{\mathfrak{p}_0}^3$ consists of matrices of the form

$$\begin{pmatrix} \cosh t & \sinh t & & & & \\ \sinh t & \cosh t & & & 0 & \\ & & I_{n-1} & & & \\ & & & \cosh t & -\sinh t & \\ & & & -\sinh t & \cosh t & \\ 0 & & & & & I_{n-1} \end{pmatrix},$$

where $t \in R$. The nilpotent subgroup U^3 corresponding to the subalgebra $\mathfrak{n}_0^3$ consists of the matrices

$$\begin{pmatrix}
1+ia+\frac{|b|^2+|f|^2}{2} & -ia-\frac{|b|^2+|f|^2}{2} & b_2\dots b_n & d & d & f_2\dots f_n \\
ia+\frac{|b|^2+|f|^2}{2} & 1-ia-\frac{|b|^2+|f|^2}{2} & b_2\dots b_n & d & d & f_2\dots f_n \\
\bar b_2 & -\bar b_2 & & f_2 & f_2 & \\
\vdots & \vdots & I_{n-1} & \vdots & \vdots & 0 \\
\bar b_n & -\bar b_n & & f_n & f_n & \\
-\bar d & \bar d & \bar f_2\dots\bar f_n & 1-ia+\frac{|b|^2+|f|^2}{2} & -ia+\frac{|b|^2+|f|^2}{2} & -\bar b_2\dots-\bar b_n \\
\bar d & -\bar d & -\bar f_2\dots-\bar f_n & ia-\frac{|b|^2+|f|^2}{2} & 1+ia-\frac{|b|^2+|f|^2}{2} & \bar b_2\dots\bar b_n \\
\bar f_2 & -\bar f_2 & & -b_2 & -b_2 & \\
\vdots & \vdots & 0 & \vdots & \vdots & I_{n-1} \\
\bar f_n & -\bar f_n & & -b_n & -b_n &
\end{pmatrix},$$

where $a\in R$, all the remaining parameters are complex, and $|b|^2=|b_2|^2+\cdots+|b_n|^2$, $|f|^2=|f_2|^2+\cdots+|f_n|^2$.

We write out the corresponding components of the decomposition for the group G_1:

$$K^1=\left\{\begin{pmatrix}1 & 0\\ 0 & k\end{pmatrix}\right\},\quad A_0^1=\left\{\begin{pmatrix}\cosh t & \sinh t & 0\\ \sinh t & \cosh t & \\ 0 & & I_{n-1}\end{pmatrix}\right\},$$

$$U^1=\left\{\begin{pmatrix}1+\frac{|b|^2}{2} & -\frac{|b|^2}{2} & b_2\dots b_n\\ \frac{|b|^2}{2} & 1-\frac{|b|^2}{2} & b_2\dots b_n\\ b_2 & -b_2 & \\ \vdots & \vdots & I_{n-1}\\ b_n & -b_n & \end{pmatrix}\right\}.$$

Here $k\in SO(n)$, $t\in R$, $b_j\in R$, $j=2,\cdots,n$ and $|b|^2=b_2^2+\cdots+b_n^2$. Further, G will denote any of the groups G_1, G_2 or G_3. We adopt a similar convention for the subgroups U, A_0, K.

1.2. *Symmetric spaces.*

We consider the set of classes $S=G/K$ as a homogeneous space. It is an irreducible symmetric space of rank 1. It is clear from the Cartan classification (see [8], Chapter 9, §4) that the cases of S under consideration, corresponding to the groups G_1, G_2 and G_3, cover all the irreducible noncompact symmetric spaces of rank 1 except for one exceptional one (type FII of dimension 16).

If $z\in S$ and $g\in G$, we let $gz\in S$ denote the image of z under the map g. For later use it is convenient for us to choose a special system of coordinates on S. Namely, it is easy to see that the action of UA_0 is simple and transitive on S; in other words, UA_0 acts transitively on S and the condition $z'=gz$, where $g\in UA_0$, determines g uniquely. For coordinates we take the parameters of the matrix elements of the groups U and A_0 in the Iwasawa decomposition (see §1.1). We first give a system of coordinates for the space S_3. In the matrix elements of the group we perform the substitution $y=e^t$. In addition, for the group U^3 we set $x_1=a$, $x_j=b_j$, $x_{n+1}=d$ and $x_{n+j}=f_j$, $j=2,\cdots,n$.

Thus the coordinates in the space S_3 are given by the vector $\{y,x\} = \{y, x_1, x_2, \cdots, x_{2n}\}$. Here $y > 0$, $x_1 \in R$ and $x_k \in C$ for $k = 2, \cdots, 2n$. We introduce coordinates $\{y, x\}$ in the spaces S_1 and S_2 in an analogous way. In the case of S_1 we perform the substitution $y = e^t$ in the group A_0^1, and in U^1 we set $x_j = b_j$, $j = 2, \cdots, n$, $x_j \in R$. In the case of S_2 we perform the substitution $y = e^t$ in A_0^2, and in U^2 we set $x_1 = a$, $x_j = b_j$, $j = 2, \cdots, n$, $x_1 \in R$, $x_j \in C$. Further, when speaking of coordinates in S, we shall use the coordinates in S_3, and assume that in the case of S_1 or S_2 certain of the variables are absent.

We consider the normalizer P of the subgroup U in G. It is easy to see that $P = UA_0M$, where the subgroup $M \subset K$, and the centralizer $Z(A_0)$ of A_0 in G is given by $Z(A_0) = A_0M$.

Definition 1. We call a transformation of the space S a *shift-similitude type transformation* if a point $z \in S$ with coordinates $\{y, x\}$ (we let $z = \{y, x\}$) is taken to a point $z' = \{y', x'\}$ such that y' differs from y by a positive multiplicative constant, and x' is a linear nonhomogeneous transform of x.

Lemma 1.1. *A transformation $g \in G$ is a shift-similitude type transformation if and only if $g \in P$.*

Proof. We first compute the centralizer $Z(A_0)$. After elementary computations we obtain

$$M^1 = \left\{\begin{pmatrix} I_2 & 0 \\ 0 & k_1 \end{pmatrix}\right\}, \quad M^2 = \left\{\begin{pmatrix} e^{i\delta} I_2 & 0 \\ 0 & e^{-\frac{2i\delta}{n-1}} k_2 \end{pmatrix}\right\},$$

where $k_1 \in SO(n-1)$ and $k_2 \in SU(n-1)$.

In the case of G_3, $M^3 \subset K^3$ consists of matrices of the form

$$\begin{pmatrix} \alpha & 0 & 0 & -\beta & 0 & 0 \\ 0 & \alpha & 0 & 0 & \beta & 0 \\ 0 & 0 & A & 0 & 0 & B \\ \beta & 0 & 0 & \bar{\alpha} & 0 & 0 \\ 0 & -\beta & 0 & 0 & \bar{\alpha} & 0 \\ 0 & 0 & -\bar{B} & 0 & 0 & \bar{A} \end{pmatrix}.$$

To prove the lemma it suffices to describe the action of the different elements of the Iwasawa decomposition for G. For $u \in U$ and $a \in A_0$ we consider the point $z \in S$ with coordinates ua. Further, let $g' = u'a'k'$, where $u' \in U$, $a' \in A_0$ and $k' \in K$. Since the action of UA_0 on S is simple and transitive, it follows that there is a unique element $k'' \in K$ such that $g'uak'' \in UA_0$. We successively consider the cases $g' = u'$, $g' = a'$ and $g' = k'$.

If $u'(a, b, d, f) \in U$, then $k'' = e$ is the unit element of G, and the coordinates of the point $z' = u'z$ equal

$$y' = y, \qquad x_1' = x_1 + a + \sum_{k=2}^{n} \left(\frac{b_k \bar{x}_k - x_k \bar{b}_k}{2i} + \frac{f_k \bar{x}_{n+k} - \bar{f}_k x_{n+k}}{2i} \right),$$

$$x'_{n+1}=x_{n+1}+d+\sum_{k=2}^{n}(b_kx_{n+k}-x_kf_k),$$

$$x'_j=x_j+b_j,\ x'_{n+j}=x_{n+j}+f_j,\ j=2,\ldots,n.$$

These are the coordinates of the transformation $u'\in G_3$; the corresponding coordinates are absent in the case of G_1 and G_2.

If $a'=a'(t)\in A_0$ and $e^t=\xi$, then $k''=e$, and

$$\begin{aligned} y'&=\xi y, & x'_j&=\xi x_j,\\ x'_1&=\xi^2x_1, & x'_{n+j}&=\xi x_{n+j},\\ x'_{n+1}&=\xi^2x_{n+1}, & j&=2,\ldots,n. \end{aligned}$$

If $g'=m'\in M$, then $y'=y$, and

$$\begin{pmatrix} x'_1\\ \operatorname{Re}x'_{n+1}\\ \operatorname{Im}x'_{n+1}\end{pmatrix}=m'_1\begin{pmatrix} x_1\\ \operatorname{Re}x_{n+1}\\ \operatorname{Im}x_{n+1}\end{pmatrix},\quad \begin{pmatrix} x'_2\\ \vdots\\ x'_n\\ x'_{n+2}\\ \vdots\\ x'_{2n}\end{pmatrix}=m'_2\begin{pmatrix} x_2\\ \vdots\\ x_n\\ x_{n+2}\\ \vdots\\ x_{2n}\end{pmatrix},$$

where m'_1 is an orthogonal matrix and m'_2 is a unitary matrix (an orthogonal matrix in the case of G_1).

Now let $k'\in K$ but $k'\notin M$. In this case the action is more complicated. The choice of the element k'' depends on which group is being considered. In the case of G_1 we take the unit element for k''; for G_2 we take k'' from the center of the group K^2:

$$k''=\begin{pmatrix} e^{i\alpha} & 0\\ 0 & e^{-\frac{i\alpha}{n}}I_n\end{pmatrix}.$$

In the case of G_3 we have $k''\in K^3$ and, in our notation, $A=I_n$ and $B=0$. The coordinates of $z'=k'z$ have the following appearance: for G_1 and G_2

$$y'=\frac{y}{|(p'(x)+y^2)d(k')+p''(x)|},$$

$$x'_j=\frac{p_j(y,x)}{|(p'(x)+y^2)d(k')+p''(x)|},$$

$j=1,\cdots,n$, where $p'(x)$ is a polynomial of degree two in x, $p''(x)$ is a first degree polynomial, $p_j(y,x)$ is a polynomial of degree two in y and x, and $d_1(k')=1-k_{11}$ and $d_2(k')=e^{i\alpha}-k_{11}e^{-i\alpha/n}$, where

$$k'_1=\begin{pmatrix} 1 & 0 & \ldots & 0\\ 0 & k_{11} & \ldots & k_{1n}\\ \vdots & \vdots & & \vdots\\ 0 & k_{n1} & \ldots & k_{nn}\end{pmatrix},\quad k'_2=\begin{pmatrix} 1 & 0 & \ldots & 0\\ 0 & k_{11} & \ldots & k_{1n}\\ \vdots & \vdots & & \vdots\\ 0 & k_{n1} & \ldots & k_{nn}\end{pmatrix}\begin{pmatrix} e^{i\alpha} & 0\\ 0 & e^{-\frac{i\alpha}{n}}I_n\end{pmatrix}$$

for G_1 and G_2, respectively.

For the group G_3, if

$$k_3' = \begin{pmatrix} \alpha & 0 & \beta & 0 \\ 0 & A & 0 & B \\ -\bar\beta & 0 & \bar\alpha & 0 \\ 0 & -\bar B & 0 & \bar A \end{pmatrix}$$

and

$$A = \begin{pmatrix} a_{11} & \dots & a_{1n} \\ \vdots & & \vdots \\ a_{n1} & \dots & a_{nn} \end{pmatrix}, \quad B = \begin{pmatrix} b_{11} & \dots & b_{1n} \\ \vdots & & \vdots \\ b_{n1} & \dots & b_{nn} \end{pmatrix},$$

then the coordinates of $z' = k'z$ equal

$$y' = \frac{y}{|(p'(x) + y^4)\, d_3(k') + p''(x)|^{1/2}},$$

$$x_j' = \frac{p_j(y, x)}{|(p'(x) + y^4)\, d_3(k') + p''(x)|}, \quad j = 1, \dots, 2n,$$

where $p'(x)$ and $p_j(y, x)$ are certain fourth degree polynomials, $p''(x)$ is a third degree polynomial, and $d_3(k') = |\alpha - a_{11}|^2 + |\beta + b_{11}|^2$. These results give us the lemma.

We shall further need an expression for the Riemann metric in our coordinates. The computation presents no difficulty in the case of S_1: $ds^2 = (y^2)^{-1}(dy^2 + \sum_{k=2}^{n} dx_k^2)$. In the case of S_2 the invariant Riemann metric can be obtained from the following considerations. S_2 is a classical symmetric domain of the first type (complex sphere). The Riemann metric is induced by the Bergman kernel, which has the following form in our coordinates:

$$c \left(\frac{\left(y^2 + \sum_{k=2}^{k} |x_2|^2 + 1 \right)^2 + 4x_1^2}{4y^2} \right)^{-(n+1)}, \quad c = \text{const},$$

and the metric equals

$$ds^2 = \frac{dy^2}{y^2} + \frac{1}{y^2} \sum_{k=2}^{n} dx_k d\bar x_k + \frac{1}{y^4} \left(dx_1 + \operatorname{Im} \sum_{k=2}^{n} \bar x_k dx_k \right)^2.$$

In these cases the action of the group $UZ(A_0)$ on S gives us a heuristic picture of the invariant Riemann metric. Proceeding from the same considerations, we may suppose that in the case of S_3 the metric has the form

$$\begin{aligned} ds^2 = \frac{dy^2}{y^2} &+ \frac{1}{y^2} \sum_{k=2}^{n} (dx_k d\bar x_k + dx_{n+k} d\bar x_{n+k}) \\ &+ \frac{1}{y^4} \left(dx_1 + \operatorname{Im} \sum_{k=2}^{n} (\bar x_k dx_k + \bar x_{n+k} dx_{n+k}) \right)^2 \\ &+ \frac{1}{y^4} \left| dx_{n+1} + \sum_{k=2}^{n} (x_{n+k} dx_k - x_k dx_{n+k}) \right|^2. \end{aligned}$$

A precise verification of the invariance of this metric relative to the action of G_3 shows that this is the case. This fact can be proved using elementary but cumbersome computations, which will not be given here.

Let ρ be the invariant distance on S. The following invariants $u(z,z')$ of pairs of points $z,z'\in S$ will be useful to us. In the spaces S_1, S_2 and S_3, respectively, we have

$$u_1(z,\ z)'=\frac{(y-y')^2+\sum_{k=2}^{n}(x_k-x'_k)^2}{4yy'},$$

$$u_2(z,\ z')=\frac{1}{4y^2y'^2}\left\{\left(y^2+y'^2+\sum_{k=2}^{n}(x_k-x'_k)^2\right)^2+4\left(\frac{1}{2i}\sum_{k=2}^{n}(\bar{x}_kx'_k-x_k\bar{x}'_k)+x'_1-x_1\right)^2\right\}-1,$$

$$u_3(z,\ z')=\frac{1}{4y^2y'^2}\left\{\left(y^2+y'^2+\sum_{k=2}^{n}((x_k-x'_k)^2+(x_{n+k}-x'_{n+k})^2)\right)^2+4\left(\frac{1}{2i}\sum_{k=2}^{n}(\bar{x}_kx'_k-x_k\bar{x}'_k+\bar{x}_{n+k}y'_{n+k}-x_{n+k}\bar{x}'_{n+k})+x'_1-x_1\right)^2+4\left|\sum_{k=2}^{n}(x_{n+k}x'_k-x'_{n+k}x_k)+x'_{n+1}-x_{n+1}\right|^2\right\}-1.$$

The invariance, i.e. the relation $u(gz,gz')=u(z,z')$ for any $g\in G$, is proved analogously to the invariance of the Riemann metric. Using the definition of ρ, it is not hard to show that $u_1=\sinh^2\frac{1}{2}\rho_1$, $u_2=\sinh^2\rho_2$ and $u_3=\sinh^2\rho_3$ for S_1, S_2 and S_3, respectively.

1.3. *The Laplace-Beltrami operator and Green's function.*

Suppose that on some Riemannian manifold in a neighborhood of a fixed point the Riemann metric is given by the formula $ds^2=\sum_{i,j}(\theta)\,d\theta_i d\theta_j$, where $\theta_1,\cdots,\theta_n$ are local coordinates.

The second order differential operator L

$$Lf=\frac{1}{\sqrt{\det h}}\sum_{k=1}^{n}\frac{\partial}{\partial\theta_k}\left(\sum_{j=1}^{n}h^{jk}\sqrt{\det h}\,\frac{\partial f}{\partial\theta_j}\right)$$

is called the Laplace-Beltrami operator. We have let $\det h$ denote the determinant of the matrix $\{h_{jk}\}$, and we have let $\{h^{jk}\}$ denote the inverse matrix of $\{h_{jk}\}$. This operator is invariant relative to isometric transformations of the manifold, and, in addition, L is formally selfadjoint with respect to the invariant measure $d\mu(\theta)=\sqrt{\det h}\,d\theta_1\cdots d\theta_n$ (see [8]).

In our case the Laplace-Beltrami operator for the Riemann metric is expressed by the following formula:

$$Lf=\frac{1}{\sqrt{\det h}}\frac{\partial}{\partial y}\left(y^2\sqrt{\det h}\,\frac{\partial f}{\partial y}\right)+y^2L''f+y^4L'f,$$

where $L'=\partial^2/\partial x_1^2+\partial^2/\partial(\operatorname{Re}x_{n+1})^2+\partial^2/\partial(\operatorname{Im}x_{n+1})^2$, and L'' is a second order operator whose derivatives have coefficients depending only on the variables $\operatorname{Re}x_k$, $\operatorname{Im}x_k$, $\operatorname{Re}x_{n+k}$ and $\operatorname{Im}x_{n+k}$, $k=2,\cdots,n$. The determinant $\det h$ depends only on the variable y, and in the spaces S_1, S_2 and S_3, respectively, it is given by: $\det h_1=1/y^{2n}$, $\det h_2=1/y^{4n+2}$ and $\det h_3=1/y^{8n+6}$. These results are obtained in dimension $n=2$ by direct computation of the matrix $\{h^{jk}\}$ and in arbitrary dimension using induction.

We replace the parameter λ in the equation

$$-L\omega-\lambda\omega=f$$

by the complex parameter s which is defined as follows: in S_1 we let $\lambda = s(n-1-s)$, in S_2 we let $\lambda=s(2n-s)/4$, and in S_3 we let $\lambda=s(4n+2-s)/4$.

Further, for the purpose of investigating Green's function for this equation, we compute the radial part of the operator L relative to the invariant $u(z,z')$ of a pair of points. In other words, if the function $\omega(z)$ on S has the form $\omega(z)=\omega(u(z,z_0))$, then, by invariance, L acts on $\omega(z)$ as a differential operator of one variable: $L\omega(z)=l\omega(u)=(u^2+u)\omega''(u)+\varphi(u)\omega'(u)$, where $\varphi(u)$ is equal to $nu+n/2$, $(n+1)u+n$ and $2(n+1)u+2n$ in S_1, S_2 and S_3, respectively. When we consider the equation $l\omega+\lambda\omega=0$, we are interested in the following solution $\omega(u,s)$ of this hypergeometric equation (see [12]):

$$\omega_1(u,s)=c_1\int_0^1 (t(1-t))^{s-\frac{n}{2}}(t-u)^{-s}\,dt,\quad \operatorname{Re}s>\frac{n}{2}-1;$$

$$\omega_1(u,s)=c_1u^{-\frac{n-2}{2}}+o\left(u^{-\frac{n-2}{2}}\right),\quad u\to 0;\quad \omega_1(u,s)=O(u^{-\operatorname{Re}s}),\ u\to\infty;$$

$$\omega_2(u,s)=c_2\int_0^1 t^{\frac{s}{2}-n}(1-t)^{\frac{s}{2}-1}(t-u)^{-\frac{s}{2}}\,dt,\quad \operatorname{Re}s>2(n-1);$$

$$\omega_2(u,s)=c_2u^{-n+1}+o(u^{-n+1}),\quad u\to 0;\quad \omega_2(u,s)=O\left(u^{-\frac{\operatorname{Re}s}{2}}\right),\quad u\to\infty;$$

$$\omega_3(u,s)=c_3\int_0^1 t^{\frac{s}{2}-2n}(1-t)^{\frac{s}{2}-2}(t-u)^{-\frac{s}{2}}\,dt,\quad \operatorname{Re}s>4n-2;$$

$$\omega_3(u,s)=c_3u^{-2n+1}+o(u^{-2n+1}),\quad u\to 0;\quad \omega_3(u,s)=O\left(u^{-\frac{\operatorname{Re}s}{2}}\right),\ u\to\infty;$$

$$c=\lim_{\zeta\to 0}u^{\frac{\dim S-2}{2}}\int_1^{\zeta}\frac{1}{q(r)}\,dr,\quad q(r)=\int_{\zeta(z,z_0)=r}d\mu(z),$$

$$c=c_j,\ j=1,2,3,\ u=u_j,\ \zeta=\zeta_j,$$

for S_1, S_2 and S_3. We have also indicated the asymptotic behavior of $\omega(u,s)$ as u tends to 0 or ∞. We note that, if we consider $\omega(u,s)=\omega(u(z,z'),s)$ on S, then this kernel behaves like the fundamental solution of the operator $-L-\lambda$ as $u\to 0$.

LEMMA 1.2. *If $f(z)$ is a function which is infinitely differentiable and bounded on S, then for $\operatorname{Re}s>\alpha$, where α is equal to $n-1, 2n$ and $4n+2$ for S_1, S_2 and S_3, respectively, the function*

$$h(z)=\int_S \omega(u(z,z'),s)f(z')\,d\mu(z')$$

has the same properties and satisfies the equation $-Lh-\lambda h=f$.

This lemma is proved by the methods of potential theory, and the proof is analogous to that of Lemma 1.1 in [4].

1.4. *The discrete group and the fundamental domain.*

Let $\Gamma \subset G$ be a discrete subgroup. We fix a point $z_0 \in S$ which is not a fixed point of any transformation $\gamma \in \Gamma$, $\gamma \neq e$, where e is the unit element of G. Further, by the fundamental domain of the group Γ in S we mean the Poincaré fundamental domain, i.e. the following set:

$$F\{z \in S \mid \rho(z, z_0) \leqq \rho(z, \gamma z_0) \quad \text{for any } \gamma \in \Gamma\}.$$

(a) We suppose that Γ has a noncompact F with finite invariant volume $\mu(F) < \infty$.

The work of Selberg [**13**] and of Garland and Raghunathan [**14**] imply the following facts concerning the structure of Γ and F.

1) The number of geodesic rays $z_\alpha(t)$, $t \in R$, such that $z_\alpha(0) = z_0$, $z_\alpha(t) \in F$, is finite for any t, $\alpha = 1, \cdots, m$. (We choose F so that this number is minimal, and we denote it by m.)

2) Let z_α be a limit point of a geodesic $z_\alpha(t)$, $t \to \infty$, $\alpha = 1, \cdots, m$. Let $\Gamma_\alpha = \{\gamma \in \Gamma \mid \gamma z_\alpha = z_\alpha\}$ be a subgroup of Γ. The fundamental domain F can be represented in the form $F = \bigcup_0^m F_\alpha$, where F_0 is a compact set, and F_α $(\alpha \neq 0)$ has the following properties:

$$F_\alpha \ni z_\alpha(t),\ t > t_0;\ F_\alpha \cap F_\beta = \varnothing,\ 0 \neq \alpha \neq \beta \neq 0,\ F_\alpha = g_\alpha \Pi_\alpha,$$

where $g_\alpha \in G$, $g_\alpha \notin \Gamma - e$; the set Π_α has the form $\Pi_\alpha = \{z = \{y, x\} \wedge S \mid y \in (a, \infty), x \in D_\alpha\}$, where $a > 0$ is sufficiently large and D_α is a compact set, $\alpha = 1, \cdots, m$.

3) The group $g_\alpha^{-1}\Gamma_\alpha g_\alpha \subset UM$, and D_α is the fundamental domain of $g_\alpha^{-1}\Gamma_\alpha g_\alpha$ in the hyperplane $y = a$ of S; $\alpha = 1, \cdots, m$.

The sets F_α, $\alpha \neq 0$, are called the parabolic vertices of the domain F. Further, we impose the following condition on the group Γ. Let g_{ij}, $i, j = 0, 1, \cdots, 2n+1$, be the matrix elements of the matrix $g \in G$ (in the case of G_1 and G_2 we have only $i, j = 0, \cdots, n$). We let $\Phi(g)$ denote the following vertex:

$$\Phi(g) = |g_{00} + g_{01} - g_{10} - g_{11}|^2 + |g_{0n+1} + g_{1n+2} - g_{0n+2} - g_{1n+1}|^2.$$

If $g \in g_\alpha^{-1}\Gamma g_\alpha$, $\alpha = 1, \cdots, m$, then we require that the condition $\Phi(g) \neq 0$ imply $\Phi(g) > \epsilon > 0$, where ϵ does not depend on g. We call this condition (b) (we recall that in the case of G_1 and G_2 the second term in the expression for $\Phi(g)$ is absent).

We note that condition (b) follows from condition (a) for $G = SL(2, R)$, which corresponds to the case G_1, $n = 2$, or G_2, $n = 1$, since (b) is equivalent to the Peterson condition (see [**6**]).

LEMMA 1.3. *Let Γ be a Borel arithmetic subgroup of an orthogonal algebraic group for which $G = G_R$ is the set of real points, and let Γ satisfy condition* (a). *Then* (b) *is fulfilled.*

PROOF. Let Γ_Z be the set of integral points of our group. Then, since all the matrix elements of $g \in \Gamma_Z$ are integers, (b) holds. We note that this condition also holds for the discrete sets $g'^{-1}\Gamma_Z g'$ and $g'\Gamma_Z$, where $g' \in G$ is fixed. Let $\Gamma' \subset \Gamma_Z$ be a discrete subgroup of Γ_Z. Condition (b) is obviously fulfilled for Γ'. Since Γ is arithmetic, there exists a subgroup Γ' of finite index in Γ_Z and a finite set $B \subset G$ such that $\Gamma \subset B\Gamma'$. This proves the lemma.

1.5. *Convergence test for a series over a discrete group* (*Selberg*).

Lemma 1.4. *Suppose that the kernel* $k(z,z') = k(u(z,z'))$ *satisfies the following conditions*:

1) $\int_S |k(z,z')|\,d\mu(z') < \infty$ *for any* $z \in S$.

2) *For any points* $z, z' \in S$, *there exist constants* δ, $c > 0$ *such that*

$$|k(z, z')| \leqslant c \int\limits_{u(z', z'')<\delta} |k(z, z'')|\, d\mu(z'').$$

Then the series $\sum_{\gamma\in\Gamma} k(z,\gamma z')$ *converges absolutely and uniformly if* z *and* z' *belong to a compact subset of* S.

Proof. We note that the discreteness of Γ implies that for fixed $\gamma \in \Gamma$ and $z' \in S$ the set $\{z \in S \mid u(\gamma z', z) < \delta\}$ can only intersect a finite number of sets of the same form but with γ' in place of γ. We denote this number by $N(z')$. Then

$$\sum_{\gamma\in\Gamma} |k(z, \gamma z')| \leqslant cN(z') \int\limits_S |k(z, z'')|\, d\mu(z'') < \infty. \tag{1}$$

The invariance of $k(z,z')$ implies that $\int_S |k(z,z'')|\,d\mu(z'')$ does not depend on z in a compact set. In addition, since z and z' play equivalent roles, it follows that (1) does not depend on z'. The proof is complete.

We let $z = z' (\operatorname{mod} \Gamma)$ denote that the equation $z = \gamma z'$ holds for some $\gamma \in \Gamma$; here $z, z' \in S$.

Corollary. *The series* $\sum_{\gamma\in\Gamma} \omega(u(z,\gamma z'), s)$ *converges uniformly on every compact set for* $z \neq z' (\operatorname{mod} \Gamma)$ *and* $\operatorname{Re} s > \alpha$, *where* α *is equal to* $n-1, 2n$ *and* $4n+2$ *for* S_1, S_2 *and* S_3, *respectively*.

The function $\omega(u,s)$ was defined in §1.3. It suffices to verify that the series $\sum_{\gamma\in\Gamma} (1 + u(z,\gamma z'))^{-s/2}$ (with $s/2$ replaced by s in the case of S_1) satisfies the conditions of Lemma 1.4. Condition 1) is verified by a simple computation. To verify the second condition we consider the following inequalities:

$$\int\limits_{u(z', z'')<\delta} (u(z, z'') + 1)^{-\sigma}\, d\mu(z'') \geqslant \int\limits_{u(z', z'')<\delta} (u(z, z') + u(z', z'') + 1)^{-\sigma}\, d\mu(z'')$$
$$\geqslant c'(u(z, z') + 1 + \delta)^{-\sigma},$$

where $\sigma > 0$ is sufficiently large and c' is the measure of the region $u(z, z_0) < \delta$. The constant c must be chosen from the condition

$$c > c' \left(\frac{u(z, z') + \delta + 1}{u(z, z') + 1}\right)^{\sigma}.$$

Obviously, c exists and does not depend on z or z'. The corollary is proved.

1.6. *Some estimates for a series over a discrete group in noncompact regions.*

Lemma 1.5. *Let* Γ' *be a discrete subset of the group* G *such that the series* $\sum_{\gamma\in\Gamma'} (1 + u(z,\gamma z'))^{-s}$ *converges absolutely and uniformly in any subregion of* S *with compact closure for* s *as in the corollary to Lemma* 1.4. *Suppose that the following condition holds for any* $g \in \Gamma'$:

$$\Phi(g) = |g_{00} + g_{01} - g_{10} - g_{11}|^2 + |g_{0n+1} + g_{1n+2} + g_{0n+2} - g_{1n+1}|^2 > \varepsilon_0 > 0,$$

where g_{ij} are the matrix elements of g and ϵ_0 does not depend on g. Then the following estimates in the cases of S_1, S_2 and S_3, respectively, hold uniformly in z and z' in the cylinder $\{y, y' \geqq h > 0; x, x' \in D\}$, where D is a compact set, h is sufficiently large, $z = \{y, x\}$ and $z' = \{y', x'\}$:

$$\sum_{\gamma\in\Gamma'} |(1+u_1(z, \gamma z'))^{-s}| \leqslant c_1 (yy')^{2(n-1)-\mathrm{Re}\, s+\epsilon}, \quad \mathrm{Re}\, s > 2(n-1);$$

$$\sum_{\gamma\in\Gamma'} \left|(1+u_2(z, \gamma z'))^{-\frac{s}{2}}\right| \leqslant c_2 (yy')^{8n-\mathrm{Re}\, s+\epsilon}, \quad \mathrm{Re}\, s > 8n;$$

$$\sum_{\gamma\in\Gamma'} \left|(1+u_3(z, \gamma z'))^{-\frac{s}{2}}\right| \leqslant c^3 (yy')^{8(2n+1)-\mathrm{Re}\, s+\epsilon}, \quad \mathrm{Re}\, s > 8(2n+1);$$

c_1, c_2 and c_3 do not depend on z or z', and $\epsilon > 0$ can be chosen arbitrarily small.

PROOF. Suppose that $g = uak \in \Gamma'$, where $u \in U$, $a \in A_0$, $k \in K$ is the Iwasawa decomposition for g. Let these matrices have the same matrix elements as in §1.1. At $a = a(t)$ we set $e^t = \xi$. A simple computation shows that

$$\Phi(g) = |g_{00} + g_{01} - g_{10} - g_{11}|^2 + |g_{0n+1} + g_{1n+2} - g_{0n+2} - g_{1n+1}|^2$$

is equal to the following in S_1, S_2 and S_3, respectively:

$$\left(\frac{1-k_{11}}{\xi}\right)^2, \quad \frac{\left|e^{i\alpha} - k_{11}e^{-\frac{i\alpha}{n}}\right|^2}{\xi^2}, \quad \frac{|\alpha - a_{11}|^2 + |\beta + b_{11}|^2}{\xi^2}.$$

We define the function $\eta(z, z'; g)$ in each of our cases as follows:

$$\eta_1(z, z'; g) = (1 + u_1(z, gz'))\, 4yy', \quad \eta_2(z, z'; g) = (1 + u_2(z, gz'))\, 4y^2y'^2,$$

$$\eta_3(z, z'; g) = (1 + u_3(z, gz'))\, 4y^2y'^2.$$

Using the action of the matrices in §1.2, we obtain

$$\eta_1(z, z'; g) = \frac{1}{\xi} |(p'(x) + y^2)\, d_1 + p''(x)|\, (u_1(z, gz') + 1)\, 4yy'(k),$$

$$\eta_2(z, z'; g) = \frac{1}{\xi^2} |(p'(x) + y^2)\, d_2 + p''(x)|^2\, (u_2(z, gz') + 1)\, 4y^2y'^2(k),$$

$$\eta_3(z, z'; g) = \frac{1}{\xi^2} |(p'(x) + y^4)\, d_3 + p''(x)|\, (u_3(z, gz') + 1)\, 4y^2y'^2(k),$$

where $p'(x)$ and $p''(x)$ are certain polynomials in x, $d_1 = 1 - k_{11}$, $d_2 = e^{i\alpha} - k_{11}e^{-i\alpha/n}$, $d_3 = |\alpha - a_{11}|^2 + |\beta + b_{11}|^2$, and $y'(k)$ is the y-coordinate of the vector kz'. We are interested in the behavior of $\eta(z, z'; g)$ as $y, y' \to \infty$. It is easy to see that

$$\eta_1(z, z'; g) = (y^2y'^2)\frac{d_1}{\xi} + o(y^2y'^2),$$

$$\eta_2(z, z'; g) = (y^4y'^4)\frac{d_2}{\xi^2} + o(y^4y'^4), \qquad y, y' \to \infty$$

$$\eta_3(z, z'; g) = (y^4y'^4)\frac{d_3}{\xi^2} + o(y^4y'^4).$$

Suppose that z, z', z_0 and z_0' lie in the cylinder indicated and that $z_0 = \{y_0, x\}$, $z_0' = \{y_0, x'\}$; y_0 is fixed. It is obvious that for sufficiently large $h > 0$ we have the inequalities

$$\eta_1(z, z'; g) \geqslant \eta_1(z_0, z_0'; g) + y^2 y'^2 \frac{d_1}{2\xi},$$

$$\eta_2(z, z'; g) \geqslant \eta_2(z_0, z_0'; g) + y^4 y'^4 \frac{d_2}{2\xi^2},$$

$$\eta_3(z, z'; g) \geqslant \eta_3(z_0, z_0'; g) + y^4 y'^4 \frac{d_3}{2\xi^2}.$$

We further limit ourselves to the case $\eta_1(z, z'; g)$, since the other cases are handled analogously. We have

$$\begin{aligned} &4yy'(1 + u_1(z, gz')) + 4y_0 y_0'(1 + u_1(z_0, gz_0')) + \frac{y^2 y'^2}{2\xi} d_1 \\ &\qquad \geqslant 4y_0 y_0'(1 + u_1(z_0, gz_0')) + \frac{y^2 y'^2}{2} \varepsilon_0. \end{aligned} \tag{2}$$

Now, using the inequality $\delta^2 a^2/2 + b^2/\delta^2 2 \geqq ab$, we remove the factor yy' in the left side of the inequality by choosing $\delta > 0$ sufficiently small. Finally, using the inequality $a^p/p + b^{p'}/p' \geqq ab$ and setting $1/p + 1/p' = 1$ for

$$p = \frac{1}{\left(n - 1 + \frac{\varepsilon}{2}\right) \frac{2}{\operatorname{Re} s}} > 1$$

with $\operatorname{Re} s > 2(n-1)$ in the right side of (2), we obtain

$$(1 + u(z, gz'))^{-\operatorname{Re} s} \leqslant c\,(yy')^{2(n-1) - \operatorname{Re} s + \varepsilon} (1 + u(z_0, gz_0))^{-n+1-\frac{\varepsilon}{2}}.$$

It remains to perform the summation and note that the series

$$\sum_{g \in \Gamma'} (1 + u(z_0, gz_0'))^{-n+1-\epsilon/2}$$

converges absolutely by the assumptions of the lemma. Using analogous reasoning, we also obtained the desired estimates in the other cases. The proof is complete.

Remark. We make the following change of variables and spectral parameter for S_1, S_2 and S_3:

$$\text{for } S_1\colon \quad y' = y^{n-1}, \quad s' = \frac{s}{n-1}, \qquad s'' = s';$$

$$\text{for } S_2\colon \quad y' = y^{2n}, \quad s' = \frac{s}{2n}, \qquad s''(1 - s'') = \frac{1}{4} s'(1 - s');$$

$$\text{for } S_3\colon \quad y' = y^{4n+2}, \quad s' = \frac{s}{4n+2}, \quad s''(1 - s'') = \frac{1}{4} s'(1 - s').$$

After this substitution in all the spaces, the measure equals $d\mu(z') = c\,dy'dx/y'^2$. We set $c = c_0^2$, where $1/c_0 = n-1, 2n$ and $4n+2$ in S_1, S_2 and S_3. The parameter λ in the equation $-L\omega - \lambda\omega = f$ becomes equal to $s''(1 - s'')/c_0^2$. Lemma 1.2 holds for $\operatorname{Re} s'' > 1$ and Lemma 1.5 holds for $\operatorname{Re} s'' > 2$. We further designate $s = s''$ and $y = y'$.

1.7. *Function spaces.*

We fix the fundamental domain F. Let $f(z)$ and $k(z, z')$ be given on F and $F \times F$, respectively. We introduce the following notation: if $F = \bigcup_0^m F_\alpha$, $F_\alpha = g_\alpha \Pi_\alpha$, $\alpha = 1, \cdots, m$, where F_0 is a compact set and F_α are the parabolic vertices, then $f_0(z)=f(z)$, $z\in F_0$; $f_2(z)=f(g_\alpha z)$, $z\in\Pi_\alpha$, $\alpha=1,\cdots,m$. We similarly define the functions $k_{\alpha\beta}(z,z')$, $\alpha,\beta=0,\cdots,m$. We shall call them the components of the function or the kernel relative to the partition $F=\bigcup_0^m F_\alpha$. We shall consider two types of function spaces on the fundamental domain.

We let $\mathfrak{H}=L_2(F,u)$ denote the Hilbert space of complex-valued functions on F which is obtained as the completion in the norm $(\int_F|f(z)|^2d\mu(z))^{1/2}$ of the set of restrictions to F of smooth functions f bounded on S such that $f(z)=f(\gamma z)$ for any $\gamma\in\Gamma$. The scalar product in $\mathfrak{H}$ is given by the formula $(f,h)=\int_F f(z)\overline{h(z)}d\mu(z)$, where $f,h\in\mathfrak{H}$.

We let $\mathfrak{B}_\mu$ denote the Banach space of complex-valued functions f which are continuous on S and such that $f(z)=f(\gamma z)$ for any $\gamma\in\Gamma$, where we consider them on F and require, in addition, that their components satisfy the condition $|f_\alpha(z)|\leqq cy^\mu$ for $\alpha=1,\cdots,m$, where $\mu\in R$ and c is a constant. The norm in this space is given by the expression

$$\|f\|_\mu=\max_{z\in F_0} f_0(z)+\sum_{\alpha=1}^{m}\sup_{z\in\Pi_\alpha} y^{-\mu}|f_\alpha(z)|.$$

In addition, we shall consider certain function spaces on the hyperplane $y=\text{const}$ of S which are analogous to $\mathfrak{H}$. These spaces will be defined in §2.

§2. The operator $\mathfrak{A}$ and its resolvent

In this section we investigate the operator $\mathfrak{A}$ which is induced in $\mathfrak{H}$ by the differential operator $-L$.

We consider the set $\mathfrak{D}$ of infinitely differentiable functions $f(z)$ which are automorphic relative to Γ, bounded on S, and such that $Lf(z)$ is bounded on S. If we fix F, then the set of restrictions of functions $f\in\mathfrak{D}$ to F is dense in $\mathfrak{H}$. We define the operator $\widetilde{\mathfrak{A}}$ by the formula

$$\widetilde{\mathfrak{A}}f=-Lf,$$

where $f\in\mathfrak{D}$ on F.

LEMMA 2.1. *The operator $\widetilde{\mathfrak{A}}$ is symmetric and nonnegative.*

PROOF. Let $f,g\in\mathfrak{D}$. We consider sequences $f_j(z)$ and $g_k(z)$, $j,k=1,2,\cdots$, of smooth automorphic functions on S such that their restrictions to F have compact support and, in addition, $f_j(z)\to f(z)$ as $j\to\infty$, $g_k(z)\to g(z)$ as $k\to\infty$ almost everywhere on F. Using Green's formula for L, the automorphicity and finiteness of $f_j(z)$ and $g_k(z)$ on F, and the formal selfadjointness of L, we obtain

$$-\int_F (Lf_j(z))\overline{g_k(z)}\,d\mu(z)=-\int_F f_j(z)\overline{(Lg_k(z))}\,d\mu(z).$$

Let $j\to\infty$. Obviously the limit in the right side exists, and the equality is preserved with f_j replaced by f. We similarly pass to the limit as $k\to\infty$. The nonnegativity of $\widetilde{\mathfrak{A}}$ is proved in the same way, and follows from the nonnegativity of the quadratic form of the operator $-L$, which, in turn, follows because the Riemann metric is positive definite. The proof is complete.

We let $\mathfrak{A}$ denote the closure of $\widetilde{\mathfrak{A}}$ in $\mathfrak{H}$.

We consider the series $\sum_{\gamma\in\Gamma}\omega(u(z,\gamma z'),s)$, where the kernel $\omega(u,s)$ was introduced in Lemma 1.2.

If we assume that $\operatorname{Re} s>1$, then by Lemma 1.4 this series converges. We denote $r(z,z';s)=\sum_{\gamma\in\Gamma}\omega(u(z,\gamma z'),s)$, and we investigate its components. We suppose that $\operatorname{Re} s>2$. It then follows from Lemmas 1.4 and 1.5 that $r_{00}(z,z';s)$ is a continuous function on $F_0\times F_0$ for the z and z' indicated, and $r_{00}(z,z';s)$ has a weak singularity as $z\to z'$. We note that the discrete sets Γg_α, $g_\alpha^{-1}\Gamma$ and $g_\alpha^{-1}\Gamma g_\beta$ satisfy condition (b) for $\alpha\neq\beta$. Consequently

$$|r_{0\alpha}(z,z';s)|=\left|\sum_{\gamma\in\Gamma g_\alpha}\omega(u(z,\gamma z'),s)\right|\leqslant cy'^{2+\varepsilon-\operatorname{Re}s},$$

$$|r_{\alpha 0}(z,z';s)|=\left|\sum_{\gamma\in g_\alpha^{-1}\Gamma}\omega(u(z,\gamma z'),s)\right|\leqslant cy^{2+\varepsilon-\operatorname{Re}s},$$

$$|r_{\alpha\beta}(z,z';s)|=\left|\sum_{\gamma\in g_\alpha^{-1}\Gamma g_\beta}\omega(u(z,\gamma z'),s)\right|\leqslant c(yy')^{2+\varepsilon-\operatorname{Re}s},$$

where $\alpha\neq\beta$, $\alpha,\beta=1,\cdots,m$; $c=c(s)$ does not depend on z or z'; and $\epsilon>0$ may be chosen arbitrarily small. We investigate the component $r_{\alpha\alpha}(z,z';s)$, $\alpha=1,\cdots,m$. It is easy to see that

$$\sum_{\substack{\gamma\in g_\alpha^{-1}\Gamma g_\alpha\\ \Phi(\gamma)=0}}\omega(u(z,\gamma z'),s)=\sum_{\gamma\in g_\alpha^{-1}\Gamma_\alpha g_\alpha}\omega(u(z,\gamma z'),s).$$

We let $w_\alpha(z,z';s)$ denote the sum on the right, and we let $v_\alpha(z,z';s)$ denote the difference:

$$v_\alpha(z,z';s)=\sum_{\gamma\in g_\alpha^{-1}\Gamma g_\alpha,\ \Phi(\gamma)\neq 0}\omega(u(t,\gamma z'),s)=r_{\alpha\alpha}(z,z';s)-w_\alpha(z,z';s).$$

The kernel $v_\alpha(z,z';s)$, $\alpha=1,\cdots,m$, is estimated using Lemma 1.5 as follows:

$$|v_\alpha(z,z';s)|\leqq c(yy')^{2+\epsilon-\operatorname{Re}s}.$$

To investigate the kernel $w_\alpha(z,z';s)$, we note that it is the Green's function of the problem in Lemma 1.2 for functions $f(z)$ which satisfy the additional automorphicity condition $f(z)=f(\gamma z)$, $\gamma\in g_\alpha^{-1}\Gamma_\alpha g_\alpha$.

We consider the ordinary differential equation

$$-\varphi''+(\mu_k^2c_0^2y^{4c_0-2}+\nu_j^2c_0^2y^{2c_0-2}-s(1-s)y^{-2})\varphi=f,$$

where $\varphi=\varphi(y)$, $f=f(y)$, $y\in(0,\infty)$; $\mu_k,\nu_j\geqq 0$; c_0 is the constant in the remark to Lemma 1.5. We construct the Green's function for this one-dimensional problem. We consider the case $\mu_k\neq 0$, $\nu_j\neq 0$; this equation reduces by a change of variables to Whittaker's equation (see [**12**]). The desired Green's function is

$$m^{k,j}(y',y';s)=\frac{(yy')^{\frac{1}{2}-c_0}}{p}\begin{cases}W_{l,l'}(\mu_k y^{2c_0})W_{-l,l'}(-\mu_k y'^{2c_0}), & y>y',\\ W_{l,l'}(\mu_k y'^{2c_0})W_{-l,l'}(-\mu_k y^{2c_0}), & y<y',\end{cases}$$

where l and l' depend on μ_k, ν_j, s and c_0, while p is the corresponding Wronskian. In the case when one of the μ_k or ν_j equals zero, the equation degenerates to a Bessel type equation. For example, if $\mu_k=0$, then

$$m^{0,j}(y, y'; s) = \frac{\sqrt{yy'}}{p} \begin{cases} H_l(i\nu_j y^{c_0}) J_l(i\nu_j y'^{c_0}), & y > y', \\ H_l(i\nu_j y'^{c_0}) J_l(i\nu_j y^{c_0}), & y < y'. \end{cases}$$

An analogous formula holds for $m^{k,0}(y, y'; s)$. Using the well-known asymptotic behavior of the Whittaker and Bessel-Hankel functions for large y and y', we obtain the estimates

$$|m^{k,j}(y, y'; s)| \leqslant c \frac{\sqrt{yy'}}{\mu_k} e^{-|y^{c_0} - y'^{c_0}|\mu_k},$$

$$|m^{k,0}(y, y'; s)| \leqslant c \frac{\sqrt{yy'}}{\mu_k} e^{-|y^{c_0} - y'^{c_0}|\mu_k},$$

$$|m^{0,j}(y, y'; s)| \leqslant c \frac{\sqrt{yy'}}{\nu_j} e^{-|y^{c_0} - y'^{c_0}|\nu_j},$$

where c does not depend on y, y', μ_k or ν_j.

Finally, if $\mu_k = \nu_j = 0$, then

$$m^{0,0}(y, y'; s) = \frac{1}{2S-1} \begin{cases} y^s y'^{1-s}, & y < y', \\ y^{1-s} y'^s, & y > y'. \end{cases}$$

We now return to our investigation of the Green's function for the problem $-Lu - \lambda u = f$ in the region $y > a$, where $u(z)$ and $f(z)$ are smooth functions which are bounded on S and are automorphic relative to $g_\alpha^{-1}\Gamma_\alpha g_\alpha$; α is fixed, $\alpha = 1, \cdots, m$. It is clear from the definition of the space S that for $z \in S$ we have $z = \{y, x\}$, $y \in (0, \infty)$, $x \in X$, where X is a Riemann space with group UM of isometric transformations. The Laplace-Beltrami operator on this space is the sum $L' + L''$, where the operators L' and L'' were defined in §1.3, and the invariant measure is Lebesgue measure. It follows from the definition of L' and L'' that $L'L'' = L''L'$; in addition, it follows from Lemma 1.1 that L' and L'' are invariant relative to the group UM. The group $g_\alpha^{-1}\Gamma_\alpha g_\alpha \subset UM$ has a compact fundamental domain D_α in X. We let $\mathfrak{H}_\alpha$ denote the Hilbert space of real functions $f(x)$, $x \in D_\alpha$, which is defined in analogy with the space $\mathfrak{H}$, but for the group $g_\alpha^{-1}\Gamma_\alpha g_\alpha$ on X; we let $\mathfrak{A}'_\alpha$ and $\mathfrak{A}''_\alpha$ denote the nonnegative selfadjoint operators induced in $\mathfrak{H}_\alpha$ by the operators $-L'$ and $-L''$, respectively. It is well known that $\mathfrak{A}'_\alpha$ and $\mathfrak{A}''_\alpha$ only have a discrete spectrum, and that, in addition, there exists a complete orthonormal system of functions $v_{kj}(x)$, $k, j = 0, 1, \cdots$, in $\mathfrak{H}_\alpha$ such that

(3) $$\mathfrak{A}'_\alpha v_{kj}(x) = \mu_k^2 v_{kj}(x), \quad \mathfrak{A}''_\alpha v_{kj}(x) = \nu_j^2 v_{kj}(x), \quad \mu_k \geqq 0, \quad \nu_j \geqq 0,$$

where the eigenfunction $v_{00}(x)$ for $\mu_0 = 0$ and $\nu_0 = 0$ is unique and equal to a constant.

We consider the series

(4) $$\sum_{\mu_k, \nu_j} m^{k,j}(y, y'; s) v_{kj}(x) v_{kj}(x'),$$

where μ_k and ν_j run through the entire spectrum of $\mathfrak{A}'_\alpha$ and $\mathfrak{A}''_\alpha$, counting multiplicity. We show that this series converges for $y \neq y'$ and coincides with the Green's function $w_\alpha(z, z'; s)$. It is easy to see that the resolvents of the operators $\mathfrak{A}'_\alpha$ and $\mathfrak{A}''_\alpha$ are integral operators with a weak singularity.

Using standard methods from the theory of integral equations with weak singularity, we find that the following series converges:

$$\sum_{\mu_k} \frac{1}{\mu_k^{\dim S-1+\epsilon}} = \sum_{k=0}^{\infty} n_k \mu_k^{-(\dim S-1+\epsilon)} < +\infty,$$

where $\dim S$ is the dimension of S, n_k is the multiplicity of the eigenvalue μ_k^2, and $\epsilon > 0$. We consider the series

$$\sum_{k=0}^{\infty} \frac{n_k}{\mu_k} e^{-\mu_k |y^{2c_0} - y'^{2c_0}|}.$$

Using our earlier estimate and summation by parts, we obtain

$$\sum_{k=0}^{\infty} \frac{n_k}{\mu_k} e^{-\mu_k |y^{2c_0}-y'^{2c_0}|} \leqslant c \frac{1}{|y^{2c_0} - y'^{2c_0}|^{\dim S-2+\epsilon}}. \tag{5}$$

In particular, this implies that (4) converges.

We now show that

$$w_\alpha(z, z'; s) = \sum_{\mu_k, \nu_j} m^{k,j}(y, y'; s)\, v_{kj}(x)\, v_{kj}(x').$$

To do this, we note that the elements g_α, $\alpha = 1, \cdots, m$, in §1.4 may be chosen so that $|D_\alpha|$, the Lebesgue measure of D_α, is equal to one. After making this choice, we normalize the system $v_{kj}(x)$ as follows:

$$\int_{D_\alpha} v_{kj}(x)\, v_{me}(x)\, dx = \delta_{km}\delta_{je},$$

so that, in particular, $v_{00}(x) = 1$.

In the equation $-Lu - \lambda u = f$ we represent $u(z)$ by a Fourier series $u(z) = \sum_{\mu_k, \nu_j} \varphi_{kj}(y)\, v_{kj}(x)$, where $\varphi_{kj}(y)$ and $L\varphi_{kj}(y)$ are bounded functions on $(0, \infty)$. This expansion holds, for example, for automorphic functions $u(z)$ (relative to $g_\alpha^{-1}\Gamma_\alpha g_\alpha$), and $f(z) = u(z) = 0$ for large y.

We have

$$\int_{\Pi_\alpha} d\mu(z) \Big\{ \sum_{\mu_k, \nu_j} v_{kj}(x)\, v_{kj}(x')\, m^{k,j}(y, y'; s) \Big\} f(z) = \int_{u(z,z')<\delta} + \int_{u(z,z')\geqslant\delta},$$

where $\delta > 0$ and $u(z, z')$ is the invariant of the pair of points.

We investigate the second integral

$$\int_{u(z,z')\geqslant\delta} \Big(\sum_{\mu_k, \nu_j} v_{kj}(x)\, v_{kj}(x')\, m^{k,j}(y, y'; s) \Big) f(z)\, d\mu(z)$$

$$= \int_{u(z,z')\geqslant\delta} \Big(\sum_{\mu_k, \nu_j} v_{kj}(x)\, v_{kj}(x')\, m^{k,j}(y, y'; s) \Big) \Big(-Lu(z) - \frac{s(1-s)}{c_0^2} u(z) \Big) d\mu(z).$$

Replacing $u(z)$ by its Fourier expansion and using the absolute convergence of the series and the integral, along with the properties of the system $v_{kj}(x)$, we obtain

$$\int_{u(z,z')\geqslant\delta} \Big(\sum_{\mu_k, \nu_j} v_{kj}(x)\, v_{kj}(x')\, m^{k,j}(y, y'; s) \Big) f(z)\, d\mu(z) =$$

$$= \sum_{\mu_k, \nu_j} v_{kj}(x') \int_{|y^{2c_0}-y'^{2c_0}|\geqslant\delta'} dy\, m^{k,j}(y, y'; s)\{-\varphi''_{kj}(y)$$
$$+ (\mu_k^2 c_0^2 y^{4c_0-2} + \nu_j^2 c_0^2 y^{2c_0-2} - s(1-s)y^{-2})\varphi_{kj}(y)\}, \quad \delta' = O(\delta).$$

It is easy to see that as $\delta \to 0$ this integral has a limit $\sum_{\mu_k,\nu_j} \varphi_{kj}(y')v_{kj}(x') = u(z')$. This follows from the definition of the Green's function $m^{k,j}(y,y';s)$.

We estimate the integrand in the first integral. It follows from (5) that

$$\left|\sum_{\mu_k, \nu_j} v_{kj}(x)v_{kj}(x')m^{k,j}(y, y'; s)\right| \leqslant cu(z, z')^{-\frac{\dim S-2+\varepsilon}{2}}.$$

The integrand is summable, and consequently the first integral disappears as $\delta \to 0$. It remains to note that our equality holds for any bounded functions $f(z)$ which are automorphic relative to $g_\alpha^{-1}\Gamma_\alpha g_\alpha$ and such that $Lf(z)$ is bounded on S. We have thereby proved that the series (4) gives the desired Green's function. It follows by uniqueness that

$$w_\alpha(z, z'; s) = \sum_{\mu_k, \nu_j} v_{kj}(x)v_{kj}(x')m^{k,j}(y, y'; s).$$

As $z \to z'$ the asymptotic behavior of $w_\alpha(z,z';s)$ does not depend on ϵ and is given by $u(z,z')^{-(\dim S-2)/2}$. We have proved the lemma.

Lemma 2.2. *The kernel $w_\alpha(z,z';s)$ admits a separation of variables in the cylinder Π_α:*

$$w_\alpha(z, z'; s) = \sum_{\mu_k, \nu_j} v_{kj}(x)v_{kj}(x')m^{k,j}(y, y'; s),$$

where the system $v_{kj}(x)$ is defined by (3), *and the $m^{k,j}(y,y';s)$ are the Green's functions of the one-dimensional problem*

$$-\varphi'' + (\mu_k^2 c_0^2 y^{4c_0-2} + \nu_j^2 c_0^2 y^{2c_0-2} - s(1-s)y^{-2})\varphi = f$$

on the semi-axis $(0, \infty)$.

These results allow us to split up the kernel $r(z,z';s)$ as follows: $r(z,z';s) = t(z,z';s) + m(z,z';s) + n(z,z';s)$, where the kernels $t(z,z';s)$ and $m(z,z';s)$ are diagonal, i.e. their only nonzero components are $t_{\alpha\alpha}(z,z';s)$ and $m_{\alpha\alpha}(z,z';s)$, $\alpha = 1, \cdots, m$, and

$$t_{\alpha\alpha}(z, z'; s) = m^{0,0}(y, y'; s), \quad m_{\alpha\alpha}(z, z'; s) = w_\alpha(z, z'; s) - m^{0,0}(y, y'; s).$$

Further, by definition, $n_{00}(z,z';s) = r_{00}(z,z';s)$, $n_{0\alpha}(z,z';s) = r_{0\alpha}(z,z';s)$, $n_{\alpha 0}(z,z';s) = r_{\alpha 0}(z,z';s)$, $n_{\alpha\beta}(z,z';s) = r_{\alpha\beta}(z,z';s)$, $n_{\alpha\alpha}(z,z';s) = v_\alpha(z,z';s)$; $\alpha = 1, \cdots, m$, $\beta = 1, \cdots, m$; $\alpha \neq \beta$.

The kernels $t(z,z';s)$, $m(z,z';s)$ and $n(z,z';s)$ are symmetric and satisfy an equation of the type $t(z,z';s) = \overline{t(z,z';s)}$.

The following results are obtained from Lemma 2.2 and are proved in the same way as the corresponding results in [**4**].

Lemma 2.3. 1) *The operator $N(s)$ with kernel $n(z,z';s)$ is completely continuous in $\mathfrak{H}$ for* $\operatorname{Re} s > 2$.

2) *The operator* $M(s)$ *with kernel* $m(z, z'; s)$ *is completely continuous in* $\mathfrak{H}$ *for* $\operatorname{Re} s > 2$.

3) *The operator* $T(s)$ *with kernel* $t(z, z'; s)$ *is bounded in* $\mathfrak{H}$ *for* $\operatorname{Re} s > \frac{1}{2}$.

We let $R(s)$ denote the integral operator with kernel $r(z, z'; s)$, $\operatorname{Re} s > 2$.

LEMMA 2.4. *The operator* $\mathfrak{A}$ *is selfadjoint in* $\mathfrak{H}$, *and its resolvent at the point* $\lambda = s(1-s)/c_0^2$ *coincides with* $R(s)$ *for* $\operatorname{Re} s > 2$.

We further let $R(s)$ denote the resolvent of $\mathfrak{A}$ at any regular point $\lambda = s(1-s)/c_0^2$.

§3. Integral equation for the kernel of the resolvent and an expansion theorem

In order to investigate the behavior of the resolvent $R(s)$ in a neighborhood of the spectrum, we construct an integral equation for the kernel $r(z, z'; s)$ which is a modification of the equation in [4] for the case of the upper half-plane. Let $\kappa > 2$ be a sufficiently large positive number. We let $\omega(s)$ denote $\{s(1-s) - \kappa(1-\kappa)\}/c_0^2$. We consider Hilbert's identity for the resolvent $R(s) = R(\kappa) + \omega(s) R(s) R(\kappa)$. There exists an integral operator (see [4]) $Q(s)$ which depends analytically on s, is bounded in $\mathfrak{H}$ for $\operatorname{Re} s > \frac{1}{2}$, and is such that the following operator equation holds in the space $\mathfrak{B}_{-1}$ for $0 < \operatorname{Re} s < 2 + \epsilon$, $\epsilon > 0$, and in $\mathfrak{H}$ for $\operatorname{Re} s > \frac{1}{2}$:

$$E = (E - \omega(s) T(\kappa))(E + \omega(s) Q(s)),$$

where E is the identity operator. The operator $Q(s)$ is given by a kernel $q(z, z'; s)$, $z, z' \in F$. The only nonzero components of $q(z, z'; s)$ are the diagonal components

$$q_{\alpha\alpha}(z, z'; s) = q(y, y'; s) = \frac{1}{2s-1} \begin{cases} \varphi(y, s) y'^{1-s}, & y < y', \\ \varphi(y', s) y^{1-s}, & y > y', \end{cases} \quad \alpha = 1, \ldots, m,$$

where

$$\varphi(y, s) = y^s + a^{2s-1} \frac{s-\kappa}{s+\kappa-1} y^{1-s}.$$

It follows from [4] that the $\varphi(y, s)$ for $s = \frac{1}{2} + i\tau$, $\tau \in R$, form a complete system of eigenfunctions for the continuous spectrum of a certain one-dimensional selfadjoint problem $\mathfrak{A}_1$ in the space of functions on the semi-axis $a \leqq y < \infty$ which are square integrable with respect to the measure $c_0^2 dy^2/y^2$, and $q(y, y'; s)$ is the kernel of the resolvent of this problem at the point $\lambda = s(1-s)/c_0^2$ for $\operatorname{Re} s > \frac{1}{2}$.

Returning to our investigation of the resolvent $R(s)$, we let V denote the sum $M(\kappa) + N(\kappa)$. From Hilbert's identity and the properties of the operator $Q(s)$ we obtain

$$R(s) = (E + \omega(s) Q(s)) R(\kappa) + \omega(s)(E + \omega(s) Q(s) V R(s)).$$

This identity may be considered as an equation for $R(s)$. We introduce the operator $B(s)$ by the following formula:

$$R(s) = Q(s) + (E + \omega(s) Q(s)) B(s) (E + \omega(s) Q(s)).$$

The operator $B(s)$ is uniquely determined by the resolvent $R(s)$. The equation for $B(s)$, which is equivalent to the equation for $R(s)$, has the form

$$B(s) = V + \omega(s) V (E + \omega(s) Q(s)) B(s). \tag{6}$$

We investigate this equation.

Lemma 3.1. *The operator* $H(s) = V(E + \omega(s) Q(s))$ *is defined in* $\mathfrak{B}_{-1}$ *for any* s, $0 < \operatorname{Re} s < 2 + \epsilon$, $\epsilon > 0$, *is completely continuous in* $\mathfrak{B}_{-1}$, *and depends analytically on* s *in the indicated strip.*

Proof. We note that if $\kappa > 2$ the operator $N(\kappa)$ is defined in $\mathfrak{B}_{-1}$ and takes this space to the space $\mathfrak{B}_{2-\kappa+\epsilon_1}$, $\epsilon_1 > 0$. It follows from Lemma 2.2 and the estimate $\int_a^\infty e^{-\alpha|y-y'|} y'^p dy' \leqq c y^p$ that the operator $M(\kappa)$ is defined in $\mathfrak{B}_{-1}$ and takes this space to $\mathfrak{B}_{-2}$. In addition, the operators $Q(s)$ and $M(\kappa)$ are orthogonal, i.e. $Q(s) M(\kappa) = M(\kappa) Q(s) = 0$. Thus, for $\kappa > 2$ sufficiently large, $H(s)$ is defined in $\mathfrak{B}_{-1}$, depends analytically on s in the strip $0 < \operatorname{Re} s < 2 + \epsilon$ and takes $\mathfrak{B}_{-1}$ to $\mathfrak{B}_{-1-\delta}$, where $\delta > 0$. It remains to note that the restricted set of equicontinuous functions in $\mathfrak{B}_{1-\delta}$ is compact in $\mathfrak{B}_{-1}$, and consequently the operator $H(s)$ is completely continuous in $\mathfrak{B}_{-1}$ for the indicated values of s.

The proof is complete.

We consider the homogeneous equation

$$v = \omega(s) H(s) v. \tag{7}$$

Lemma 3.2. *If* $s \neq \frac{1}{2}$, $\operatorname{Re} s \geqq \frac{1}{2}$, *is a singular point of the equation* (7), *i.e. a value of* s *for which* (7) *has a nontrivial solution* $v \in \mathfrak{B}_{-1}$, *then* $\lambda = s(1-s)/c_0^2$ *is an eigenvalue of the operator* $\mathfrak{A}$. (*We note that the singular points of* (7) *are situated discretely.*)

Lemma 3.3. *If* $\psi \in \mathfrak{H}$ *is such that* $\mathfrak{A}\psi = \lambda\psi$, $\lambda \geqq 0$, *then the function* $v = (E = \omega(s) T(\kappa))\psi \in \mathfrak{B}_{-1}$ *and satisfies equation* (7), *where* s *is chosen by the condition* $\lambda = s(1-s)/c_0^2$, $\operatorname{Re} s \geqq \frac{1}{2}$.

These lemmas are proved in the same way as the corresponding lemmas in [4].

Lemma 3.4. *There exists a solution of equation* (6) *in the strip* $0 < \operatorname{Re} s < 2 + \epsilon$, $\epsilon > 0$, *which is representable in the form* $B(s) = M'(s) + N'(s)$, *where* $M'(s)$ *and* $N'(s)$ *are certain operators specified below, and* $B(s)$ *is bounded in the space* $\mathfrak{B}_{-1}$ *and depends meromorphically on* s *in the strip.*

Proof. We perform the following transformations, after which our equation becomes an equation with a free term having a kernel whose components are continuous functions. Let l be a natural number such that the kernel of $H^l(s) V$ has continuous components. This will be fulfilled for sufficiently large l, namely for $l \geqq [\dim S/2] + 1$, where $\dim S$ is the dimension of S, since the operator V only has weak singularity.

We set

$$\begin{aligned} B_1(s) &= B(s) - V, \\ B_2(s) &= B_1(s) - \omega(s) H(s) V, \\ &\vdots \\ B_l(s) &= B_{l-1}(s) - \omega^{l-1}(s) H^{l-1}(s) V. \end{aligned}$$

Obviously, $B_l(s) = B(s) - V - \omega(s)H(s)V - \cdots - \omega^{l-1}(s)H^{l-1}(s)V$ satisfies the equation $B_l(s) = \omega^l(s)H^l(s)V + \omega(s)H(s)B_l(s)$, whose free term $\widetilde{B}(s) = \omega^l(s)H^l(s)V$ with kernel $\widetilde{b}(z,z';s)$ has continuous components, and, in addition $|\widetilde{b}_{\alpha\beta}(z,z';s)| \leq c(yy')^{-1}$. We write the equation for $B_l(s)$ in terms of kernels:

$$b_l(z, z'; s) = \widetilde{b}(z, z'; s) + \omega(s)\int_F h(z, z''; s)\, b_l(z'', z'; s)\, d\mu(z''),$$

where $h(z,z';s)$ is the kernel of the operator $H(s)$. For fixed z this equation may be considered an equation with completely continuous operator in the space $\mathfrak{B}_{-1}$ depending analytically on s. Consequently the desired solution exists as a meromorphic function and is connected with the solution to equation (6) by the formula $B(s) = B_l(s) + V + \omega(s)H(s)V + \cdots + \omega^{l-1}(s)H^{l-1}(s)V$. Further, combining the terms in the right side, we find that $B(s) = M'(s) + N'(s)$, where we have let $M'(s)$ denote the polynomial $M'(s) = M(\kappa) + \omega(s)M^2(\kappa) + \cdots + \omega^{l-1}(s)M^l(\kappa)$. It is obvious that $M'(s)$ depends analytically on s and $N'(s)$ depends meromorphically on s, where its kernel satisfies the estimate $|n'_{\alpha\beta}(z,z';s)| \leq c(yy')^{-1}$. The proof is complete.

These results imply the following theorem.

Theorem 3.1. *For $z \neq z'$ the kernel of the resolvent $r(z,z';s)$ is a meromorphic function of s on the entire s-plane, with singularities not depending on z or z'. The poles for which $\operatorname{Re} s \geq \frac{1}{2}$ are all located on the line $\operatorname{Re} s = \frac{1}{2}$ or in the interval $\frac{1}{2} \leq s \leq 1$ of the real axis. All of these poles, with the possible exception of the pole at the point $s = \frac{1}{2}$, are simple poles.*

Before formulating the expansion theorem, we construct a system of eigenfunctions for the continuous spectrum of the operator $\mathfrak{A}$. We consider the set of functions

$$\psi^\beta(z, s) = (E + \omega(s)(E + \omega(s)Q(s))B(s))\varphi^\beta(z, s), \quad \beta = 1, \ldots, m,$$

where $\varphi^\beta_\alpha(z,s) = \delta_{\alpha\beta}\varphi(y,s)$, $\alpha = 1, \cdots, m$, $\varphi^\beta_0(z,s) = 0$, and $\varphi(y,s)$ is from the definition of the operator $Q(s)$.

Lemma 3.5. 1) *The functions $\psi^\alpha(z,s)$ are meromorphic on the entire s-plane, with singularities not depending on z, and with poles located in the region $\operatorname{Re} s < \frac{1}{2}$ and in the interval $\frac{1}{2} \leq s \leq 1$. These functions are analytic in a neighborhood of the line $\operatorname{Re} s = \frac{1}{2}$, with the possible exception of the point $s = \frac{1}{2}$. In addition, the functions $\psi^\alpha(z,s) \in \mathfrak{B}_\mu$, $\mu = \max(\operatorname{Re} s, 1 - \operatorname{Re} s)$, and satisfy the equation $R(\kappa)\psi^\alpha(z,s) = (1/\omega(s))\psi^\alpha(z,s)$, $\alpha = 1, \cdots, m$.*

2) *The functions $\psi^\alpha(z,s)$ have the following asymptotic behavior in a neighborhood of the parabolic vertices:*

$$\psi^\alpha_\beta(z, s) = y^s\delta_{\alpha\beta} + S_{\alpha\beta}(s)\, y^{1-s} + O(1) \quad (y \to \infty),$$

where

$$S_{\alpha\beta}(s) = a^{2s-1}\frac{s - \varkappa}{s + \varkappa - 1}\delta_{\alpha\beta} + \frac{\omega^2(s)}{2s - 1}\int_F\int_F \varphi^\alpha(z, s)\, n'(z, z'; s)\, \varphi^\beta(z', s)\, d\mu(z)\, d\mu(z'),$$

and $n'(z,z';s)$ is the kernel of the operator $N'(s)$. The matrix $S(s) = \{S_{\alpha\beta}(s)\}_{\alpha,\beta=1}^m$ is symmetric, is meromorphic on the s-plane with poles in the region $\operatorname{Re} s < \frac{1}{2}$ and in the interval $\frac{1}{2} \leq s \leq 1$, and is analytic in a neighborhood of the line $\operatorname{Re} s = \frac{1}{2}$, with the possible exception of the point $s = \frac{1}{2}$.

3) *The following functional equations hold*:

$$\sum_{\alpha=1}^{m} S_{\gamma\alpha}(s)\, S_{\alpha\beta}(1-s)=\delta_{\gamma\beta},\quad \psi^{\beta}(z,\ s)=\sum_{\alpha=1}^{m}\psi^{\alpha}(z,\ 1-s)\, S_{\alpha\beta}(s).$$

Proof. The assertions of the lemma are proved in the same way as the corresponding assertions in [4] and follow from the definition of the functions $\psi^{\beta}(z,s)$ and the properties of the operator $B(s)$. We shall prove 3). Hilbert's identity for $R(s)$ implies

$$B(s)-B(s')=\frac{1}{c_0^2}(s(1-s)-s'(1-s'))B(s)(E+\omega(s)Q(s))(E+\omega(s')Q(s'))B(s'),$$

which is valid as an operator equation for any nonsingular values of s and s'. In particular, for $s'=1-s$

$$B(s)-B(1-s)=\omega^2(s)\, B(s)\,(Q(s)-Q(1-s))\, B(1-s).$$

We hence obtain

$$\sum_{\gamma=1}^{m} S_{\alpha\gamma}(s)\, S_{\alpha\beta}(1-s)=\sum_{\gamma=1}^{m}\Bigg(a^{2s-1}\frac{s-\varkappa}{s+\varkappa-1}\delta_{\alpha\gamma}+\frac{\omega^2(s)}{2s-1}$$
$$\times\int_F\int_F \varphi^{\alpha}(z,\ s)\, n'(z,\ z';\ s)\,\varphi^{\gamma}(z',\ s)\, d\mu(z)\, d\mu(z')\Bigg)$$
$$\times\Bigg(a^{-2s+1}\frac{s+\varkappa-1}{s-\varkappa}\delta_{\gamma\beta}-\frac{\omega^2(s)}{2s-1}\int_F\int_F \varphi^{\gamma}(z,\ 1-s)\, n'(z,\ z';(1-s))\,\varphi^{\beta}(z',\ 1-s)$$
$$\times\, d\mu(z)\, d\mu(z')\Bigg)=\delta_{\alpha\beta}+\frac{\omega^2(s)}{2s-1}\int_F\int_F \varphi^{\alpha}(z,\ s)\,(b(z,\ z';\ s)-b(z,\ z';\ 1-s))$$
$$\times\,\varphi^{\beta}(z',\ 1-s)\, d\mu(z)\, d\mu z'-\frac{\omega^4(s)}{(2s-1)^2}\int_F\int_F\int_F\int_F \varphi^{\alpha}(z,\ s)\, n'(z,\ z';\ s)$$
$$\times\left(\sum_{\gamma=1}^{m}\varphi^{\gamma}(t,s)\,\varphi^{\gamma}(t',1-s)\right) n'(t',\ z';\ 1-s)\,\varphi^{\beta}(z',\ 1-s)\, d\mu(z)$$
$$\times\, d\mu(t)\, d\mu(t')\, d\mu(z')=\delta_{\alpha\beta}.$$

In these computations we have used the following formulas, which are easily verified:

$$B(s)\,\varphi^{\beta}(z,\ s)=N'(s)\,\varphi^{\beta}(z,\ s),\qquad \beta=1,\ \ldots,\ m,$$

and

$$q(z,\ z';\ s)-q(z,\ z';\ 1-s)=\frac{1}{2s-1}\sum_{\beta=1}^{m}\varphi^{\beta}(z,\ s)\,\varphi^{\beta}(z',\ 1-s),$$

$$\varphi(y,\ s)=\varphi(y,\ 1-s)\, a^{2s-1}\frac{s-\varkappa}{s+\varkappa-1}.$$

We verify the second equation in 3):

$$\sum_{\alpha=1}^{m}\psi^{\alpha}(z,\ 1-s)\, S_{\alpha\beta}(s)=\psi^{\beta}(z,\ 1-s)\, a^{2s-1}\frac{s-\varkappa}{s+\varkappa-1}$$
$$+\,\omega^2(s)\,(Q(s)-Q(1-s))\, N'(s)\,\varphi^{\beta}(z,\ s)+\omega^3(s)\, N'(1-s)$$
$$\times\,(Q(s)-Q(1-s))\, N'(s)\,\varphi^{\beta}(z,\ s)+\omega^4(s)\, Q(1-s)\, N'(1-s)$$
$$\times\,(Q(s)-Q(1-s))\, N'(s)\,\varphi^{\beta}(z,\ s)\ =$$

$$= \psi^{\beta}(z, 1-s)\, a^{2s-1} \frac{s-\varkappa}{s+\varkappa-1}$$
$$+ \omega^2(s)(Q(s)-Q(1-s))\, N'(s)\, \varphi^{\beta}(z, s) + \omega(s)(N'(s)-N'(1-s))\, \varphi^{\beta}(z, s)$$
$$+ \omega^2(s)\, Q(1-s)(N'(s)-N'(1-s))\, \varphi^{\beta}(z, s) = \psi^{\beta}(z, s).$$

The proof is complete.

We let $\mathfrak{H}_0$ denote the Hilbert space of Lebesgue square integrable vector-valued functions $\xi(t) = \{\xi_1(t), \cdots, \xi_m(t)\}$ on the semi-axis $[0, \infty)$ with scalar product

$$(\xi, \eta)_0 = \frac{1}{2\pi} \sum_{\alpha=1}^{m} \int_0^{\infty} \xi_\alpha(t)\, \overline{\eta_\alpha(t)}\, dt, \quad \text{where } \xi, \eta \in \mathfrak{H}_0;$$

the bar denote complex conjugation, and E_0 denotes the identity operator in $\mathfrak{H}_0$. We define the following selfadjoint operator $\mathfrak{A}_0$ on the corresponding dense set in $\mathfrak{H}$: $\mathfrak{A}_0\xi(t) = c_0^{-2}(\frac{1}{4}+t^2)\xi(t)$. Further, we let $\mathfrak{P}$ denote the projection onto the subspace of the discrete spectrum of the operator $\mathfrak{A}$ in $\mathfrak{H}$.

Theorem 3.2. *The transformation* $\mathfrak{N}$ *defined by*

$$(\mathfrak{N}f)_\alpha(t) = \xi_\alpha(t) = \int_F \psi^{\alpha}\left(z, \frac{1}{2}+it\right) f(z)\, d\mu(z), \alpha = 1, \ldots, m, f \in \mathfrak{H},$$

maps $\mathfrak{H}$ *isometrically onto all of* $\mathfrak{H}_0$, *and*

(8) $$\mathfrak{N}\mathfrak{N}^* = E_0, \quad \mathfrak{N}^*\mathfrak{N} = E - \mathfrak{P}, \quad \mathfrak{N}h(\mathfrak{A}) = h(\mathfrak{A}_0)\mathfrak{N},$$

where h *is any bounded measurable function on* $[0, \infty)$.

This theorem is an expansion theorem for eigenfunctions of the operator $\mathfrak{A}$. The proof, which is by the methods of perturbation theory of a continuous spectrum, does not depend on the specific nature of the spaces S_1, S_2 and S_3 as opposed to the upper half-plane, and is identical to the proof of Theorem 4.1 in [**4**].

Remark. When proving Lemma 2.2 we imposed a certain condition on the set D_α, namely: the Lebesgue measure of D_α, which we denoted $|D_\alpha|$, is equal to one. The general case reduces to this one using a change of variables. Theorem 3.2 is stated and proved analogously in the general case, except that in place of the function $\psi^\alpha(z,s)$ we must consider the normalized eigenfunctions, i.e. $\psi^\alpha(z,s)/\sqrt{|D_\alpha|}$.

§4. The Selberg trace formula

Lemma 4.1. *If* p *is a natural number such that* $p \geqq p_0 = [\dim S/4]+1$, *then* $R^{2p}(s) - Q^{2p}(s)$ *is a nuclear operator in* $\mathfrak{H}$ *for complex* s *in the region* $\operatorname{Re} s > \frac{1}{2}$, $s \notin [\frac{1}{2}, 1]$.

Proof. We suppose that $\operatorname{Re} s > 4$. We have $R(s) = T(s) + M(s) + N(s)$, where the operators $T(s)$, $M(s)$ and $N(s)$ were defined in §2.

We let $P(s)$ denote $T(s) - Q(s)$. We show that $P(s)$ is a nuclear operator in $\mathfrak{H}$. Using the explicit expression for $T(s)$ and $Q(s)$, we find that the kernel $p(z,z';s)$ of the operator $P(s)$ only has nonzero diagonal components $p_{\alpha\alpha}(z,z';s) = c(s)y^{1-s}y'^{1-s}$, $\alpha = 1, \cdots, m$. It therefore suffices to prove that $P(s)$ is nuclear in the subspace $\{f \in \mathfrak{H} \mid f_\beta = 0, \beta \neq \alpha\}$, where $\beta, \alpha = 1, \cdots, m$.

It is obvious that this kernel has all possible partial derivatives with respect to z and z' up to order $[\dim S/2]+1$, inclusive, and these kernels are Hilbert-Schmidt kernels (in Lebesgue measure). Applying the well-known criteria for nuclearity of integral operators (see [15]), we easily find that $P(s)$ is nuclear in $\mathfrak{H}$. We now consider the difference $T^{2p}(s) - Q^{2p}(s) = (Q(s)+P(s))^{2p} - Q(s)^{2p}$. This difference is obviously a polynomial in $P(s)$ and $Q(s)$ with $P(s)$ appearing in every term. Consequently this difference is a nuclear operator, since $Q(s)$ is a bounded operator. Returning to $R^{2p}(s) - Q^{2p}(s)$, we note that the operator $(M(s)+N(s))^{2p}$ is nuclear in $\mathfrak{H}$. In fact, $M(s)$ and $N(s)$ are operators with weak singularity, and $(M(s)+N(s))^p$ is a Hilbert-Schmidt operator for the indicated values of p.

To prove the lemma for $\operatorname{Re} s > 4$ it suffices to prove the nuclearity of the operator $\widetilde{N}(s)$, whose kernel $n(z,z';s)$ is defined as follows:

$$\tilde{n}_{\alpha\beta}(z, z'; s) = n_{\alpha\beta}(z, z'; s), \ \tilde{n}_{00}(z, z'; s) = 0,$$

$\alpha = 0, \cdots, m$; $\beta = 0, \cdots, m$, except for $\alpha = \beta = 0$. For s as indicated the kernel $\tilde{n}(z,z';s)$ can be represented in the form of a series over a discrete group (see §2), and this series can be differentiated term by term with respect to z and z' an arbitrary number of times while preserving the estimate

$$|D\tilde{n}_{\alpha\beta}(z, z'; s)| \leqslant c\,(yy')^{2+\varepsilon-\operatorname{Re} s}.$$

We have let D denote an arbitrary partial derivative. Applying the above convergence test, we obtain the lemma for $\operatorname{Re} s > 4$. It remains to note that Hilbert's identity for $R(s)$ and $Q(s)$ implies that $R^{2p}(s) - Q^{2p}(s)$ is nuclear at any common regular point, i.e. in the region $\operatorname{Re} s > \frac{1}{2}$, $s \notin [\frac{1}{2}, 1]$. The proof is complete.

If $r > 0$, we define the integral operator K_r by the following formula:

$$K_r f(z) = -\frac{1}{2\pi}\sum_{\alpha=1}^{m}\int_F\int_0^r\int_0^{r'}\Big\{\psi^\alpha\Big(z, \frac{1}{2}+it\Big)\overline{\psi^\alpha\Big(z', \frac{1}{2}+it\Big)} - \varphi^\alpha\Big(z, \frac{1}{2}+it\Big)\overline{\varphi^\alpha\Big(z', \frac{1}{2}+it\Big)}\Big\} f(z')\,dt\,dr'\,d\mu(z').$$

LEMMA 4.2. *K_r is nuclear in $\mathfrak{H}$.*[1)]

The proof of this lemma is analogous to that of the corresponding assertion in [5].

If $h(\lambda)$ is a function on the semi-axis $0 \leqq \lambda < \infty$, we shall also denote this function by $h(r)$, where $\lambda = (\frac{1}{4}+r^2)/c_0^2$.

THEOREM 4.1. *Let $h(\lambda)$ be a twice continuously differentiable function on the semi-axis $0 \leqq \lambda < \infty$, such that*

$$h(r) = O(1+|r|^2)^{-2p_0}, \quad h''(r) = O(1+|r|^2)^{-2p_0-1}, \quad \textit{where } p_0 = [\dim S/4]+1.$$

In addition, for fixed z suppose that the function $h(r)\,|\psi^\alpha(z, \frac{1}{2}+ir)|$ is continuous at zero. Then the following equation is fulfilled:

---

1) For simplicity we assume that the functions $\psi^\alpha(\cdot, \frac{1}{2}+it)$ are analytic at zero.

$$(9)\qquad \sum_j h(\lambda_j) = Tr\,(h(\mathfrak{A}) - h(\mathfrak{A}_1)) + \int_0^\infty h''(r)\,TrK_r dr;$$

λ_j on the right runs through all eigenvalues of the operator $\mathfrak{A}$ counting multiplicity.[2)]

Proof. We first consider $h(\lambda) = (\lambda - \lambda_0)^{-2p_0}$, where $\lambda_0 \notin (0, \infty)$. If $f \in \mathfrak{H}$, then

$$((h(\mathfrak{A}) - h(\mathfrak{A}_1))f, f) = \int_{-\infty}^{+\infty} h(\lambda)\, d((E_\lambda - E'_\lambda)f, f),$$

where we have let E_λ and E'_λ denote the resolution of the identity for $\mathfrak{A}$ and $\mathfrak{A}_1$, respectively. Continuing the equation, we obtain

$$\begin{aligned}((h(\mathfrak{A}) - h(\mathfrak{A}_1))f, f) &= -\int_0^\infty h'(\lambda)\int_0^\lambda d((E_{\lambda'} - E'_{\lambda'})f, f)\,d\lambda \\ &= -\int_0^\infty h'(\lambda)(E^s_\lambda f, f)\,d\lambda + \int_0^\infty h''(r)\frac{1}{2\pi}\sum_{\alpha=1}^m \int_0^r\int_0^{r'}\Big\{\Big|\int_F \psi^\alpha\Big(z, \frac{1}{2}+it\Big)f(z)\,d\mu(z)\Big|^2 \\ &\quad - \Big|\int_F \varphi^\alpha\Big(z, \frac{1}{2}+it\Big)f(z)\,d\mu(z)\Big|^2\Big\}\,dt\,dr'\,dr = -\int_0^\infty h'(\lambda)(E^s_\lambda f, f)\,d\lambda \\ &\quad - \int_0^\infty h''(r)(K_r f, f)\,dr,\end{aligned}\qquad (10)$$

where we have let E^s_λ denote the projection onto the eigensubspace of the discrete spectrum of $\mathfrak{A}$ in the interval $[0,\lambda]$. Formula (10) is obtained by integrating by parts, using the following fact from the theory of selfadjoint operators:

$$\int_a^b (dE_\lambda f, f) = \frac{1}{2\pi i}\lim_{\varepsilon\to 0}\int_a^b ((R(\lambda + i\varepsilon) - R(\lambda - i\varepsilon))f, f)\,d\lambda,$$

where $R(\lambda)$ and E_λ are the resolvent and resolution of the identity for any selfadjoint operator. We have also used the identities

$$r(z, z'; s) - r(z, z'; 1-s) = \frac{1}{2s-1}\sum_{\alpha=1}^m \psi^\alpha(z, s)\psi^\alpha(z', 1-s),$$

$$q(z, z'; s) - q(z, z'; 1-s) = \frac{1}{2s-1}\sum_{\alpha=1}^m \varphi^\alpha(z, s)\varphi^\alpha(z', 1-s)$$

at nonsingular points.

Since, in particular, $h(\mathfrak{A}) - h(\mathfrak{A}_1)$ is nuclear on the subspace of the continuous spectrum for $\mathfrak{A}$ (Lemma 4.1), it follows that the operators $\int_0^\infty h'(\lambda)E^s_\lambda d\lambda$ and $\int_0^\infty h''(r)K_r dr$ are nuclear in $\mathfrak{H}$. It remains to note that E^s_λ is a nuclear operator in $\mathfrak{H}$, as the projection onto a finite-dimensional subspace. This,

---

[2)] We are allowing a common abuse of notation, where $h(\mathfrak{A}_1)$ denotes the operator constructed in the same was as $Q(s)$ is from $q(y, y'; s)$.

together with Lemma 4.2, implies the theorem. To prove the general case we consider a basis $\{f_i\}$ in the space $\mathfrak{H}$. For the matrix trace of the operator $h(\mathfrak{A}) - h(\mathfrak{A}_1)$ formula (10) easily implies the estimate

$$\sum_i (((h(\mathfrak{A}) - h(\mathfrak{A}_1)) f_i, f_i) \leqslant \sum_i (((\mathfrak{A} - \lambda_0)^{-2p_0} - (\mathfrak{A}_1 - \lambda_0)^{-2p_0}) f_i, f_i) < \infty.$$

It then follows that $h(\mathfrak{A}) - h(\mathfrak{A}_1)$ is nuclear in $\mathfrak{H}$, as asserted in the theorem. The proof is complete.

The last assertion which we shall make in this section is connected with Eisenstein series. We show that the functions $\psi^\alpha(z,s)$ realize an analytic continuation of Eisenstein series onto the entire s-plane. If $\operatorname{Re} s > 1$, we consider the Eisenstein series

$$J^\alpha(z, s) = \sum_{\gamma \in \Gamma_\alpha \backslash \Gamma} y^s(\gamma g_\alpha z), \quad \alpha = 1, \ldots, m,$$

where $y(z)$ is the y-coordinate of the vector z. We show that for such s this series converges under our assumptions concerning the discrete group Γ.

In fact, let $\gamma g_\alpha = ua(t)k$ be the components of the Iwasawa decomposition (see §1). A theorem of Godement (see [**16**]) implies that the series converges if we prove that the inequality $t < d$ holds when γ runs through $\Gamma_\alpha \backslash \Gamma$, where d is a sufficiently large positive number not depending on γ. This follows from condition (b) and the proof of Lemma 1.5.

It is obvious that $J^\alpha(z,s)$ is automorphic relative to Γ and an eigenfunction of the operator $-L$ with eigenvalue $s(1-s)/c_0^2$. In addition, the Eisenstein series has the following asymptotic behavior in a neighborhood of the parabolic vertices:

$$J^\alpha_\beta(z,s) = y^s \delta_{\alpha\beta} + O(1), \qquad \alpha, \beta = 1, \cdots, m.$$

This implies that for $\operatorname{Re} s > 1$ the difference $J^\alpha(z,s) - \psi^\alpha(z,s)$ is a bounded eigenfunction with the same eigenvalue, and, since $\mathfrak{A}$ is selfadjoint, we have $J^\alpha(z,s) = \psi^\alpha(z,s)$.

§5. Some examples of computations of the Selberg trace formula

In this section we perform a concrete computation of formula (9); more precisely, we compute the additional contribution to the trace formula compared with the compact case. We limit ourselves to the space $S = S_1$ and to discrete groups Γ of two different types. In the first example we consider discrete groups Γ whose subgroups Γ_α do not contain nontrivial elements of finite order. In the second example we compute the additional contribution for the group $SL(2, Z + iZ) \subset SL(2, C)$, where Z is the ring of integers.

Throughout this section we use the coordinate system for $S = S_1$ in §1.2. We recall that in this coordinate system the invariant measure has the form $d\mu(z) = dy dx_2 \cdots dx_n / y^n$, and the invariant of a pair of points $u(z,z')$ is chosen to be equal to

$$u(z, z') = \frac{(y - y')^2 + (x_2 - x_2')^2 + \ldots + (x_n - x_n')^2}{yy'}.$$

In addition,

$$Lf = y^n \frac{\partial}{\partial y}\left(\frac{1}{y^{n-2}} \frac{\partial f}{\partial y}\right) + y^2 \Delta_x f, \text{ where } \Delta_x = \frac{\partial^2}{\partial x_2^2} + \cdots + \frac{\partial^2}{\partial x_n^2}.$$

5.1. *The Selberg map.*

We consider the ring of differential operators on S which are invariant relative to the action of the group $G=G_1$. It follows from the definition of S that this ring has a single generator, and L may be taken to be this generator. If K is the invariant integral operator on S with kernel $k(z,z')$, it follows from [1] that $Kf=h(\lambda)f$, where f is an eigenfunction: $Lf=\lambda f$. Here $h(\lambda)$ depends only on K and is determined by a sufficiently large "representative" family of eigenfunctions $f(z)$. On the other hand, S is a two-point homogeneous space (see [8]), so that $k(z,z')=k(u(z,z'))$. By the Selberg map we mean the one-to-one correspondence $h(\lambda)\to k(u)$ which is defined in [1] in a general situation.

In our case, if we choose y^s as the "representative" family of eigenfunctions, then

$$\int_S k(u(z,\ z'))\, y'^s d\mu(z) = h(s(n-1-s))\, y^s,\quad \lambda = s(n-1-s).$$

Assuming that $h(\lambda)$ is an analytic function (below we impose precise conditions on $h(\lambda)$) and performing elementary computations, we obtain

$$\begin{aligned}
&k(t) = \frac{(-1)^k}{\pi^k} Q^{(k)}(\omega)\Big|_{\omega=t}, && n = 2k+1,\\
&Q(\omega) = \pi^k \int_\omega^\infty dt_1 \int_{t_1}^\infty dt_2 \ldots \int_{t_{k-1}}^\infty dt k(t), &&\\
&k(t) = \frac{(-1)^k}{\pi^k} \int_t^\infty \frac{Q^{(k)}(\omega)\, d\omega}{\sqrt{\omega - t}}, && n = 2k,\\
&Q(\omega) = \pi^{k-1} \int_\omega^\infty dt_1 \int_{t_1}^\infty dt_2 \ldots \int_{t_{k-1}}^\infty \frac{k(t)\, dt}{\sqrt{t - t_{k-1}}}, &&
\end{aligned} \tag{11}$$

where $Q(w)=Q(e^u+e^{-u}-2)=g(u)$ and

$$h(r) = \int_{-\infty}^{+\infty} e^{iru} g(u)\, du,\quad g(u) = \frac{1}{2\pi}\int_{-\infty}^{+\infty} e^{-iru} h(r)\, dr.$$

We have let $h(r)$ denote the function $h(((n-1)/2)^2+r^2)$.

We suppose that $h(r)$ is an even analytic function in the strip $|\operatorname{Im} r| < (n-1)/2+\epsilon$, $\epsilon>0$, and $h(r)=O(1+|r|^2)^{-2p_0}$, $p_0=[n/4]+1$; in addition, for fixed z the functions $h(r)|\psi^\alpha(z,(n-1)/2+ir)|^2$ are continuous at zero. We show that in this case the $k(t)$ in (11) exists, the series $\sum_{\gamma\in\Gamma} k(u(z,\gamma z'))$ converges and is the kernel of the integral operator K in the space $\mathfrak{H}$, and $K=h(\mathfrak{A})$. In addition, we have the formula (9). To prove this, deforming the contour of integration in the region $|\operatorname{Im} t| < (n-1)/2+\epsilon$, $\epsilon>0$, we find that the function $g(u)=(1/2\pi)\int_{-\infty}^{+\infty} e^{-iru} h(r)\, dr$ satisfies

$$g(u) = O\Big(\exp\Big(-\Big(\frac{n-1}{2}+\varepsilon\Big)|u|\Big)\Big).$$

It is then easy to show that $k(t)=O(t^{-n+1-\epsilon})$, and consequently, by the corollary to Lemma 1.4, the series $\sum_{\gamma\in\Gamma} k(u(z,\gamma z'))$ converges and induces the operator K in $\mathfrak{H}$. It follows from the definition of $\mathfrak{A}$ and (11) that $K = h(\mathfrak{A})$. Finally, from Theorem 4.1 we obtain the new assertion.

5.2. *Computation of the contribution from Eisenstein series.*

Let F be the fundamental domain of Γ in S, and let Y be a sufficiently large positive number. We introduce $F_Y = \bigcup_{\alpha=1}^m F_\alpha^Y$, where $F_\alpha^Y = g_\alpha \Pi_\alpha^Y$, $\alpha = 1, \cdots, m$ (see §1),

$$\Pi_\alpha^Y = \{z \in S,\ z = \{y,\ x\} \mid y \in [a,\ Y],\ x \in D_\alpha\}.$$

We consider formula (9):

$$\sum_j h(\lambda_j) = \mathbf{Tr}(h(\mathfrak{A}) - h(\mathfrak{A}_1)) - \int_0^\infty h''(r) \frac{1}{2\pi} \sum_{\alpha=1}^m \int_F \int_0^r \int_0^{r'} \left(\left| \psi^\alpha\left(z,\ \frac{n-1}{2} + it\right)\right|^2 - \left| \varphi^\alpha\left(z,\ \frac{n-1}{2} + it\right)\right|^2 \right) dt\,dr'\,d\mu(z)\,dr.$$

We introduce the function subspace

$$\mathfrak{H}_Y = \{f \in \mathfrak{H} \mid f(z) = 0,\ z \in F - F_Y\}.$$

We set

$$h_Y(\mathfrak{A}) = P_Y h(\mathfrak{A}) P_Y,\quad h_Y(\mathfrak{A}_1) = P_Y h(\mathfrak{A}_1) P_Y,$$

where P_Y is the projection onto $\mathfrak{H}_Y$. We note that $h_Y(\mathfrak{A})$ and $h_Y(\mathfrak{A}_1)$ are nuclear operators in $\mathfrak{H}$. We have

$$\sum_j h(\lambda_j) = \lim_{Y\to\infty} \left\{ \mathbf{Tr}\,(h_Y(\mathfrak{A}) - h_Y(\mathfrak{A}_1)) - \int_0^\infty h''(r) \frac{1}{2\pi} \times \sum_{\alpha=1}^m \int_0^r \int_0^{r'} \int_{F_Y} \left(\left|\psi^\alpha\left(z,\ \frac{n-1}{2} + it\right)\right|^2 - \left|\varphi^\alpha\left(z,\ \frac{n-1}{2} + it\right)\right|^2\right) d\mu(z)\,dt\,dr'\,dr \right\}. \tag{12}$$

It is obvious that $\mathbf{Tr}(h_Y(\mathfrak{A}) - h_Y(\mathfrak{A}_1)) = \mathbf{Tr}\,h_Y(\mathfrak{A}) - \mathbf{Tr}\,h_Y(\mathfrak{A}_1)$. In addition,

$$\frac{1}{2\pi} \int_0^\infty h(r) \sum_{\alpha=1}^m \int_{F_Y} \left|\varphi^\alpha\left(z,\ \frac{n-1}{2} + ir\right)\right|^2 d\mu(z)\,dr = \mathbf{Tr}\,h_Y(\mathfrak{A}_1).$$

Further, $\mathbf{Tr}\,h_Y(\mathfrak{A}) = \int_{F_Y} \sum_{\gamma\in\Gamma} k(u(z, \gamma z))\,d\mu(z)$, where $k(u)$ is the Selberg map for the function $h(\lambda)$. Thus, after we integrate by parts, formula (12) takes the form

$$\sum_j h(\lambda_j) = \lim_{Y\to\infty} \left\{ \int_{F_Y} \sum_{\gamma\in\Gamma} k(u(z,\ \gamma z))\,d\mu(z) - \frac{1}{2\pi} \sum_{\alpha=1}^m \int_0^\infty h(r) \int_{F_Y} \left|\psi^\alpha\left(z,\ \frac{n-1}{2} + ir\right)\right|^2 d\mu(z)\,dr \right\}. \tag{13}$$

In this subsection we investigate the second term in the limit in (13). We consider Green's formula for the operator L:

$$\int_{F_Y} \{\psi^\alpha(z,\ s) L\psi^\alpha(z,\ s') - \psi^\alpha(z,\ s') L\psi^\alpha(z,\ s)\}\,d\mu(z) = \int_{\partial F_Y} \frac{1}{y^{n-2}} \left\{ \psi^\alpha(z,\ s) \frac{\partial \psi^\alpha(z,\ s')}{\partial \nu} - \psi^\alpha(z,\ s') \frac{\partial\psi^\alpha(z,\ s)}{\partial\nu} \right\} dS = . \tag{14}$$

Since the formula only contains functions which are smooth and automorphic on S, it follows that F_Y may be considered a manifold whose boundary is $\bigcup_1^m g_\alpha D_\alpha = \partial F_Y$, where ν denotes the exterior normal direction. We continue (14):

$$= \sum_{\beta=1}^{m} \int_{D_\beta} dx_2 \ldots dx_n \frac{1}{Y^{n-2}} \left\{ \psi_\beta^\alpha(z, s) \frac{\partial \psi_\beta^\alpha(z, s')}{\partial y} - \psi_\beta^\alpha(z, s') \frac{\partial \psi_\beta^\alpha(z, s)}{\partial y} \right\}_{y=Y}.$$

On the other hand, we easily see that

$$(s(n-1-s) - s'(n-1-s'))\psi^\alpha(z, s)\psi^\alpha(z, s') = \psi^\alpha(z, s) L\psi^\alpha(z, s') - \psi^\alpha(z, s') L\psi^\alpha(z. s).$$

We thereby obtain

$$\int_{F_Y} \psi^\alpha(z, s)\psi^\alpha(z, s')\, d\mu(z) = \frac{1}{s(n-1-s) - s'(n-1-s')} \times \sum_{\beta=1}^{m} \int_{D_\beta} (\psi_\beta^\alpha(z, s) \frac{\partial \psi_\beta^\alpha(z, s')}{\partial y} - \psi_\beta^\alpha(z, s') \frac{\partial \psi_\beta^\alpha(z, s)}{\partial y} \bigg|_{y=Y} \frac{dx_2 \ldots dx_n}{Y^{n-2}}.$$

We are interested in the behavior of the right side for large Y. Replacing $\psi^\alpha(z,s)$ by its asymptotic behavior in Lemma 3.5 and setting $s = (n-1)/2 + ir$ and $s' = (n-1)/2 - ir'$, $r, r' \in R$, we obtain

$$\sum_{\alpha=1}^{m} \int_{F_Y} \psi^\alpha(z, s)\overline{\psi^\alpha(z, s')}\, d\mu(z) = \sum_{\alpha=1}^{m}\sum_{\beta=1}^{m} \Big\{ \frac{i}{r-r'} (S_{\alpha\beta}(s) S_{\alpha\beta}(s') Y^{i(r'-r)} - \delta_{\alpha\beta} Y^{i(r-r')}) + \frac{i}{r+r'} (\delta_{\alpha\beta} S_{\alpha\beta}(s) Y^{-i(r+r')} - \delta_{\alpha\beta} S_{\alpha\beta}(s') Y^{i(r+r')}) \Big\} + \alpha(Y, r, r'),$$
$$\alpha(Y, r, r') = o(1) \quad (Y \to \infty).$$

We perform the transformations

$$\frac{1}{2\pi} \sum_{\alpha=1}^{m} \int_0^\infty h(r) \int_{F_Y} \left| \psi^\alpha\left(z, \frac{n-1}{2} + ir\right) \right|^2 d\mu(z)\, dr = \frac{1}{4\pi} \lim_{r' \to r} \sum_{\alpha=1}^{m} \int_{-\infty}^{+\infty} h(r) \times \int_{F_Y} \psi^\alpha\left(z, \frac{n-1}{2} + ir\right) \overline{\psi^\alpha\left(z, \frac{n-1}{2} + ir'\right)}\, d\mu(z)\, dr$$

(15)
$$= \frac{1}{4\pi} \sum_{\alpha=1}^{m}\sum_{\beta=1}^{m} \lim_{r' \to r} \left\{ \int_{-\infty}^{+\infty} h(r) \left\{ \frac{i}{r-r'} \left(S_{\alpha\beta}\left(\frac{n-1}{2} + ir\right) S_{\alpha\beta}\left(\frac{n-2}{2} - ir'\right) \times Y^{i(r'-r)} - \delta_{\alpha\beta} Y^{i(r-r')} \right) + \frac{i}{(r+r')} \left(\delta_{\alpha\beta} S_{\alpha\beta}\left(\frac{n-1}{2} + ir\right) Y^{-i(r+r')} - \delta_{\alpha\beta} S_{\alpha\beta}\left(\frac{n-1}{2} - ir'\right) Y^{i(r+r')} \right) + \alpha(Y, r, r') \right\} dr \right\}.$$

It is possible to pass to the limit as $r' \to r$ because the integral is uniformly convergent for fixed Y.

Because of the convergence of the integral

$$\int_0^\infty h''(r) \int_0^r \int_0^{r'} \int_F \left(\left| \psi^\alpha\left(z, \frac{n-1}{2} + it\right) \right|^2 - \left| \varphi^\alpha\left(z, \frac{n-1}{2} + it\right) \right|^2 \right) d\mu(z)\, dt\, dr'\, dr,$$

the fact that the matrix $S(s)$, $s=(n-1)/2+ir$, $r\in R$, is unitary, and also the formula obtained from (14) by replacing the functions $\psi^\alpha(z,s)$ by $\varphi^\alpha(z,s)$, it follows that the integral $\int_0^\infty h(r)\alpha(Y,r,r')$ converges uniformly in r' and Y. This implies that

$$\int_0^\infty h(r)\,\alpha(Y,\ r,\ r')\,dr=o(1)\quad (Y\to\infty).$$

We investigate the expression

$$\sum_{\alpha=1}^m\sum_{\beta=1}^m\frac{i}{r-r'}\left(S_{\alpha\beta}\left(\frac{n-1}{2}+ir\right)S_{\alpha\beta}\left(\frac{n-1}{2}-ir'\right)Y^{i(r-r')}-\delta_{\alpha\beta}Y^{i(r-r')}\right).$$

Adding and subtracting

$$\sum_{\alpha=1}^m\sum_{\beta=1}^m\frac{i}{r-r'}S_{\alpha\beta}\left(\frac{n-1}{2}+ir'\right)S_{\alpha\beta}\left(\frac{n-1}{2}-ir'\right)Y^{i(r'-r)},$$

we reduce it to the form

$$Y^{i(r'-r)}\sum_{\alpha=1}^m\sum_{\beta=1}^m\left\{-S_{\alpha\beta}\left(\frac{n-1}{2}-ir'\right)\frac{S_{\alpha\beta}\left(\frac{n-1}{2}+ir\right)-S_{\alpha\beta}\left(\frac{n-1}{2}+ir'\right)}{i\,(r-r')}\right\}$$
$$+\frac{im}{r-r'}\{Y^{i(r'-r)}-Y^{i(r-r')}\}.$$

We now suppose that the integral

$$\int_{-\infty}^{+\infty}h(r)\,Tr\left[S^{-1}\left(\frac{n-1}{2}+ir\right)S'\left(\frac{n-1}{2}+ir\right)\right]dr$$

is absolutely convergent. We further introduce the absolute convergence of this integral from (13). We note that, independently of this, absolute convergence can be proved in terms of nonstationary scattering theory, since $S((n-1)/2+ir), r\in R$, is, to within a trivial factor, a subharmonic scattering operator for the perturbation $\mathfrak{A}-\mathfrak{A}_1$ (see [**17**]).

Substituting the expression

$$Y^{i(r'-r)}\sum_{\alpha=1}^m\sum_{\beta=1}^m\left(-S_{\alpha\beta}\left(\frac{n-1}{2}-ir'\right)\frac{S_{\alpha\beta}\left(\frac{n-1}{2}+ir\right)-S_{\alpha\beta}\left(\frac{n-1}{2}+ir'\right)}{i\,(r-r')}\right)$$

in (15) and passing to the limit in r', we obtain the contribution

$$-\frac{1}{4\pi}\int_{-\infty}^{+\infty}h(r)\,Tr\left(S^{-1}\left(\frac{n-1}{2}+ir\right)S'\left(\frac{n-1}{2}+ir\right)\right)dr+o(1)\quad (Y\to\infty).$$

The second term $(im/(r-r'))(Y^{i(r'-r)}-Y^{i(r-r')})$ gives the following contribution to (15) as $r'\to r$:

$$mg(0)\ln Y+o(1)\quad (Y\to\infty),$$

where $g(u)$ is the function in (11).

Passing to the limit as $r'\to r$ in (15), we separate the resulting integral into two terms:

$$\frac{1}{4\pi}\sum_{\alpha=1}^{m}\int_{-\infty}^{+\infty} h(r)\frac{i}{2r}S_{\alpha\alpha}\left(\frac{n-1}{2}+ir\right)(Y^{-2ir}-Y^{2ir})\,dr$$
$$+\frac{1}{4\pi}\sum_{\alpha=1}^{m}\int_{-\infty}^{+\infty} h(r)\frac{i}{2r}Y^{2ir}\left\{S_{\alpha\alpha}\left(\frac{n-1}{2}+ir\right)-S_{\alpha\alpha}\left(\frac{n-1}{2}-ir\right)\right\}dr.$$

Applying the Fourier integral formula, we find that the contribution we are looking for is equal to

$$\frac{1}{4}h(0)\,Tr\left(S\left(\frac{n-1}{2}\right)\right)+o(1)\quad (Y\to\infty).$$

Finally, bringing together all the contributions, we obtain

$$\frac{1}{2\pi}\sum_{\alpha=1}^{m}\int_{0}^{\infty} h(r)\int_{F_Y}\left|\psi^{\alpha}\left(z,\ \frac{n-1}{2}+ir\right)\right|^2 d\mu(z)\,dr=mg(0)\ln Y$$
$$-\frac{1}{4\pi}\int_{-\infty}^{+\infty} h(r)\,\mathrm{Tr}\left(S^{-1}\left(\frac{n-1}{2}+ir\right)S'\left(\frac{n-1}{2}+ir\right)\right)dr \tag{16}$$
$$+\frac{1}{4}h(0)\,\mathrm{Tr}\left(S\left(\frac{n-1}{2}\right)\right)+o(1)\quad (Y\to\infty).$$

5.3. *Computation of the contribution from parabolic elements* (*first example*). We consider the integral $\int_{F_Y}\sum_{\gamma\in\Gamma}k(u(z,\gamma z))\,d\mu(z)$ in (13). Using the absolute convergence of the integral and the series for fixed Y, by a change of variables we reduce this expression to the form

$$\int_{F_Y}\sum_{\gamma\in\Gamma}k(u(z,\ \gamma z))\,d\mu(z)=\sum_{\{\gamma\}_\Gamma}\ \int_{\bigcup_{\gamma_0\in\Gamma^\gamma\setminus\Gamma}\gamma_0 F_Y}k(u(z,\ \gamma z))\,d\mu(z).$$

We have let $\{\gamma\}_\Gamma$ denote the conjugacy class in Γ with representative γ, and we have let Γ^γ denote the centralizer of γ in Γ. The element γ runs through a set of representatives of all the classes $\{\gamma\}_\Gamma$, i.e. the elliptic, hyperbolic, loxodromic, unit and parabolic classes. By a parabolic class we mean a conjugacy class $\{\gamma\}_\Gamma$ with representative $\gamma\neq e$, $\gamma\in\Gamma_\alpha$ for some $\alpha=1,\cdots,m$. We let Ω_Y^γ denote the union $\Omega_Y^\gamma=\bigcup_{\gamma_0\in\Gamma^\gamma\setminus\Gamma}\gamma_0 F_Y$. Our subsequent goal is to compute the following expression in terms of $h(\lambda)$:

$$\sum_{\substack{\{\gamma\}_\Gamma\\ \gamma\in\Gamma_\alpha,\ \gamma\neq e}}\ \int_{\Omega_Y^\gamma}k(u(z,\ \gamma z))\,d\mu(z). \tag{17}$$

In this subsection we suppose that $g_\alpha^{-1}\Gamma_\alpha g_\alpha\subset U$, $\alpha=1,\cdots,m$. It is easy to see that this assumption concerning Γ is fulfilled, for example, for the congruence subgroups of degree $\geqq 3$ of the group $SL(n+1,Z)\cap G$. In addition, for convenience of computation we make the inessential assumption that $g_\alpha^{-1}\Gamma_\alpha g_\alpha$ does not depend on $\alpha=1,\cdots,m$, and we introduce the notation $\Gamma_\infty=g_\alpha^{-1}\Gamma_\alpha g_\alpha$, $\Pi=\Pi_\alpha$ and $D=D_\alpha$.

If $\gamma\in\Gamma_\infty$, it obviously follows from the commutativity of U that $\{\gamma\}_\Gamma=\Gamma_\infty-e$, $\Gamma^\gamma=\Gamma_\infty$ and $\Omega_Y^\gamma=\Omega_Y$ does not depend on γ.

Thus (17) takes the form

$$\sum_{\alpha=1}^{m} \sum_{\substack{\gamma \in \Gamma_\alpha \\ \gamma \neq e}} \int\limits_{\substack{\cup \gamma_0 F_Y \\ \gamma_0 \in \Gamma_\alpha \setminus \Gamma}} k(u(z, \gamma z))\, d\mu(z). \tag{18}$$

We transform the domain of integration in (18) as follows. We note that

$$\bigcup_{\gamma_0 \in \Gamma_\alpha \setminus \Gamma} \gamma_0 F = \Omega(\alpha)$$

is a fundamental domain of the group Γ_α in S. Further,

$$\bigcup_{\gamma_0 \in \Gamma_\alpha \setminus \Gamma} \gamma_0 F_Y = \{\Omega(\alpha) - (F - F_Y)\} - \bigcup_{\substack{\gamma_0 \in \Gamma_\alpha \setminus \Gamma \\ \gamma_0 \neq e}} \gamma_0 (F - F_Y).$$

In accordance with this, the integral (18) splits into two terms. We investigate the second. Let Γ^* denote the group $g_\alpha^{-1} \Gamma g_\alpha$ and let F^* denote the region $g_\alpha^{-1} F$. It is obvious that F^* is a fundamental domain for Γ^* in S. By the invariance of the kernel $k(u(z, z'))$ we obtain

$$\sum_{\alpha=1}^{m} \sum_{\substack{\gamma \in \Gamma_\alpha \\ \gamma \neq e}} \int\limits_{\substack{\cup \gamma_0 (F - F_Y) \\ \gamma_0 \in \Gamma_\alpha \setminus \Gamma \\ \gamma_0 \neq e}} k(u(z, \gamma z))\, d\mu(z) = \sum_{\alpha=1}^{m} \sum_{\substack{\gamma \in \Gamma_\infty \\ \gamma \neq e}} \int\limits_{\substack{\cup \gamma_0 (F^* - F_Y^*) \\ \gamma_0 \in \Gamma_\infty \setminus \Gamma^* \\ \gamma_0 \neq e}} k(u(z, \gamma z))\, d\mu(z).$$

We note that if

$$z = \{y, x\} \in \bigcup_{\substack{\gamma_0 \in \Gamma_\infty \setminus \Gamma^* \\ \gamma_0 \neq e}} \gamma_0 (F^* - F_Y^*),$$

then the variable y is uniformly bounded from above as $Y \to \infty$, i.e. $0 < y < A$, where $A = \text{const}$ does not depend on Y. We further show that

$$\sum_{\substack{\gamma \in \Gamma_\infty \\ \gamma \neq e}} \int_0^A k(u(z, \gamma z)) \frac{dy}{y^n}$$

is absolutely convergent. This implies that

$$\sum_{\alpha=1}^{m} \sum_{\substack{\gamma \in \Gamma_\alpha \\ \gamma \neq e}} \int\limits_{\substack{\cup \gamma_0 (F - F_Y) \\ \gamma_0 \in \Gamma_\alpha \setminus \Gamma \\ \gamma_0 \neq e}} k(u(z, \gamma z))\, d\mu(z) = o(1) \quad (Y \to \infty),$$

since the measure of the domain of integration approaches zero. Thus the nonvanishing contribution from the two terms in (18) may be written in the form

$$\sum_{\alpha=1}^{m} \sum_{\substack{\gamma \in \Gamma_\alpha \\ \gamma \neq e}} \int\limits_{\Omega(\alpha) - (F - F_Y)} k(u(z, \gamma z))\, d\mu(z).$$

Choosing $\Omega(\alpha)$ in the form $\Omega(\alpha) = g_\alpha \{z \in S \mid y \in (0, \infty), x \in D\}$, from the invariance of the kernel we obtain

$$\sum_{\alpha=1}^{m} \sum_{\substack{\gamma \in \Gamma_\alpha \\ \gamma \neq e}} \int\limits_{\Omega(\alpha) - (F - F_Y)} k(u(z, \gamma z))\, d\mu(z) = m \sum_{\substack{\gamma \in \Gamma_\infty \\ \gamma \neq e}} \int_0^Y \int_D k(u(z, \gamma z)) \frac{dx\,dy}{y^n} + o(1) \quad (Y \to \infty). \tag{19}$$

We let Γ'_∞ denote $\Gamma_\infty - e$. If $\gamma \in \Gamma'_\infty$ and if $z = \{y, x\} \in S$, then $\gamma z = \{y, x + \xi\}$, where $\xi \neq 0$ is an $(n-1)$-dimensional vector. We let $|x|^2$ denote $x_2^2 + \cdots + x_n^2$. We have

$$\sum_{\gamma \in \Gamma'_\infty} \int_0^Y \int_D k(u(z, \gamma z))\, d\mu(z) = \sum_{\xi \neq 0} \int_0^Y k\left(\frac{|\xi|^2}{y^2}\right) \frac{dy}{y^n} = \sum_{\xi \neq 0} \frac{1}{2} \frac{1}{|\xi|^{n-1}} \int_{\frac{|\xi|^2}{Y^2}}^{\infty} (\sqrt{t})^{n-3} k(t)\, dt,$$

since $|D| = 1$. We order the set Γ'_∞ according to nondecreasing $|\xi|$, i.e. $|\xi^{(i+1)}| \geqq |\xi^{(i)}|$. Hence

$$\sum_{\xi \neq 0} \int_0^Y k\left(\frac{|\xi|^2}{y^2}\right) \frac{dy}{y^n} = \frac{1}{2} \sum_{i=1}^{\infty} \frac{1}{|\xi^{(i)}|^{n-1}} \int_{\frac{|\xi^{(i)}|^2}{Y^2}}^{\infty} (\sqrt{t})^{n-3} k(t)\, dt. \tag{20}$$

We prove the following assertion:

$$\lim_{N \to \infty} \sum_{i=1}^{N} \left\{ \frac{1}{|\xi^{(i)}|^{n-1}} - \frac{2\pi^{\frac{n-1}{2}}}{\Gamma\left(\frac{n-1}{2}\right)} \ln \frac{|\xi^{(i+1)}|}{|\xi^{(i)}|} \right\} < +\infty.$$

To prove this we choose a D containing zero, and we consider

$$\sum_{i=1}^{N} \int_D \frac{1}{|\gamma_i x|^{n-1}}\, dx = \int_{\bigcup_{i=1}^{N} \gamma_i D} \frac{dx}{|x|^{n-1}}.$$

Setting $R_1 = |\xi^{(N)}| + \operatorname{diam} D$, $R_2 = |\xi^{(N)}| - \operatorname{diam} D$ and letting K_R denote the ball in R^{n-1} with center at zero and radius R, we obtain

$$\int_{K_{R_2}} \frac{dx}{|x|^{n-1}} \leqslant \int_{\bigcup_{i=1}^{N} \gamma_i D} \frac{dx}{|x|^{n-1}} \leqslant \int_{K_{R_1}} \frac{dx}{|x|^{n-1}}$$

for large N. It is obvious that

$$\int_{K_{R_1} - K_{R_2}} \frac{dx}{|x|^{n-1}} = \frac{2\pi^{\frac{n-1}{2}}}{\Gamma\left(\frac{n-1}{2}\right)} \ln \frac{R_1}{R_2} \xrightarrow[N \to \infty]{} 0.$$

In addition,

$$\int_{K_{R_1 - \operatorname{diam} D}} \frac{dx}{|x|^{n-1}} = \frac{2\pi^{\frac{n-1}{2}}}{\Gamma\left(\frac{n-1}{2}\right)} \ln (\xi^{(N)}).$$

Finally,

$$\sum_{i=1}^{m} \int_D \left\{ \frac{1}{|\gamma_i x|^{n-1}} - \frac{1}{|\xi^{(i)}|^{n-1}} \right\} dx < +\infty,$$

which proves our assertion.

We now note that the decreasing of $k(t) = O(t^{-n+1-\epsilon})$ for large t implies that (20) converges absolutely for fixed Y. To investigate (19) further, we perform the following transformations:

$$\frac{1}{2}\sum_{i=1}^{N}\frac{1}{|\xi^{(i)}|^{n-1}}\int_{\frac{|\xi^{(i)}|^2}{Y^2}}^{\infty}(\sqrt{t})^{n-3}k(t)\,dt = \frac{1}{2}\sum_{i=1}^{N}\left\{\frac{1}{|\xi^{(i)}|^{n-1}} - \frac{2\pi^{\frac{n-1}{2}}}{\Gamma\left(\frac{n-1}{2}\right)}\ln\frac{|\xi^{(i+1)}|}{|\xi^{(i)}|}\right\}$$

$$\times\int_{\frac{|\xi^{(i)}|^2}{Y^2}}^{\infty}(\sqrt{t})^{n-3}k(t)\,dt + \frac{\pi^{\frac{n-1}{2}}}{\Gamma\left(\frac{n-1}{2}\right)}\sum_{i=1}^{N}\ln\frac{|\xi^{(i+1)}|}{|\xi^{(i)}|}\int_{\frac{|\xi^{(i)}|^2}{Y^2}}^{\infty}(\sqrt{t})^{n-3}k(t)\,dt. \tag{21}$$

The first series converges uniformly in Y as $N \to \infty$, so that, passing to the limit as $N \to \infty$, we obtain the contribution

$$\frac{c_1(n)}{2}\int_0^{\infty}(\sqrt{t})^{(n-3)}k(t)\,dt + o(1) \qquad (Y \to \infty),$$

where

$$c_1(n) = \lim_{n\to\infty}\sum_{i=1}^{N}\left\{\frac{1}{|\xi^{(i)}|^{n-1}} - \frac{2\pi^{\frac{n-1}{2}}}{\Gamma\left(\frac{n-1}{2}\right)}\ln\frac{|\xi^{(i+1)}|}{|\xi^{(i)}|}\right\}.$$

We investigate the second term in (21):

$$\frac{\pi^{\frac{n-1}{2}}}{\Gamma\left(\frac{n-1}{2}\right)}\sum_{i=1}^{\infty}\ln\frac{|\xi^{(i+1)}|}{|\xi^{(i)}|}\int_{\frac{|\xi^{(i)}|^2}{Y^2}}^{\infty}k(t)(\sqrt{t})^{n-3}\,dt = \frac{\pi^{\frac{n-1}{2}}}{\Gamma\left(\frac{n-1}{2}\right)}\sum_{i=1}^{\infty}\ln|\xi^{(i+1)}|$$

$$\times\int_{\frac{|\xi^{(i)}|^2}{Y^2}}^{\frac{|\xi^{(i+1)}|^2}{Y^2}}k(t)(\sqrt{t})^{n-3}\,dt - \frac{\pi^{\frac{n-1}{2}}}{\Gamma\left(\frac{n-1}{2}\right)}\ln|\xi^{(1)}|\int_{\frac{|\xi^{(1)}|^2}{Y^2}}^{\infty}(\sqrt{t})^{n-3}k(t)\,dt.$$

It is easy to see that

$$\frac{\pi^{\frac{n-1}{2}}}{\Gamma\left(\frac{n-1}{2}\right)}\sum_{i=1}^{\infty}\ln|\xi^{(i+1)}|\int_{\frac{|\xi^{(i)}|^2}{Y^2}}^{\frac{|\xi^{(i+1)}|^2}{Y^2}}(\sqrt{t})^{n-3}k(t)\,dt = \frac{1}{2}\frac{\pi^{\frac{n-1}{2}}}{\Gamma\left(\frac{n-1}{2}\right)}\sum_{i=1}^{\infty}\ln\frac{|\xi^{(i+1)}|^2}{Y^2}$$

$$\times\int_{\frac{|\xi^{(i)}|^2}{Y^2}}^{\frac{|\xi^{(i+1)}|^2}{Y^2}}(\sqrt{t})^{n-3}k(t)\,dt + \frac{\pi^{\frac{n-1}{2}}}{\Gamma\left(\frac{n-1}{2}\right)}\sum_{i=1}^{\infty}\ln Y\int_{\frac{|\xi^{(i)}|^2}{Y^2}}^{\frac{|\xi^{(i+1)}|^2}{Y^2}}(\sqrt{t})^{n-3}k(t)\,dt. \tag{22}$$

We note that the intervals $[|\xi^{(i)}|^2/Y^2, |\xi^{(i+1)}|^2/Y^2]$ form a partition of the semi-axis for $i=1,2,\cdots$. Hence the first term in (22) may be considered as an integral sum as $Y\to\infty$, and its contribution is equal to

$$\frac{1}{2}\frac{\pi^{\frac{n-1}{2}}}{\Gamma\left(\frac{n-1}{2}\right)}\int_0^\infty \ln t\,(\sqrt{t})^{n-3}\,k(t)\,dt+o(1)\qquad (Y\to\infty).$$

The second term in (22) obviously gives the contribution

$$\frac{\pi^{\frac{n-1}{2}}}{\Gamma\left(\frac{n-1}{2}\right)}\ln Y\int_0^\infty (\sqrt{t})^{n-3}\,k(t)\,dt+o(1)\quad (Y\to\infty).$$

Combining all of these results, we obtain

$$\sum_{\substack{\{\gamma\}_\Gamma\\ \gamma\in\Gamma_\alpha,\ \gamma\neq e}}\int_{\Omega_{Y}(\alpha)} k(u(z,\ \gamma z))\,d\mu(z)=\frac{m\pi^{\frac{n-1}{2}}}{\Gamma\left(\frac{n-1}{2}\right)}\ln\ Y\int_0^\infty (\sqrt{t})^{n-3}\,k(t)\,dt$$

(23)

$$+\frac{m}{2}\frac{\pi^{\frac{n-1}{2}}}{\Gamma\left(\frac{n-1}{2}\right)}\int_0^\infty \ln t\,(\sqrt{t})^{n-3}\,k(t)\,dt+\frac{mc(n)}{2}\int_0^\infty (\sqrt{t})^{n-3}\,k(t)\,dt+o(1)\ \ (Y\to\infty),$$

where

$$c(n)=c_1(n)-\frac{2\pi^{\frac{n-1}{2}}}{\Gamma\left(\frac{n-1}{2}\right)}\ln|\xi^{(1)}|.$$

We compute all the integrals in this formula. Let $n=2k+1$, $k=1,2,\cdots$. We consider

$$\int_0^\infty (\sqrt{t})^{n-3}\,k(t)\,dt=\int_0^\infty t^{k-1}k(t)\,dt.$$

From (11) we see that

$$k(t)=\frac{(-1)^k}{\pi^k}\,Q^{(k)}(\omega)\Big|_{\omega=t}.$$

Substituting this expression, integrating by parts $k-1$ times, and using the fact that $Q^{(k-1)}(t)$ decreases faster than t^{k+1} as $t\to\infty$, we obtain

$$\int_0^\infty t^{k-1}k(t)\,dt=\frac{(k-1)!}{\pi^k}\,g(0).$$

We similarly compute the integral

$$\int_0^\infty \ln t\,t^{k-1}k(t)\,dt=\frac{(-1)^k}{\pi^k}(-1)^{k-1}\int_0^\infty (k-1)!\left(\ln t+1+\tfrac{1}{2}+\ldots+\tfrac{1}{k-1}\right)dQ(t).$$

We denote

$$c_2(k)=\begin{cases} 0, & k=1,\\ 1+\frac{1}{2}+\dots+\frac{1}{k-1}. \end{cases}$$

We note that $c_2(k) = \Psi(k)+\gamma$ (see [18]), where $\Psi(k)$ is the Euler function $\Psi(z) = d\ln\Gamma(z)/dz$ and γ is Euler's constant. Consequently

$$(24)\qquad \int_0^\infty \ln t\, t^{k-1}k(t)\,dt=\frac{(k-1)!}{\pi^k}(\Psi(k)+\gamma)g(0)-\frac{(k-1)!}{\pi^k}\int_0^\infty \ln t d(Q(t)).$$

We consider the second term in (24):

$$\int_0^\infty \ln t dQ(t)=\lim_{\varepsilon\to 0}\int_\varepsilon^\infty \ln t dQ(t)=\lim_{\varepsilon\to 0}\left\{-\ln\varepsilon Q(\varepsilon)-\int_\varepsilon^\infty \frac{Q(t)}{t}\,dt\right\}.$$

We compute the second integral:

$$\int_\varepsilon^\infty \frac{Q(t)}{t}\,dt=\int_\delta^\infty \frac{g(u)(e^u-e^{-u})}{e^u+e^{-u}-2}\,du=\frac{1}{2\pi}\int_\delta^\infty\int_{-\infty}^{+\infty}\frac{e^{\frac{u}{2}}+e^{-\frac{u}{2}}}{e^{\frac{u}{2}}-e^{-\frac{u}{2}}}e^{-iru}h(r)\,drdu$$

$$(25)\qquad =\frac{1}{2\pi}\int_\delta^\infty\int_{-\infty}^{+\infty}\frac{1+e^{-u}}{1-e^{-u}}e^{-iru}h(r)\,drdu=\frac{1}{2\pi}\int_\delta^\infty\int_{-\infty}^{+\infty}\left(1+\frac{2e^{-u}}{1-e^{-u}}\right)e^{-iru}h(r)\,drdu$$

$$=\frac{1}{2\pi}\int_\delta^\infty\int_{-\infty}^{+\infty}\left(e^{-iru}+\frac{2e^{-u}}{u}-\frac{2e^{-u}}{u}+\frac{2e^{-u}}{1-e^{-u}}e^{-iru}\right)h(r)\,drdu.$$

In (25) we have used (11) and have set $\delta=e^\epsilon+e^{-\epsilon}-2$. Continuing the computations, we divide the last integral in (25) into two parts:

$$\frac{1}{2\pi}\int_\delta^\infty\int_{-\infty}^{+\infty}\left(e^{-iru}+\frac{2e^{-u}}{u}\right)h(r)\,drdu-\frac{1}{\pi}\int_{-\infty}^{+\infty}\int_\delta^\infty\left(\frac{e^{-u}}{u}-\frac{e^{-u(1+ir)}}{1-e^{-u}}\right)h(r)\,dudr.$$

We investigate the first term:

$$\frac{1}{2\pi}\int_\delta^\infty\int_{-\infty}^{+\infty}\left(e^{-iru}+\frac{2e^{-u}}{u}\right)h(r)\,drdu=\int_\delta^\infty g(u)\,du+g(0)\int_\delta^\infty\frac{2e^{-u}}{u}\,du$$

$$=\int_\delta^\infty g(u)\,du-2g(0)\ln\delta e^{-\delta}+2g(0)\int_\delta^\infty \ln ue^{-u}du.$$

We now note that δ^2 and ϵ are infinitesimal of the same order. Hence

$$\int_0^\infty \ln td(Q(t))=\lim_{\varepsilon\to 0}\left\{-\int_\delta^\infty g(u)\,du-2g(0)\int_\delta^\infty \ln ue^{-u}du\right.$$

$$\left.+\frac{1}{\pi}\int_{-\infty}^{+\infty}\int_\delta^\infty\left(\frac{e^{-u}}{u}-\frac{e^{-u(1+ir)}}{1-e^{-u}}\right)h(r)\,dudr.\right.$$

Passing to the limit and taking into account that (see [19])

$$\int_0^\infty \left(\frac{e^{-u}}{u} - \frac{e^{-u(1+ir)}}{1-e^{-u}}\right) du = \Psi(1+ir),$$

$$\int_0^\infty \ln u e^{-u} du = -\gamma,$$

we obtain

$$\int_0^\infty \ln t dQ(t) = -\frac{1}{2} h(0) + 2g(0)\gamma \frac{1}{\pi} \int_{-\infty}^{+\infty} \Psi(1+ir) h(r)\, dr.$$

We combine these results for $n = 2k+1$ into one formula:

$$\sum_{\substack{\{\gamma\}_\Gamma \\ \gamma\in\Gamma_\alpha,\, \gamma\neq e}} \int_{\mathfrak{Q}_Y(\alpha)} k(u(z, \gamma z)\, d\mu(z) = mg(0) \ln Y + \frac{m}{2}(\Psi(k) - \gamma) g(0)$$

(25′)

$$+\frac{m}{4} h(0) - \frac{m}{2\pi} \int_{-\infty}^{+\infty} \Psi(1+ir) h(r) dr + \frac{(k-1)!}{\pi^k} \frac{m}{2} c(n) g(0) + o(1) \quad (Y \to \infty).$$

Now let $n = 2k$. We have

$$\int_0^\infty (\sqrt{t})^{n-3} k(t)\, dt = \frac{(-1)^k}{\pi^k} \int_0^\infty (\sqrt{t})^{2k-3} \int_t^\infty \frac{dQ^{(k-1)}(\omega)}{\sqrt{\omega - t}} dt$$

$$= \frac{(-1)^k}{\pi^k} \int_0^\omega dt (\sqrt{t})^{2k-3} \int_0^\infty \frac{dQ^{(k-1)}(\omega)}{\sqrt{\omega - t}} = \frac{(-1)^k}{\pi^k} \int_0^\omega t^{k-\frac{3}{2}} (\omega - t)^{-\frac{1}{2}} dt$$

$$\times \int_0^\infty dQ^{(k-1)}(\omega) = \frac{(-1)^k}{\pi^k} \frac{\Gamma\left(k - \frac{1}{2}\right)\Gamma\left(\frac{1}{2}\right)}{\Gamma(k)} \int_0^\infty \omega^{k-1} dQ^{(k-1)}(\omega).$$

In these computations we have used simple properties of $\Gamma(z)$. Computing as in the case $n = 2k+1$, we obtain

$$\int_0^\infty (\sqrt{t})^{n-3} k(t)\, dt = \frac{1}{\pi^k} \Gamma\left(k - \frac{1}{2}\right) \Gamma\left(\frac{1}{2}\right) g(0).$$

We consider the integral

$$\int_0^\infty \ln t (\sqrt{t})^{n-3} k(t)\, dt = \frac{(-1)^k}{\pi^k} \int_0^\infty \ln t\, t^{k-\frac{3}{2}} dt \int_t^\infty \frac{dQ^{(k-1)}(\omega)}{\sqrt{\omega - t}}$$

$$= \frac{(-1)^k}{\pi^k} \int_0^\omega dt \frac{\ln t\, t^{k-\frac{3}{2}}}{\sqrt{\omega - t}} \int_0^\infty dQ^{(k-1)}(\omega).$$

We further note that

$$\int_0^\omega \frac{\ln t\, t^{k-\frac{3}{2}}}{\sqrt{\omega - t}} dt = \omega^{k-1} \int_0^1 \ln t\, t^{k-\frac{3}{2}} (1-t)^{-\frac{1}{2}} dt + \omega^{k-1} \ln \omega \int_0^1 t^{k-\frac{2}{2}} (1-t)^{-\frac{1}{2}} dt.$$

It is well known (see [18]) that

$$\int_0^1 \ln t\, t^{k-\frac{3}{2}}(1-t)^{-\frac{1}{2}}\,dt = \frac{\Gamma\left(k-\frac{1}{2}\right)\Gamma\left(\frac{1}{2}\right)}{\Gamma(k)}\left(\Psi\left(k-\frac{1}{2}\right)-\Psi(k)\right),$$

$$\int_0^1 t^{k-\frac{3}{2}}(1-t)^{-\frac{1}{2}}\,dt = \frac{\Gamma\left(k-\frac{1}{2}\right)\Gamma\left(\frac{1}{2}\right)}{\Gamma(k)}.$$

These results allow us to reduce this case to the case $n = 2k+1$. In fact,

$$\int_0^\infty \ln t\,(\sqrt{t})^{n-3}k(t)\,dt = \frac{(-1)^k}{\pi^k}\frac{\Gamma\left(k-\frac{1}{2}\right)\Gamma\left(\frac{1}{2}\right)}{\Gamma(k)}\left\{\left(\Psi\left(k-\frac{1}{2}\right)\right.\right.$$

$$\left.\left.-\Psi(k)\right)\int_0^\infty \omega^{k-1}dQ^{(k-1)}(\omega) + \int_0^\infty \ln\omega\,\omega^{k-1}dQ^{(k-1)}(\omega)\right\}.$$

We hence obtain

$$\int_0^\infty \ln t\,(\sqrt{t})^{n-3}k(t)\,dt = \frac{\Gamma\left(k-\frac{1}{2}\right)\Gamma\left(\frac{1}{1}\right)}{\pi^k}\left\{g(0)\left(\Psi\left(k-\frac{1}{2}\right)-\Psi(k)\right)\right.$$

$$\left.+(\Psi(k)-\gamma)g(0)+\frac{h(0)}{2}-\frac{1}{\pi}\int_{-\infty}^{+\infty}h(r)\Psi(1+ir)\,dr\right\}$$

$$=\frac{\Gamma\left(k-\frac{1}{2}\right)}{\pi^{k-\frac{1}{2}}}\left\{g(0)\left(\Psi\left(k-\frac{1}{2}\right)-\gamma\right)+\frac{1}{2}h(0)-\frac{1}{\pi}\int_{-\infty}^{+\infty}h(r)\Psi(1+ir)\,dr\right\}.$$

Finally, (23) for $n = 2k$ takes the form

$$\sum_{\substack{\{\gamma\}_\Gamma \\ \gamma\in\Gamma_\alpha,\ \gamma\neq e}}\int_{\Omega_Y^\alpha} k(u(z,\gamma z))\,d\mu(z) = m\ln Y + \frac{m}{2}\left(\Psi\left(k-\frac{1}{2}\right)-\gamma\right)g(0)$$

$$(26)\qquad +\frac{m}{4}h(0)-\frac{m}{2\pi}\int_{-\infty}^{+\infty}\Psi(1+ir)h(r)\,d(r)$$

$$+\frac{\Gamma\left(k-\frac{1}{2}\right)}{\pi^{k-\frac{1}{2}}}\frac{m}{2}c(n)g(0)+o(1)\qquad (Y\to\infty).$$

Comparing this formula with (25′), we see that (23) does not depend on parity. Finally, for any n

$$\sum_{\substack{\{\gamma\}_\Gamma \\ \gamma\in\Gamma_\alpha,\ \gamma\neq e}}\int_{\Omega_Y^\alpha} k(u(z,\gamma z))\,d\mu(z) = mg(0)\ln Y + \frac{m}{2}\left(\Psi\left(\frac{n-1}{2}\right)-\gamma\right)g(0)$$

$$(27)\qquad +\frac{m}{4}h(0)-\frac{m}{2\pi}\int_{-\infty}^{+\infty}\Psi(1+ir)h(r)\,dr+\frac{\Gamma\left(\frac{n-1}{2}\right)}{\pi^{\frac{n-1}{2}}}\frac{m}{2}c(n)g(0)+o(1).$$

Returning to (13), we write it in the form

$$\sum_j h(\lambda_j) = \lim_{Y\to\infty}\Bigg\{ \sum_{\substack{\{\gamma\}_\Gamma \\ \gamma \text{ nonparab}}} \int_{\Omega_Y^\gamma} k(u(z,\ \gamma z))\, d\mu(z) + \sum_{\substack{\{\gamma\}_\Gamma \\ \gamma \text{ parab}}} \int_{\Omega_Y^\gamma} k(u(z,\ \gamma z))\, d\mu(z) \tag{28}$$
$$- \frac{1}{2\pi} \sum_{\alpha=1}^{m} \int_0^\infty h(r) \int_{F_Y} \left|\psi^\alpha\left(z,\ \frac{n-1}{2}+ir\right)\right|^2 d\mu(z)\, dr\Bigg\}.$$

We now note that

$$\sum_{\{\gamma\}_\Gamma} \int_{\Omega_Y^\gamma} k(u(z,\ \gamma z))\, d\mu(z) = \sum_{\{\gamma\}_\Gamma} \sum_{\gamma_0 \in \Gamma^\gamma\backslash\Gamma} \int_{F_Y} k(u(z,\ \gamma_0^{-1}\gamma\gamma_0 z))\, d\mu(z).$$

It follows from Lemma 1.5 that the integral operator with kernel

$$\sum_{\{\gamma\}_\Gamma \text{ nonparab}} \sum_{\gamma_0 \in \Gamma^\gamma\backslash\Gamma} k(u(z,\ \gamma_0^{-1}\gamma\gamma_0 z')), \quad z,\ z' \in F$$

is nuclear in $\mathfrak{H}$.

Thus the first term in (28) has a limit as $Y \to \infty$, equal to

$$\sum_{\{\gamma\}_\Gamma \text{ nonparab}} \sum_{\gamma_0 \{\Gamma^\gamma\backslash\Gamma} \int_{\gamma_0 F} k(u(z,\ \gamma z))\, d\mu(z) < +\infty.$$

Since we have independently computed the asymptotic behavior of all the terms in (28), it follows from the existence of the limit that, in particular, the following integral converges:

$$\int_{-\infty}^{+\infty} h(r)\, Tr\left(S^{-1}\left(\frac{n-1}{2}+ir\right) S'\left(\frac{n-1}{2}+ir\right)\right) dr.$$

The full trace formula finally takes the form

$$\sum_j h(\lambda_j) = \sum_{\substack{\{\gamma\}_\Gamma \\ \gamma \text{ nonparab}}} \int_{\Omega^\gamma} k(u(z,\ \gamma z))\, d\mu(z) + \frac{1}{4\pi} \int_{-\infty}^{+\infty} h(r)\, \mathrm{Tr}\left(S^{-1}\left(\frac{n-1}{2}+ir\right)\right.$$
$$\left.\times S'\left(\frac{n-1}{2}+ir\right)\right) dr + \frac{m}{2}\left(\Psi\left(\frac{n-1}{2}\right) - \gamma + \frac{\Gamma\left(\frac{n-1}{2}\right)}{\pi^{\frac{n-1}{2}}} c(n)\right) g(0) \tag{29}$$
$$- \frac{m}{2\pi} \int_{-\infty}^{+\infty} h(r)\, \Psi(1+ir)\, dr + \frac{h(0)}{4}\left(m - \mathrm{Tr}\, S\left(\frac{n-1}{2}\right)\right).$$

5.4. *Computation of the contribution from parabolic elements* (*second example*). In this subsection we compute (17) for a special discrete group Γ. We realize the space $S = S_1$ for $n = 3$ as $SL(2, C)/SU(2)$. It is not hard to compute the action of $g \in SL(2, C)$ on S: if

$$g = \begin{pmatrix} a & b \\ c & d \end{pmatrix}, \qquad z = \{y,\ x\} \in S, \quad x = x_1 + ix_2;$$

then $z' = gz$ has coordinates

$$y' = \frac{y}{|cx+d|^2 + |c|^2 y^2}, \qquad x' = \frac{(ax+b)\overline{(cx+d)} + a\bar{c}y^2}{|cx+d|^2 + |c|^2 y^2}.$$

We set

$$G = SL(2,\ C)\Big/\left\{\pm\begin{pmatrix}1&0\\0&1\end{pmatrix}\right\}$$

and for $\Gamma\subset G$ we choose the discrete group

$$SL(2,\ Z+iZ)\Big/\left\{\pm\begin{pmatrix}1&0\\0&1\end{pmatrix}\right\},$$

where Z is the ring of integers. We shall write the elements $g\in G$ as matrices, where we identify g and $-g$.

The group Γ has one maximal parabolic subgroup Γ_∞ up to conjugation in Γ, and this subgroup is given in Γ by the condition $c=0$. In accordance with this, the fundamental domain F for Γ has one parabolic vertex. For example, F may be chosen as follows:

$$F=\left\{z\in S\mid x_1^2+x_2^2+y^2>1,\ \ x_1>x_2,\ \ x_1<\tfrac{1}{2},\ \ x_2>-\tfrac{1}{2}\right\}.$$

We compute the conjugacy classes $\{\gamma\}_\Gamma$, $\gamma\in\Gamma_\infty$, $\gamma\neq e$. Let

$$\gamma=\begin{pmatrix}1&n_1+in_2\\0&1\end{pmatrix},\qquad n_1,\ n_2\in Z.$$

The conjugacy class with representative γ consists of γ and

$$\begin{pmatrix}1&-n_1-in_2\\0&1\end{pmatrix}.$$

The other conjugacy classes have the following representatives in Γ_∞:

$$\gamma_1=\begin{pmatrix}i&0\\0&-i\end{pmatrix},\quad \gamma_2=\begin{pmatrix}i&1\\0&-i\end{pmatrix},\quad \gamma_3=\begin{pmatrix}i&-i\\0&-i\end{pmatrix},\quad \gamma_4=\begin{pmatrix}i&1-i\\0&-i\end{pmatrix}.$$

It is easy to see that the centralizers of these representatives are as follows: in the first case Γ^γ consists of

$$\begin{pmatrix}1&m_1+im_2\\0&1\end{pmatrix},$$

where $m_1, m_2\in Z$; in the other cases

$$\Gamma^1=\Gamma^{\gamma_1}=\left\{\begin{pmatrix}1&0\\0&1\end{pmatrix},\ \begin{pmatrix}i&0\\0&-i\end{pmatrix},\ \begin{pmatrix}0&1\\-1&0\end{pmatrix},\ \begin{pmatrix}0&i\\i&0\end{pmatrix}\right\},$$

$$\Gamma^2=\Gamma^{\gamma_2}=\left\{\begin{pmatrix}1&0\\0&1\end{pmatrix},\ \begin{pmatrix}i&1\\0&-i\end{pmatrix},\ \begin{pmatrix}i&0\\2&-i\end{pmatrix},\ \begin{pmatrix}-1&i\\2i&1\end{pmatrix}\right\},$$

$$\Gamma^3=\Gamma^{\gamma_3}=\left\{\begin{pmatrix}1&0\\0&1\end{pmatrix},\ \begin{pmatrix}i&-i\\0&-i\end{pmatrix},\ \begin{pmatrix}1&-1\\2&-1\end{pmatrix},\ \begin{pmatrix}i&0\\2i&-i\end{pmatrix}\right\},$$

$$\Gamma^4=\Gamma^{\gamma_4}=\left\{\begin{pmatrix}1&0\\0&1\end{pmatrix},\ \begin{pmatrix}i&1-i\\0&-i\end{pmatrix},\ \begin{pmatrix}i&0\\1+i&-i\end{pmatrix},\ \begin{pmatrix}1&-1-i\\1-i&-1\end{pmatrix}\right\}.$$

We consider the series

$$\sum_{\substack{\{\gamma\}_\Gamma\\ \gamma\in\Gamma_\infty,\ \gamma\neq e}}\ \int\limits_{\substack{\cup\gamma_0 F_\gamma\\ \gamma_0\in\Gamma^\gamma\setminus\Gamma}} k(u(z,\ \gamma z))\,d\mu(z). \tag{30}$$

Let

$$\gamma=\begin{pmatrix}1 & n_1+in_2\\ 0 & 1\end{pmatrix}, \quad n_1,\ n_2\in Z.$$

We note that, for such γ, γ and γ^{-1} give the same contribution to (30), so that it is easy to see that the contribution to (30) from all the conjugacy classes of type γ is equal to half the contribution from the elements of the groups Γ'_∞ in Example 1. In other words,

$$\sum_{\substack{\{\gamma\}_\Gamma\\ \gamma=\begin{pmatrix}1 & n_1+in_2\\ 0 & 1\end{pmatrix}\\ \gamma\neq e}} \int_{\substack{\cup\gamma_0F_Y\\ \gamma_0\in\Gamma^\gamma\setminus\Gamma}} k(u(z,\gamma z))\,d\mu(z)=\frac{1}{2}g(0)\ln Y+\frac{1}{4}(\Psi(1)-\gamma)g(0) \tag{31}$$
$$+\frac{1}{8}h(0)-\frac{1}{4\pi}\int_{-\infty}^{+\infty}\Psi(1+ir)h(r)\,dr+\frac{\Gamma(1)}{\pi}\frac{c(3)}{4}g(0)+o(1)\qquad (Y\to\infty).$$

It remains for us to investigate the sum

$$\sum_{j=1}^{4}\int_{\substack{\cup\gamma_0F_Y\\ \gamma_0\in\Gamma^j\setminus\Gamma}} k(u(z,\gamma_jz))\,d\mu(z). \tag{32}$$

We note that this sum is equal to

$$\sum_{j=1}^{4}\sum_{\gamma_0\in\Gamma^j\setminus\Gamma}\int_{F_Y} k(u(z,\gamma_0^{-1}\gamma_j\gamma_0z))\,d\mu(z)$$
$$=\frac{1}{4}\sum_{j=1}^{4}\sum_{\gamma_0\in\Gamma}\int_{F_Y} k(u(z,\gamma_0^{-1}\gamma_j\gamma_0z))\,d\mu(z)=\frac{1}{4}\sum_{j=1}^{4}\int_{\substack{\cup\gamma_0F_Y\\ \gamma_0\in\Gamma}} k(u(z,\gamma_jz))\,d\mu(z).$$

We set $S^Y=\{z\in S\,|\,y\leq Y\}$. We introduce polar coordinates r and φ in the hyperplane $y=\text{const}$ of the space S. We consider the region $B_1^Y\subset S$ lying inside the closed surface given by the equation $r=y\sqrt{1/yY-1}$. We note that the region B_1^Y is the image of the set $S-S^Y$ under the mapping

$$\begin{pmatrix}0 & 1\\ -1 & 0\end{pmatrix}.$$

From the definition of F it is easy to see that the set

$$S^Y-B_1^Y-\bigcup_{\gamma_0\in\Gamma}\gamma_0F_Y$$

lies in the region $\{y<1,\ r>y\sqrt{1/yY_0-1}\}$, where Y_0 is a sufficienttly large fixed positive number. It is also obvious that

$$\mu(S^Y-B_1^Y-\bigcup_{\gamma_0\in\Gamma}F_Y)=o(1)\qquad (Y\to\infty).$$

It remains to note that

$$\int_0^1\frac{dy}{y^3}\int_{r>y\sqrt{\frac{1}{Y_0y}-1}} dr\,k\left(\frac{4r^2}{y^2}\right)$$

converges absolutely. It hence follows that

$$\int\limits_{\substack{\cup\gamma_0 F_Y\\ \gamma_0\in\Gamma}} k(u(z,\ \gamma_1 z))\,d\mu(z) = \int\limits_{S^Y - B_1^Y} k(u(z,\ \gamma_1 z))\,d\mu(z) + o(1) \qquad (Y\to\infty). \tag{33}$$

We compute the integral on the right in (33):

$$\int\limits_{S^Y - B_1^Y} k(u(z,\ \gamma_1 z))\,d\mu(z) = 2\pi\int\limits_{\frac{1}{Y}}^{Y}\frac{dy}{y^3}\int\limits_0^\infty r\,dr\,k\left(\frac{4r^2}{y^2}\right) + 2\pi\int\limits_0^{\frac{1}{Y}}\frac{dy}{y^3}\int\limits_{y\sqrt{\frac{1}{yY}-1}}^{\infty} r\,dr\,k\left(\frac{4r^2}{y^2}\right).$$

After a natural change of variables, we transform the first integral to the form

$$\frac{\pi}{4}\int\limits_{\frac{1}{Y}}^{Y}\frac{dy}{y}\int\limits_0^\infty k(t)\,dt = \frac{1}{2}\ln Y\,g(0),$$

where the function $g(u)$ is from the Selberg map (11). Integrating by parts, we find that the second integral equals

$$-\frac{\ln 2}{2}g(0) + \frac{\pi}{4}\int\limits_0^\infty \ln(t+4)\,k(t)\,dt.$$

Consequently

$$\int\limits_{\substack{\cup\gamma_0 F_Y\\ \gamma_0\in\Gamma}} k(u(z,\ \gamma_1 z))\,d\mu(z) = \frac{1}{2}g(0)\ln Y - \frac{\ln 2}{2}g(0) + \frac{\pi}{4}\int\limits_0^\infty \ln(t+4)k(t)\,dt. \tag{34}$$

The remaining integrals in (32) are investigated in the same way. Here we indicate the regions B_j^Y, $j=2,3,4$, which are analogous to B_1^Y. In the case γ_2, $x = i/2 + re^{i\varphi}$, the boundary of B_2^Y is given by the equation $r = y\sqrt{1/4yY-1}$. For γ_3, $x = 1/2 + re^{i\varphi}$, the boundary of B_3^Y is given by the equation $r = y\sqrt{1/4yY-1}$. Finally, in the case γ_4, $x = (1+i)/2 + re^{i\varphi}$, the equation of the boundary has the form $r = y\sqrt{1/2yY-1}$. The corresponding contributions to (32) are equal as $Y\to\infty$ to

$$\int\limits_{\substack{\cup\gamma_0 F_Y\\ \gamma_0\in\Gamma}} k(u(z,\ \gamma_2 z))\,d\mu(z) = \frac{1}{2}g(0)\ln Y + \frac{\pi}{4}\int\limits_0^\infty \ln(t+4)\,k(t)\,dt + o(1),$$

$$\int\limits_{\substack{\cup\gamma_0 F_Y\\ \gamma_0\in\Gamma}} k(u(z,\ \gamma_3 z))\,d\mu(z) = \frac{1}{2}g(0)\ln Y + \frac{\pi}{4}\int\limits_0^\infty \ln(t+4)\,k(t)\,dt + o(1),$$

$$\int\limits_{\substack{\cup\gamma_0 F^Y\\ \gamma_0\in\Gamma}} k(u(z,\ \gamma_4 z))\,d\mu(z) = \frac{1}{2}g(0)\ln Y - \frac{\ln 2}{4}g(0) + \frac{\pi}{4}\int\limits_0^\infty \ln(t+4)\,k(t)\,dt + o(1).$$

Adding these results, we obtain

$$\sum_{j=1}^{4}\int\limits_{\substack{\cup\gamma_0 F_Y\\ \gamma_0\in\Gamma^j\setminus\Gamma}} k(u(z,\ \gamma_j z))\,d\mu(z) = \tag{35}$$

$$= \frac{1}{2} g(0) \ln Y - \frac{3}{16} \ln 2\, g(0) + \frac{\pi}{4} \int_0^\infty k(t) \ln(t+4)\, dt + o(1) \quad (Y \to \infty).$$

We compute the integral

$$\frac{\pi}{4} \int_0^\infty k(t) \ln(t+4)\, dt = -\frac{1}{4} \int_0^\infty \ln(t+4)\, dQ(t) = \frac{1}{2} \ln 2\, g(o)\, \frac{1}{4} \int_0^\infty \frac{Q(t)}{t+4}\, dt.$$

Replacing $Q(t)$ by its expression in terms of the function $h(r)$ in the Selberg map (11), we obtain

$$\int_0^\infty \frac{Q(t)}{t+4}\, dt = \frac{1}{2\pi} \int_0^\infty \int_{-\infty}^{+\infty} \frac{1-e^{-u}}{1+e^{-u}} e^{-iru} h(r)\, dr\, du$$

$$= \frac{1}{2\pi} \int_0^\infty \int_{-\infty}^{+\infty} \left(1 - \frac{2e^{-u}}{1+e^{-u}}\right) e^{-iru} h(r)\, dr\, du = \frac{h(0)}{2} = \frac{1}{\pi} \int_0^\infty \int_{-\infty}^{+\infty} \frac{e^{-(1+ir)u}}{1+e^{-u}} h(r)\, dr\, du.$$

Continuing the computations, we note (see [18]) that

$$\int_0^\infty \frac{e^{-(1+ir)u}}{1+e^{-u}}\, du = \frac{1}{2} \left\{ \Psi\left(1 + \frac{ir}{2}\right) - \Psi\left(\frac{1+ir}{2}\right) \right\}.$$

Finally, we obtain

$$\frac{\pi}{4} \int_0^\infty k(t) \ln(t+4)\, dt = \frac{\ln 2}{2} g(0) + \frac{h(0)}{8} + \frac{1}{8\pi} \int_{-\infty}^{+\infty} \left(\Psi\left(\frac{1+ir}{2}\right) - \Psi\left(1 + \frac{ir}{2}\right) \right) h(r)\, dr.$$

Substituting this result in (35) and taking (11) into account, we have

$$\sum_{\substack{\{\gamma\}_\Gamma \\ \gamma \in \Gamma_\infty,\ \gamma \neq e}} \int_{\substack{\bigcup_{\gamma_0} F_Y \\ \gamma_0 \in \Gamma^\gamma \backslash \Gamma}} k(u(z, \gamma z))\, d\mu(z) = g(0) \ln Y + \left(\frac{5}{16} \ln 2 - \frac{\gamma}{2} + \frac{c(3)}{4\pi}\right) g(0)$$

(36)
$$+ \frac{h(0)}{4} + \frac{1}{8\pi} \int_{-\infty}^{+\infty} \left(\Psi\left(\frac{1+ir}{2}\right) - \Psi\left(1 + \frac{ir}{2}\right) \right) h(r)\, dr$$

$$- \frac{1}{4\pi} \int_{-\infty}^{+\infty} \Psi(1+ir)\, h(r)\, dr + o(1) \quad (Y \to \infty).$$

To derive the full trace formula we must compute the contribution from the normalized Eisenstein series $\psi(z,s)/\sqrt{|D|}$, which, it is not hard to see, coincides with (16) for $m = 1$, and we must perform the summation in (13) and add to (36).

The final trace formula has the form

$$\sum_j h(\lambda_j) = \sum_{\substack{\{\gamma\}_\Gamma \\ \gamma \text{ nonparab}}} \int_{\Omega\gamma} k(u(z, \gamma z))\, d\mu(z) + \frac{1}{4\pi} \int_{-\infty}^{+\infty} h(r) \frac{S'(1+ir)}{S(1+ir)}\, dr$$

$$+ \left(\frac{5}{16} \ln 2 - \frac{\gamma}{2} + \frac{1}{4\pi} c(3)\right) g(0) + \frac{h(0)}{4}(1 - S(1))$$

$$+ \frac{1}{8\pi} \int_{-\infty}^{+\infty} h(r) \left\{ \Psi\left(\frac{1+ir}{2}\right) - \Psi\left(1 + \frac{ir}{2}\right) - 2\Psi(1+ir) \right\} dr.$$

Bibliography

[1] A. Selberg, *Harmonic analysis and discontinuous groups in weakly symmetric Riemannian spaces with applications to Dirichlet series,* J. Indian Math. Soc. **20** (1956), 47-87. MR **19**, 531.

[2] ———, *Discontinuous groups and harmonic analysis,* Proc. Internat. Congress Math. (Stockholm, 1962), Inst. Mittag-Leffler, Djursholm, 1963, pp. 177-189. MR **31** #372.

[3]* R. P. Langlands, *On functional equations, satisfied by Eisenstein series,* preprint, 1965.

[4] L. D. Faddeev, *Expansion in eigenfunctions of the Laplace operator on the fundamental domain of a discrete group on the Lobačevskiĭ plane,* Trudy Moskov. Mat. Obšč. **17** (1967), 323-350 = Trans. Moscow Math. Soc. **17** (1967), 357-386. MR **38** #5062.

[5] A. B. Venkov, V. L. Kalinin and L. D. Faddeev, *Nonarithmetic deduction of the Selberg trace formula for the upper half-plane,* Zap. Naučn. Sem. Leningrad. Otdel. Mat. Inst. Steklov. (LOMI) **37** (1973), 3-38 = J. Soviet Math. (to appear).

[6] H. Petersson, *Zur analytischen Theorie der Grenzkreisgruppen.* I, Math. Ann. **115** (1937), 23-67.

[7] A. B. Venkov, *Expansion in automorphic eigenfunctions of the Laplace operator and the Selberg trace formula in the space $SO_0(n,1)/SO(n)$,* Dokl. Akad. Nauk SSSR **200** (1971), 266-268 = Soviet Math. Dokl. **12** (1971), 1363-1366. MR **45** #452.

[8] S. Helgason, *Differential geometry and symmetric spaces,* Pure and Appl. Math., vol. 12, Academic Press, New York, 1962. MR **26** #2986.

[9] Harish-Chandra, *Representations of a semisimple Lie group on a Banach space.* I, Trans. Amer. Math. Soc. **75** (1953), 185-243. MR **15**, 100.

[10] R. Takahashi, *Sur les représentations unitaires des groupes de Lorentz généralisés,* Bull. Soc. Math. France **91** (1963), 289-433. MR **31** #3544.

[11] Mitsuo Sugiura, *Conjugate classes of Cartan subalgebras in real semisimple Lie algebras,* J. Math. Soc. Japan **11** (1959), 374-434. MR **26** #3827.

[12] E. Kamke, *Differentialgleichungen. Lösungsmethoden und Lösungen.* Teil. 1: *Gewöhnliche Differentialgleichungen,* 3rd ed., Math. und ihre Anwendungen in Physik und Technik, Band 18, Geest & Portig, Leipzig, 1944. MR **9**, 33.

[13] A. Selberg, *Recent developments in the theory of discontinuous groups of motions of symmetric spaces,* Proc. Fifteenth Scandinavian Congress (Oslo, 1968), Lecture Notes in Math., vol. 118, Springer-Verlag, Berlin, 1970, pp. 99-120. MR **41** #8595.

[14] H. Garland and M. S. Raghunathan, *Fundamental domains for lattices in (R-) rank 1 semisimple Lie groups,* Ann. of Math. (2) **92** (1970), 279-326. MR **42** #1943.

[15] W. F. Stinespring, *A sufficient condition for an integral operator to have a trace,* J. Reine Angew. Math. **200** (1958), 200-207. MR **20** #5431.

[16] R. Godement, *Introduction à la théorie de Langlands,* Séminaire Bourbaki. Vol. 1966/67, Exposé 321, Benjamin, New York, 1968. MR **37** #10.

[17] M. G. Kreĭn, *Some new studies in the theory of perturbations of self-adjoint operators,* First Math. Summer School, part I, "Naukova Dumka", Kiev, 1964, pp. 103-187. (Russian) MR **32** #2919.

[18] I. M. Ryžik and I. S. Gradšteĭn, *Tables of integrals, sums, series and products,* 3rd ed., GITTL, Moscow, 1951; English and German transl., VEB Deutscher Verlag, 1957. MR **14**, 643; **22** #3120.

[19] E. T. Whittaker and G. N. Watson, *A course of modern analysis,* Cambridge Univ. Press, New York, 1927; 4th ed., reprint, 1962. MR **31** #2375.

[20] Harish-Chandra, *Automorphic forms on semisimple Lie groups,* Lecture Notes in Math., no. 62, Springer-Verlag, Berlin and New York, 1968. MR **38** #1216.

Translated by
N. KOBLITZ

---

* *Editor's note.* A brief account of this work has been published (*Eisenstein series,* in Algebraic Groups and Discontinuous Subgroups (Proc. Sympos. Pure Math., vol. 9, Boulder, Colo., 1965), Amer. Math. Soc., Providence, R. I., 1966, pp. 235-252. MR **40** #2784).

Trudy Mat. Inst. Steklov.
125 (1973)

Proc. Steklov Inst. Math.
125 (1973)

ON THE SOLVABILITY OF THE DIRICHLET PROBLEM FOR SOME CLASSES OF SECOND ORDER ELLIPTIC SYSTEMS

A. V. IVANOV

Abstract. The classes of quasilinear elliptic systems of second order are distinguished and examined from the point of view of solvability of the Dirichlet problem. Systems of the first class have the form $A^s_{ij}(x,u^t,u^s_x)u^s_{ij} = B^s(x,u^t,u^s_x)$, $s = 1,\cdots,N$. The existence theorems proved for these systems embrace the corresponding results for the case of a single elliptic equation, obtained by the author in previous articles. The second class includes systems of the form $A_{ij}(x,u^t,|u_x|)u^s_{ij} = B^s(x,u^t,u^t_x)$, $s = 1,\cdots,N$, where $|u_x| = (\sum_{t,l}|u^t_l|^2)^{1/2}$. The existence theorems proved for these systems extend the familiar results of Ladyženskaja and Ural′ceva on the solvability of the Dirichlet problem for uniformly elliptic systems with common principal parts.

Bibliography: 6 items.

We consider in a bounded domain $\Omega \subset E_n$ a system of quasilinear elliptic equations of the form

$$A^s_{ij}(x, u^t, u^t_x)\, u^s_{ij} = B^s(x, u^t, u^t_x), \quad s = 1, \ldots, N, \tag{0.1}$$

where

$$x = (x_1, \ldots, x_n),\ f(x, u^t, u^t_x) \equiv f(x, u^1, \ldots, u^t, \ldots, u^N, u^1_x, \ldots, u^t_x, \ldots, u^N_x),$$
$$u^t_x = (u^t_1, \ldots, u^t_n),\ u^t_k = \frac{\partial u^t}{\partial x_k},\quad u^s_{ij} = \frac{\partial^2 u^s}{\partial x_i \partial x_j},\quad A^s_{ij}u^s_{ij} = \sum_{i,\, j=1}^{n} A^s_{ij}u^s_{ij}.$$

The ellipticity of the system (0.1) lies in the fact that, with arbitrary $\xi \neq 0$ and arbitrary $s = 1,\cdots,N$

$$A^s_{ij}\xi_i\xi_j > 0. \tag{0.2}$$

We denote by $\lambda^s = \lambda^s(x,u^t,p)$ and $\Lambda^s = \Lambda^s(x,u^t,p)$ the minimum and maximum eigenvalues respectively of the matrix $\| A^s_{ij}(x,u^t,p^t)\|$; obviously, for any ξ,

$$\lambda^s\xi^2 \leqslant A^s_{ij}\xi_i\xi_j \leqslant \Lambda^s\xi^2. \tag{0.3}$$

In the case $n > 2$ the existence theorem for the solution of the Dirichlet problem was proved by O. A. Ladyženskaja and N. N. Ural′ceva [1] for uniformly elliptic systems (with common principal parts) of the form

$$A_{ij}(x, u^t)\, u^s_{ij} = B^s(x, u^t, u^t_x), \quad s = 1, \ldots, N. \tag{0.4}$$

In the present article, the structure of the elliptic systems for which the solvability of the Dirichlet problem is proved is generalized in a number of ways.

In Chapter I of the article we distinguish certain classes of (in general nonuniformly) elliptic systems with different matrices $\|A^s_{ij}\|$ of the form

$$A^s_{ij}(x, u^t, u^s_x)\, u^s_{ij} = B^s(x, u^t, u^s_x), \quad s = 1, \ldots, N. \tag{0.5}$$

AMS (MOS) subject classifications (1970). Primary 35J25, 35J55, 35J60.

The existence theorems proved for these systems embrace fully the corresponding results for the case of a single elliptic equation

$$A_{ij}(x, u, u_x)u_{ij} = B(x, u, u_x), \tag{0.6}$$

previously obtained by the author in [**2**–**4**].

In Chapter II of the article we examine certain classes of (in general nonuniformly) elliptic systems with common matrices $\|A^s_{ij}\| = \|A_{ij}\|$ of the form

$$A_{ij}(x, u^t, |u_x|)u^s_{ij} = B^s(x, u^t, u^t_x), \quad s = 1, \ldots, N, \tag{0.7}$$

which include as a particular case systems of the form (0.4). The theorems on the solvability of such systems embrace in particular the above-mentioned result of Ladyženskaja and Ural′ceva.

When proving the solvability of the Dirichlet problem for the elliptic systems mentioned, an important role is played by the result of Ladyženskaja and Ural′ceva on estimation of the Hölder norm of the gradient of the solution of a single uniformly elliptic equation. By means of this estimate, the proof of the existence theorem is in essence reduced (see §§3 and 6 below) to proving a priori estimates for $\max_{\bar\Omega}(|v^s| + |v^s_x|)$ for the solutions of the family of problems

$$A^s_{ij}v^s_{ij} = \tau B^s \text{ in } \Omega, \quad v^s = \tau\varphi^s \text{ on } \partial\Omega, \quad s = 1, \ldots, N, \quad \tau \in [0, 1]. \tag{0.8}$$

In the present article the main attention has to focus on estimates of $\max_{\partial\Omega}|v_x|$ and $\max_{\bar\Omega}|v_x|$. It may be noted that the a priori estimate of $\max_{\partial\Omega}|v_x|$ obtained in §§1 and 4 is for more general systems than (0.5) or (0.7). When proving the estimate for $\max_{\partial\Omega}|v_x|$ we generalize and develop certain ideas put forward by Ladyženskaja and Ural′ceva in [**1**] and by Serrin in [**5**]. When obtaining the a priori estimate for $\max_\Omega|v_x|$ (see §§2 and 5) we extend some ideas used in our own articles [**2**–**4**]. In certain cases an estimate is also obtained for $\max_\Omega|v|$ when proving the existence theorems. In such cases the relevant existence theorem is of an unconditional type (see Theorems 3.3 and 6.3).

A distinctive feature of the present article, as in the case of [**2**–**4**], is that the admissible growth of the right sides of the systems (0.1) is as a rule described in terms of majorizing functions $\mathscr{E}^s_1 = A^s_{ij}p^s_ip^s_j$ and $\mathscr{E}^s_2 = \operatorname{tr}\|A^s_{ij}\| \cdot |p^s|$ and their sums. We know from negative results obtained for elliptic equations (see, for example, [**5**]) that a faster growth of the right sides than $\mathscr{E}_1 + \mathscr{E}_2$ is inadmissible for solvability of the Dirichlet problem, no matter what the sufficiently smooth initial data. Some of the conditions obtained in the present article on the growth of the coefficients thus link up with limiting conditions.

I. SYSTEMS WITH UNEQUAL MATRICES $\|A^s_{ij}\|$

§1. Estimates for $\max_{\partial\Omega;\, s=1,\ldots,N}|u^s_x|$

When obtaining estimates for $\max_{\partial\Omega;\, s=1,\ldots,N}|u^s_x|$ (here and in §4), use will be made of the following familiar results.

LEMMA 1.1. (SERRIN [5]). *Let the function* $u(x) \in C^2(\Omega) \cap C^1(\overline{\Omega})$ *satisfy in the domain* Ω *the elliptic equation*

$$a_{ij}(x, u, u_x) u_{ij} - b(x, u, u_x) = 0 \tag{1.1}$$

(*the coefficients* a_{ij} *and* b *are continuous functions of their arguments and the quadratic form* $a_{ij}\xi_i\xi_j$ *is positive definite, i.e.* $a_{ij}\xi_i\xi_j > 0$ *for all* $\xi \neq 0$). *Let the (barrier) function* $\omega(x) \in C^2(\Omega) \cap C^1(\overline{\Omega})$ *satisfy in* Ω *the inequality*

$$a_{ij}(x, \omega + c, \omega_x)\omega_{ij} - b(x, \omega + c, \omega_x) \leqslant 0 \tag{1.2}$$

with an arbitrary constant $c \geqq 0$. *Then, if* $u(x) \leqq \omega(x)$ *on the boundary* $\partial\Omega$ *of* Ω, *we have* $u(x) \leqq \omega(x)$ *throughout* Ω.

We shall now assume that the boundary $\partial\Omega$ of Ω belongs to the class C^3. A function $d = d(x)$ can then be defined (see [5], p. 421), representing the distance from the point x to the boundary $\partial\Omega$, in a domain N adjoining the boundary surface $\partial\Omega$; and in the domain N (characterized by the inequality $d(x) < \delta_0$, where δ_0 is determined solely by the boundary $\partial\Omega$) the function $d(x)$ belongs to the class C^2.

LEMMA 1.2 (SERRIN [5]). *Let the boundary* $\partial\Omega$ *of* Ω *belong to the class* C^3. *Let*

$$\omega(x) = \varphi(x) + h(d), \tag{1.3}$$

where $\varphi(x) \in C^3(\overline{\Omega})$, $h(d) \in C^2(0,\delta) \cap C[0,\delta]$, $d = d(x)$, $\delta \leqq \delta_0$, *and* $h' > 0$. *An upper bound for*

$$a_{ij}(x, \omega + c, \omega_x)\omega_{ij} - b(x, \omega + c, \omega_x) \tag{1.4}$$

is then given by

$$F\frac{h''}{h'^2} + K\,\mathbf{tr}\,\|a_{ij}\|h' + a_{ij}\varphi_{ij} - b, \tag{1.5}$$

where $F = a_{ij}(p - p_0)_i(p - p_0)_j$, $p = p_0 + \nu h'$, ν *is the unit inward normal to the surface* $\partial\Omega$ *at the point* $y = y(x)$ *nearest to the point* $x \in N$, $a_{ij} = a_{ij}(x, \omega + c, p)$, $b = b(x, \omega + c, p)$, $K = \sup_{i=1,\cdots,n-1} k_i(y)$ *are the principal curvatures of the surface* $\partial\Omega$ *at the point* $y \in \partial\Omega$.

If the domain Ω *is convex, an upper bound for* (1.4) *is also given by*

$$F\frac{h'' + Kh'}{h'^2} - k\,\mathbf{tr}\,\|a_{ij}\| + a_{ij}\varphi_{ij} - b, \tag{1.5'}$$

where $k = \inf_{i=1,\cdots,n-1} k_i(y)$, *and* $k \geqq 0$.

THEOREM 1.1. *Let the coefficients* $A^s_{ij}(x, u^t, p^t)$ *and* $B^s(x, u^t, p^t)$ *be continuous functions of their arguments and satisfy the conditions*

$$|B^s(x, u^t, p^t)| \leqslant \psi(|p^s|)|p^s|\mathscr{E}^s_1 + \delta(|p^s|)\mathscr{E}^s_2, \quad s = 1, \ldots, N, \tag{1.6}$$

where $\mathscr{E}^s_1 = A^s_{ij}p^s_ip^s_j$, $\mathscr{E}^s_2 = \mathrm{tr}\,A^s|p^s|$, $A^s = \|A^s_{ij}\|$, $\psi(\rho)$ *is a positive decreasing function satisfying the condition* $\int^{\infty}(1/\rho^2\psi(\rho))d\rho = \infty$, *and* $\delta(\rho)$ *is a positive decreasing function such that* $\delta(\rho) \to 0$ *as* $\rho \to \infty$. *Let* $u(x) = \{u^t(x)\}_1^N$ *be a solution of the Dirichlet problem*

$$A^s_{ij}(x, u^t, u^t_x)u^s_{ij} = B^s(x, u^t, u^t_x) \text{ in } \Omega, \quad s = 1, \ldots, N;$$
$$u^s = \varphi^s \text{ on } \partial\Omega, \quad s = 1, \ldots, N, \tag{1.7}$$

such that $u^s \in C^2(\Omega) \cap C^1(\overline{\Omega})$, $s = 1, \cdots, N$; $\max_{\overline{\Omega}; s=1,\cdots,N} |u^s| \leq m$. *Finally, let the domain* Ω *be strictly convex, and let its boundary* $\partial\Omega$ *and the functions* φ^s *(defined in* $\overline{\Omega}$*) belong to the class* C^3. *Then* $\max_{\partial\Omega; s=1,\cdots,N} |u^s_n|$, *where n is the inward normal to* $\partial\Omega$ *at the point* $y = y(x) \in \partial\Omega$, *is bounded by a quantity* M_0:

$$\max_{\substack{\partial\Omega \\ s=1,\ldots,N}} |u^s_n| \leqslant M_0, \tag{1.8}$$

which depends only on m, the functions $\delta(\rho)$ *and* $\psi(\rho)$ *of condition* (1.6), *the norm* $\|\varphi^s\|_{C^2(\overline{\Omega})}$, *and on* k^{-1} *and K, where*

$$k = \inf_{\substack{y \in \partial\Omega \\ i=1,\ldots,n-1}} k_i(y), \quad K = \sup_{\substack{y \in \partial\Omega \\ i=1,\ldots,n-1}} k_i(y)$$

and $k_i(y)$ *are the principal curvatures of the surface* $\partial\Omega$ *at the point* $y \in \partial\Omega$.

PROOF. We fix an arbitrary $s = 1, \cdots, N$ and consider the corresponding equation

$$A^s_{ij}(x, u^t, u^t_x)u^s_{ij} = B^s(x, u^t, u^t_x) \tag{1.9}$$

as an equation of the form

$$a_{ij}(x, u^s, u^s_x)u^s_{ij} = b(y, u^s, u^s_x), \tag{1.10}$$

where $a_{ij} = A^s_{ij}$ and $b = B^s$. Obviously a_{ij} and b depend continuously on their arguments. We introduce a function

$$\omega^s(x) = \varphi^s(x) + h(d) \tag{1.11}$$

in the domain $N \subset \Omega$, defined by the condition $d(x) < \delta$, where $h(d)$ and δ are the same as in Lemma 2. By Lemma 2, with an arbitrary constant $c > 0$

$$a_{ij}(x, \omega^s + c, \omega^s_x)\omega^s_{ij} - b(x, \omega^s + c, \omega^s_x)$$
$$\leqslant F^s \frac{h'' + Kh'}{h'^2} - k \operatorname{tr} \|a_{ij}\| h' + a_{ij}\varphi^s_{ij} - b, \tag{1.12}$$

where $F^s = a_{ij}(p^s - p^s_0)_i(p^s - p^s_0)_j$, $p^s = p^s_0 + \nu h'$, ν is the unit inward normal to $\partial\Omega$ at the point $y \in \partial\Omega$, $a_{ij} = a_{ij}(x, \omega^s + c, p^s)$, and $b = b(x, \omega^s + c, p^s)$. It may be assumed without loss of generality that $\operatorname{tr}\|a_{ij}\| = \operatorname{tr} A^s = 1$. We then have

$$|a_{ij}\varphi^s_{ij}| \leqslant c_\varphi, \quad c_\varphi = \max_{s=1,\ldots,N} \|\varphi^s\|_{C^2(\Omega)}. \tag{1.13}$$

Recalling condition (1.6), we see that

$$|b| \leqslant \psi(|p^s|)|p^s|\mathscr{E}^s_1 + \delta(|p^s|)|p^s|, \tag{1.14}$$

where $\mathscr{E}^s_1 = a_{ij}p^s_i p^s_j$. Obviously

$$h' - c_\varphi \leqslant |p^s| \leqslant h' + c_\varphi, \tag{1.15}$$

where we shall assume that $h' > c_\varphi$. In addition, it may easily be shown that

$$\mathscr{E}^s_1 \leqslant 2F^s + 4c^2_\varphi. \tag{1.16}$$

It then follows from (1.12) – (1.16) that

$$a_{ij}(x, \omega^s + c, \omega^s_x)\omega^s_{ij} - b(x, \omega^s + c, \omega^s_x) \leqslant F^s \frac{h'' + Kh'}{h'^2} - kh' + c_\varphi + \psi(|p^s|)$$

$$\times (2F^s + 4c_\varphi^2) + \delta(|p^s|)|p^s| = F^s h'\left\{\frac{h''}{h'^3} + \frac{K}{h'^2} + \frac{2\psi(h' - c_\varphi)}{h'}(h' + c_\varphi)\right\} \tag{1.17}$$

$$+ h'\left\{-k + \frac{c_\varphi}{h'} + 4c_\varphi^2 \frac{\psi(|p^s|)|p^s|}{h'} + \frac{\delta(|p^s|)|p^s|}{h'}\right\}.$$

Noting that $|p^s| \leqq h' + c_\varphi \leqq 2h'$, we obtain from this

$$\begin{aligned} a_{ij}\omega^s_{ij} - b \leqslant F^s h'&\left\{\frac{h''}{h'^3} + \left[\frac{K}{h'^2} + 4\psi(h' - c_\varphi)\right]\right\} \\ &+ h'\left\{-k + \frac{c_\varphi}{h'} + 8c_\varphi^2\psi(h' - c_\varphi) + 2\delta(h' - c_\varphi)\right\}. \end{aligned} \tag{1.18}$$

We shall assume that $h' \geqq \alpha$, where $\alpha > c_\varphi$ is so large that the second summand is nonpositive. The α here depends only on c_φ, k, and the functions $\psi(\rho)$ and $\delta(\rho)$. Then

$$a_{ij}\omega^s_{ij} - b \leqslant F^s h'\left\{\frac{h''}{h'^3} + \left[\frac{K}{h'^2} + 4\psi(h' - c_\varphi)\right]\right\} \equiv F^s h'\left\{\frac{h''}{h'^3} + \Psi(h')\right\}. \tag{1.19}$$

It follows from the conditions imposed on the function ψ that also $\int^\infty (1/\rho^2\Psi(\rho))d\rho = \infty$. But a function $h(d) = h(d, q)$ was constructed under this condition in [5] such that

$$\frac{h''}{h'^3} + \Psi(h') = 0, \qquad h(0) = 0,\ h(\delta) = q,\ h'(d) \geqslant \alpha, \tag{1.20}$$

where $q > c_\varphi > \max_{\bar\Omega}|\varphi| + m$. Then on the boundary of the domain N, i.e., at points x at which $d(x) = 0$ or $d(x) = \delta$, we obviously have

$$u^s(x) \leqslant \omega^s(x). \tag{1.21}$$

Applying Lemma 1, we now find that

$$u^s(x) \leqslant \omega^s(x) \text{ for all } x \in N. \tag{1.22}$$

From this we have

$$\left.\frac{\partial u^s}{\partial n}\right|_{\partial\Omega} \leqslant \left|\frac{\partial \omega^s}{\partial n}\right|\Big|_{\partial\Omega}. \tag{1.23}$$

Since $-u^s(x)$ is a solution of the completely analogous equation of the form (1.1), whose coefficients satisfy just the same conditions as above, we have

$$-\left.\frac{\partial u^s}{\partial n}\right|_{\partial\Omega} \leqslant \left|\frac{\partial \omega^s}{\partial n}\right|\Big|_{\partial\Omega}. \tag{1.24}$$

It obviously follows from (1.23) and (1.24) that

$$\left|\frac{\partial u^s}{\partial n}\right|\Big|_{\partial\Omega} \leqslant \left|\frac{\partial \omega^s}{\partial n}\right|\Big|_{\partial\Omega}, \tag{1.25}$$

which is what we had to prove.

As a particular consequence of Theorem 1.1 we have the following new result for a single elliptic equation.

THEOREM 1.1′. *Let the coefficients $A_{ij}(x,u,p)$ and $B(x,u,p)$ depend continuously on their arguments and satisfy the condition*

$$|B(x, u, p)| \leqslant \psi(|p|)|p|\mathscr{E}_1 + \delta(|p|)\mathscr{E}_2, \tag{1.26}$$

where $\mathscr{E}_1 = A_{ij}p_ip_j$, $\mathscr{E}_2 = \operatorname{tr} A|p|$, $A = \|A_{ij}\|$, and $\psi(\rho)$ and $\delta(\rho)$ are the same functions as in Theorem 1.1. Let $u(x)$ be a solution of the Dirichlet problem

$$\begin{gathered} A_{ij}(x, u, u_x)u_{ij} = B(x, u, u_x) \text{ in } \Omega, \\ u = \varphi \text{ on } \partial\Omega, \end{gathered} \tag{1.27}$$

belonging to the class $C^2(\Omega) \cap C^1(\overline{\Omega})$, and $\max_{\overline{\Omega}}|u| \leq m$. Finally, let the domain Ω be strictly convex, and let its boundary $\partial\Omega$ and the function φ (continued throughout the closed domain $\overline{\Omega}$) belong to the class C^3. Then $\max_{\partial\Omega}|u_n|$ (where n is the inward normal to $\partial\Omega$ at the point $y \in \partial\Omega$) has a bound which depends only on m, the functions $\delta(\rho)$ and $\psi(\rho)$ of condition (1.26), the norms $\|\varphi\|_{C^2(\overline{\Omega})}$, and on k^{-1} and K, where k and K are the same quantities as in Theorem 1.1.

We now assume that the following extra condition, in addition to the conditions of Theorem 1.1, is satisfied:

$$\mathscr{E}_1^s \geqslant \frac{\operatorname{tr} A^s}{\psi(|p^s|)}, \quad \int^{\infty} \frac{d\rho}{\rho^2\psi(\rho)} = \infty, \quad s = 1, \ldots, N. \tag{1.28}$$

Then, obviously,

$$\mathscr{E}_2^s = \operatorname{tr} A^s|p^s| \leqslant \psi(|p^s|)|p^s|\mathscr{E}_1^s. \tag{1.29}$$

It can be assumed in this case that the inequality (1.6) holds with $\delta(|p^s|) \equiv 0$. But then (with fixed s) we can show that

$$a_{ij}\omega^s_{ij} - b \leqslant 0 \tag{1.30}$$

in complete analogy with the case of a single elliptic equation considered in [**5**], pp. 431-434, where, under the extra condition (1.29), an inequality of the form (1.30) is obtained with arbitrary $k = \inf k_i(y)$ (i.e. in the case of an arbitrary, in general nonconvex, domain with smooth boundary (belonging to the class C^3)). The required estimate for $\max_{\partial\Omega}|u^s_n|$ follows from the inequality (1.30) by means of Lemma 1.

THEOREM 1.2. *Let all the conditions of Theorem 1.1 be satisfied, with the exception of the condition that Ω be convex, and in addition let the extra condition (1.28) hold. Then $\max_{\partial\Omega}|u_n|$ is bounded by a quantity which depends only on m, the functions $\delta(\rho)$ and $\psi(\rho)$ of condition (1.26), on the norms $\|\varphi\|_{C^2(\overline{\Omega})}$, and on K.*

NOTE. Obviously, together with a bound for $\max_{\partial\Omega}|u_n|$, where n is the inward normal to $\partial\Omega$, we also have a bound for $\max_{\partial\Omega}|u_x|$, since a bound is known for $\max_{\partial\Omega}|u_\tau|$, where τ is any tangential vector to $\partial\Omega$ (or else we know the value of $u(x) = \varphi(x)$ on $\partial\Omega$).

§2. Estimates for $\max_{\overline{\Omega}}|u_x|$

We consider here a system of the form

$$A^s_{ij}(x, u^t, u^s_x)u^s_{ij} = B^s(x, u^t, u^s_x). \tag{2.1}$$

We introduce the following notation. Let

$$(A^s)^\tau = A^s_{ij}\tau_i\tau_j \tag{2.2}$$

and, given a function $D = D(x, u^t, u^t_x)$, let the expression $\delta^s(D)$ denote

$$\delta^s(D) = \frac{\partial D}{\partial u^t} u^t_k \frac{u^s_k}{|u^s_x|} + \frac{\partial D}{\partial x_k}\frac{u^s_k}{|u^s_x|} \tag{2.3}$$

(summation over all possible indices t and k is assumed in (2.3)).

We take the sth equation of the system (2.1) and apply to it the operator $u^s_k\partial/\partial x_k$. Then, using the notation (2.3) and introducing the function

$$v^s = \sum_{k=1}^{n} (u^s_k)^2, \tag{2.4}$$

we obtain the identity

$$\frac{1}{2} A^s_{ij} v^s_{ij} = A^s_{ij} u^s_{ik} u^s_{jk} + \frac{1}{2}\left(\frac{\partial B^s}{\partial p^s_e} - \frac{\partial A^s_{ij}}{\partial p^s_e} u^s_{ij}\right) v^s_e + \sqrt{v^s}\,\delta^s(B^s - A^s_{ij}u^s_{ij}). \tag{2.5}$$

We shall first prove

THEOREM 2.1. *Let the coefficients of the system (2.1) be continuously differentiable with respect to all their arguments and satisfy the following conditions: as $|p^s| \to \infty$*

$$\begin{gathered} |(\sqrt{(A^s)^\tau})_x| + |p^s|\,|(\sqrt{(A^s)^\tau})_u| \leqslant \delta_1(|p^s|)\sqrt{\frac{\mathscr{E}^s_1}{\omega^s}}, \\ |B^s - B^s_{p^s}p^s| \leqslant \mu\mathscr{E}^s_1, \quad |B^s_x| + |p^s|\,|B^s_u| \leqslant \delta_2(|p^s|)\frac{\mathscr{E}^s_1}{\omega^s}, \\ \omega^s = \omega^s(|p^s|) = \sup_{x\in\bar\Omega,\ |u^t|\leqslant m,\ \sigma^s = \frac{p^s}{|p^s|},\ |\tau|=1} \frac{|(\sqrt{(A^s)^\tau})_{p^s}p^s|^2\,|p^s|^2}{\mathscr{E}^s_1}, \end{gathered} \tag{2.6}$$

where $\mathscr{E}^s_1 = \sum_{i,j} A^s_{ij}p^s_i p^s_j$, and $\delta_i(\rho) \to 0$ as $\rho \to \infty$. Then, for any solution $u(x) = \{u^s(x)\}_1^N$ of the system (2.1) such that $u^s \in C^3(\Omega) \cap C^1(\bar\Omega)$, $\max_{\bar\Omega;\,s=1,\dots,N}|u^s| \leqq m$, and $\max_{\partial\Omega;\,s=1,\dots,N}|u^s_x| \leqq M_0$, we have the bound

$$\max_{\bar\Omega;\ s=1,\dots,N} |u^s_x| \leqslant M \tag{2.7}$$

with a constant M which depends only on m, M_0, and the functions $\delta_i(\rho)$ and constants μ of conditions (2.6).

PROOF. We multiply (2.5) by the function $f(v^s)$ (where $f(\rho) > 0$ for $\rho > 0$) and introduce the function $w^s = \int_0^{v^s} f(\tau)\,d\tau$. We obviously have

$$w^s_i = f v^s_i, \quad w^s_{ij} = f v^s_{ij} + f' v^s_i v^s_j. \tag{2.8}$$

We then obtain from (2.5) the identity

$$\begin{aligned} \frac{1}{2} A^s_{ij} w^s_{ij} &= \frac{1}{2} f' A^s_{ij} v^s_i v^s_j + f A^s_{ij} u^s_{ki} u^s_{kj} \\ &\quad + \frac{1}{2}\left[B^s_{p^s_l} - (A^s_{ij})_{p^s_l} u^s_{ij}\right] w^s_l + f\sqrt{v^s}\,\delta^s(B^s - A^s_{ij}u^s_{ij}). \end{aligned} \tag{2.9}$$

We now introduce a function $\bar w^s$ such that

$$w^s = z(u)\,\bar w^s, \quad z(u) > 0, \tag{2.10}$$

while

$$w_i^s = z' u_i \bar{w}^s + z \bar{w}_i^s,$$

(2.11)
$$w_{ij}^s = z'' u_i u_j \bar{w}^s + z' u_{ij} \bar{w}^s + z' u_i \bar{w}_j^s + z' u_j \bar{w}_i^s + z \bar{w}_{ij}^s.$$

It then follows from (2.9) that

(2.12)
$$z A_{ij}^s \bar{w}_{ij}^s + \sum_k b_k \bar{w}_k^s = -z'' \mathcal{E}_1^s \bar{w}^s + \frac{f'}{f^2} z'^2 \mathcal{E}_1^s (\bar{w}^s)^2 + 2f A_{ij}^s u_{ki}^s u_{kj}^s + z' (B_{p_l^s}^s p_l^s - B^s) \bar{w}^s - z' (A_{ij}^s)_{p_l^s} u_l^s u_{ij}^s \bar{w}^s + 2f \sqrt{v^s}\, \delta^s (B^s - A_{ij}^s u_{ij}^s).$$

We can now obtain, just as in the case of a single elliptic equation (see [4], formula (2.12)), the bound

(2.13)
$$-z' (A_{ij}^s)_{p_l^s} u_l^s u_{ij}^s \bar{w}^s \geqslant -f A_{ij}^s u_{ki}^s u_{kj}^s - c_0 \frac{z'^2}{f} \frac{\omega^s}{v^s} \mathcal{E}_1^s (\bar{w}^s)^2,$$

where ω can be regarded as a function of v^s. Let $f(v^s)$ be chosen in such a way as to satisfy the conditions

(2.14)
$$\frac{f'}{f} = 2c_0 \frac{\omega^s}{v^s}, \quad f(\rho) > 0 \quad \text{as} \quad \rho > 0.$$

It then follows from (2.11)–(2.13) that

(2.15)
$$z A_{ij}^s \bar{w}_{ij}^s + \sum_k b_k \bar{w}_k^s \geqslant \left\{ -z'' + \frac{f' \int_0^{v^s} f(\tau)\, d\tau}{f^2} \frac{z'^2}{z} \right\} \mathcal{E}_1^2 \bar{w}^s \cdot + f A_{ij}^s u_{ki}^s u_{kj}^s + z' (B_{p_l^s}^s u_l^s - B^s) \bar{w}^s + f \sqrt{v^s}\, \delta^s (B^s - A_{ij}^s u_{ij}^s).$$

We now find a bound for $f\sqrt{v^s}\delta^s(A_{ij}^s)u_{ij}^s$. We have

(2.16)
$$\left| f \sqrt{v^s}\, \delta^s (A_{ij}^s) u_{ij}^s \right| = \left| \sqrt{f} \sqrt{\tilde{A}_{ii}^s}\, \tilde{u}_{ii}^s \right| \left| \sqrt{f} \frac{\delta^s (\tilde{A}_{ii}^s)}{\sqrt{\tilde{A}_{ii}^s}} \sqrt{v^s} \right|,$$

where $\| \tilde{u}_{ij}^s \| = T^* \| u_{ij}^s \| T$, $\| \widetilde{A}_{ij}^s \| = T^* \| A_{ij}^s \| T$, and the matrix $\| \tilde{u}_{ij}^s \|$ is diagonal. Applying Cauchy's inequality and returning to the old quantities u_{ij}^s, we can obtain a bound for the right side of (2.16) in terms of the sum

(2.17)
$$f A_{ij}^s u_{ki}^s u_{kj}^s + v^s f \frac{[\delta^s (\tilde{A}_{ii}^s)]^2}{\tilde{A}_{ii}^s}.$$

We now observe that, in view of conditions (2.6),

(2.18)
$$| \delta^s ((A^s)^\tau) | \leqslant \delta_4 (| p^s |) \left[1 + \frac{(M')^s}{| p^s |} \right] \sqrt{\frac{\mathcal{E}_1^s (A^s)^\tau}{\omega^s}},$$

where $\varphi(t) \to 0$ as $t \to \infty$ and $(M') = \max_{\Omega; t \neq s} | u_x^t |$. We then obviously have

(2.19)
$$\sqrt{v^s}\, f \delta^s (A_{ij}^s) u_{ij}^s \geqslant -f A_{ij}^s u_{ki}^s u_{kj}^s - \frac{v^s f}{\omega^s} \delta_4^2 (v^s) \left\{ 1 + \frac{[(M')^s]^2}{v^s} \right\} \mathcal{E}_1^s.$$

It may easily be shown (see [4], formula (2.17)) that

(2.20)
$$f \leqslant 4c_0 \frac{\omega^s}{v^s} w^s.$$

It then follows from (2.15), (2.19) and (2.20) that

(2.21)
$$z A_{ij}^s \bar{w}_{ij}^s + \sum_k b_k \bar{w}_k^s \geqslant \left\{ -z'' + \frac{f' \int_0^{v^s} f(\tau)\, d\tau}{2f^2} \right\} \mathcal{E}_1^s \bar{w}^s$$

$$+ z' (B^s_{p^s} p^s - B^s) \bar{w}^s + f \sqrt{v^s} \delta^s (B^s) - c\delta_4^2 (v^s) z \left\{1 + \frac{[(M')^s]^2}{v^s}\right\} \mathcal{E}_1^s \bar{w}^s.$$

Recalling conditions (2.6) on the growth of B^s and (2.20), we obtain from this

$$(2.22)\quad \begin{aligned} zA^s_{ij}\bar{w}^s_{ij} + \sum_k b_k \bar{w}^s_k &\geqslant \left\{ -z'' + \frac{f' \int_0^{v^s} f(\tau)\,d\tau}{f^2} - \frac{z'^2}{z} \right\} \mathcal{E}_1^s \bar{w}^s \\ &- c(|z'| + \delta_5(v^s) z) \mathcal{E}_1^s \bar{w}^s + \delta_6(v^s) z \left\{ 1 + \frac{(M')^s}{\sqrt{v^s}} + \frac{[(M')^s]^2}{v^s} \right\} \mathcal{E}_1^s \bar{w}^s \\ &\geqslant \left\{ -z'' - c|z'| - \delta_7(v^s) z \left\{ 1 + \frac{(M')^s}{\sqrt{v^s}} + \frac{[(M')^s]^2}{v^s} \right\} \right\} \mathcal{E}_1^s \bar{w}^s. \end{aligned}$$

It will be assumed that an s has been chosen from the set $1, \cdots, N$ for which

$$(2.23)\quad \max_{\bar\Omega} |u^s_x| = M = \max_{\substack{\bar\Omega \\ t=1,\ldots,N}} |u^t_x|.$$

Clearly $(M^s)' \leq M$. We shall show that the expression $1 + M/\sqrt{v^s} + M^2/v^s$ is bounded by a known quantity at the point $x_0 \in \Omega$ at which $\bar{w}^s$ attains its maximum. Let c_0 be a fixed number greater than 1. If $v^s(x_0) \geqq M^2/c_0$, we have

$$(2.24)\quad 1 + \frac{M}{\sqrt{v^s}} + \frac{M^2}{v^s}\bigg|_{x=x_0} \leqslant 1 + \sqrt{c_0} + c_0;$$

while if $v^s(x_0) \leqq M^2/c_0$, then

$$(2.25)\quad w^s(x_0) \leqslant \int_0^{\frac{M^2}{c_0}} f(\tau)\,d\tau \leqslant \frac{M_w^2}{c_0},$$

where we use the notation $M_w^2 = \max_\Omega w^s(x)$, since the function $w(v) = \int_0^v f(\tau)d\tau$ increases more rapidly (not more slowly) than the linear function $g(v) \equiv kv$, for which $g(v/c) = g(v)/c$. However, with a sufficiently large c_0, (2.25) is impossible. For, setting

$$(2.26)\quad z = 1 + e^{\alpha m} - e^{\alpha u},$$

we have

$$(2.27)\quad w^s(x_0) = z(x_0)\bar{w}^s(x_0) \geqslant \bar{w}^s(x_0) = M^2_{\bar{w}}$$

and

$$(2.28)\quad M_w^2 = \max_{\bar\Omega} w^s \leqslant (1 + e^{\alpha m}) M^2_{\bar{w}}.$$

It then follows from (2.25) – (2.28) that

$$(2.29)\quad M_w^2 \leqslant (1 + e^{\alpha m}) M^2_{\bar{w}} \leqslant (1 + e^{\alpha m}) w^s(x_0) \leqslant \frac{1 + e^{\alpha m}}{c_0} M_w^2,$$

which is impossible, if say $c_0 = 2(1 + e^{\alpha m})$. Hence at the point x_0 at which $\bar{w}^s(x)$ attains its maximum we must necessarily have

$$(2.30)\quad v^s(x_0) \geqslant \frac{M^2}{c_0}, \quad c_0 = 2(1 + e^{\alpha m}),$$

whence it follows that

$$c\left(1+\frac{M}{\sqrt{v^s}}+\frac{M^2}{v^s}\right)\Big|_{x=x_0}\leqslant c_1. \tag{2.31}$$

It then follows from (2.22) and (2.30) that at the point $x = x_0$

$$zA^s_{ij}\overline{w}^s_{ij}+\sum b_k\overline{w}^s_k\geqslant[-z''-c\,|\,z'\,|-c_1\delta_6(v^s)\,z]\,\mathcal{E}^s_1\overline{w}_s. \tag{2.32}$$

Since $z' = -\alpha e^{\alpha u}$ and $z'' = -\alpha^2 e^{\alpha u}$, we have

$$-z''-c\,|\,z'\,|-c_1\delta_6(v^s)\,z\geqslant(\alpha^2-c\alpha)\,e^{\alpha u}-c_1\delta_6(v^s)\,(1+e^{\alpha m}). \tag{2.33}$$

On choosing a sufficiently large α and assuming that $v^s(x_0)$ is sufficiently large, say $v^s(x_0) \geqq L$, where L is a known number, determined by the constants c_1 and m and the function δ_6, we get

$$(\alpha^2-c\alpha)\,e^{\alpha u}-c_1\delta_6(v^s)\,(1+e^{\alpha m})\,|_{x=x_0}>0. \tag{2.34}$$

In view of (2.31), this inequality contradicts the fact that at the point x_0 where $\overline{w}^s(x)$ has its maximum we must have

$$zA^s_{ij}\overline{w}^s_{ij}+\sum b_k\overline{w}^s_k\leqslant 0.$$

Consequently

$$\max_{\overline{\Omega}}\overline{w}^s\leqslant\max\left\{\max_{\partial\Omega}\overline{w}^s,\ \max_{\substack{x\in\Omega\\ v(x)\leqslant L}}\overline{w}^s\right\}=\max\left\{\max_{\partial\Omega}\frac{w^s}{z},\ \max_{\Omega,\,v\leqslant L}\frac{w^s}{z}\right\}. \tag{2.35}$$

On now noting that $z \geqq 1$ and $w = \int_0^v f(\tau)d\tau$, we find from (2.35) that

$$\max_{\overline{\Omega}}w^s\leqslant\max_{\overline{\Omega}}z\,\max_{\overline{\Omega}}\overline{w}^s\leqslant(1+e^{\alpha m})\max\left\{\int_0^{M_0^2}f(\tau)\,d\tau,\ \int_0^L f(\tau)\,d\tau\right\}, \tag{2.36}$$

where $M_0^2 = \max_{\partial\Omega} v$. The required bound for $\max_{\overline{\Omega}} v$ is obtained in an obvious way from (2.36). Theorem 2.1 is proved.

Let us now prove

THEOREM 2.2. *Let the coefficients of the system* (2.1) *be continuously differentiable with respect to all their arguments and satisfy the following conditions: as* $|p^s| \to \infty$

$$\begin{gathered}|(\sqrt{(A^s)^\tau})_p|\leqslant\mu\frac{\sqrt{\operatorname{tr}A^s}}{|p^s|},\\ |(A^s)^\tau_x|+|p^s|\,|(A^s)^\tau_u|\leqslant\delta_1(|p^s|)\,(A^s)^\tau,\\ |B^s_{p^s}|\leqslant\delta_2(|p^s|)\frac{\mathcal{E}^s_2}{|p^s|},\qquad|B^s_x|+|p^s|\,|B^s_u|\leqslant\delta_3(|p^s|)\,\mathcal{E}^s_2,\end{gathered} \tag{2.37}$$

where $\delta_i(\rho) \to 0$ *as* $\rho \to \infty$. *Then for any solution* $u(x) = \{u^s(x)\}_1^N$ *of system* (2.1) *such that*

$$u^s(x)\in C^3(\Omega)\cap C^1(\overline{\Omega}),\qquad\max_{\substack{\overline{\Omega}\\ s=1,\dots,N}}|u^s|\leqslant m,\qquad\max_{\substack{\partial\Omega\\ s=1,\dots,N}}|u^s_x|\leqslant M_0,$$

we have

$$\max_{\substack{\overline{\Omega}\\ s=1,\dots,N}}|u^s_x|\leqslant M, \tag{2.38}$$

where M depends only on m, M_0, the diameter of the domain Ω, the functions $\delta_i(\rho)$ and the constant μ of conditions (2.37).

Proof. Consider the identity (2.5). In precisely the same way as in [**2**], formula (22), we first obtain the inequality

$$-\frac{1}{2}\frac{\partial A^s_{ij}}{\partial p^l_s}u^s_{ij}v^s_l \geqslant -\frac{1}{2}A^s_{ij}u^s_{ki}u^s_{kj} - c\ \mathrm{tr}\ A^s\frac{(v^s_x)^2}{v^s}. \tag{2.39}$$

In fact, we have

$$\begin{aligned}\frac{\partial A^s_{ij}}{\partial p^s_l}u^s_{ij}v^s_l = \frac{\partial \tilde{A}^s_{ii}}{\partial p^s_l}\tilde{u}^s_{ii}v^s_l = \left(\sqrt{\tilde{A}^s_{ii}}\tilde{u}^s_{ii}\right)\left(\frac{(\tilde{A}^s_{ii})_{p^s_l}v^s_l}{\sqrt{\tilde{A}^s_{ii}}}\right)\\ = \left(\sqrt{\tilde{A}^s_{ii}}\tilde{u}^s_{ii}\right)\left(\left(\sqrt{\tilde{A}^s_{ii}}\right)_{p^s_l}v^s_l\right),\end{aligned} \tag{2.40}$$

where the matrix $\|\tilde{u}^s_{ij}\| = T^*\|u^s_{ij}\|T$ is diagonal, $\|\tilde{A}^s_{ij}\| = T^*\|A_{ij}\|T$, and $T = \|\tau_{ij}\|$. Since $\tilde{A}^s_{ii} = A^s_{kl}\tau_{ki}\tau_{li} = (A^s)^{\tau^i}$, where τ^i is the ith column of the matrix T (and $|\tau^i| = 1$), we find, on applying the Cauchy inequality to (2.40) and using the condition $|(\sqrt{(A^s)^\tau})_{p^s}| \leq \mu\sqrt{\mathrm{tr}\,A^s}/|p^s|$, that

$$\begin{aligned}\frac{1}{2}\left|\frac{\partial A^s_{ij}}{\partial p^s_l}u^s_{ij}v^s_l\right| \leqslant \frac{1}{2}\tilde{A}^s_{ii}(\tilde{u}^s_{ii})^2 + c\frac{\mathrm{tr}\ A^s}{v^s}(v^s_x)^2\\ \leqslant \frac{1}{2}A^s_{ij}u^s_{ki}u^s_{kj} + c\frac{\mathrm{tr}\ A^s}{v^s}(v^s_x)^2,\end{aligned} \tag{2.41}$$

of which (2.39) is an obvious consequence. Now consider the term $\sqrt{v^s}\delta^s(A^s_{ij}u^s_{ij})$. We first observe that conditions (2.37) imply the inequality

$$|\delta^s((A^s)^\tau)| \leqslant \delta_4(|p^s|)\left(1 + \frac{(M')^s}{|p^s|}\right)(A^s)^\tau, \tag{2.42}$$

where $(M')^s = \max_{\bar{\Omega};t\neq s}|u^t_x|$ and $\delta_4(\rho) \to 0$ as $\rho \to \infty$. Then, noting that

$$\delta^s(A^s_{ij}u^s_{ij}) = \delta^s(\tilde{A}^s_{ij}\tilde{u}^s_{ij}) = \delta^s(\tilde{A}^s_{ii})\tilde{u}^s_{ii}, \tag{2.43}$$

where $\|\tilde{u}^s_{ij}\|$ and $\|\tilde{A}^s_{ij}\|$ are the same as above, we have

$$\begin{aligned}|\sqrt{v^s}\delta^s(A^s_{ij}u^s_{ij})| = \left(\sqrt{\tilde{A}^s_{ii}}\tilde{u}^s_{ii}\right)\left(\sqrt{\frac{v^s}{\tilde{A}^s_{ii}}}\delta^s(A^s_{ii})\right) \leqslant \frac{1}{2}\tilde{A}^s_{ii}(\tilde{u}^s_{ii})^2 + \frac{1}{2}v^s\frac{\delta^s(\tilde{A}^s_{ii})}{\tilde{A}^s_{ii}}\\ \leqslant \frac{1}{2}\tilde{A}^s_{ii}(\tilde{u}^s_{ii})^2 + \delta_5(v^s)\left(1 + \frac{[(M')^s]^2}{v^s}\right)\sum_{i=1}^n \tilde{A}^s_{ii}v^s\\ = \frac{1}{2}A^s_{ij}u^s_{ki}u^s_{kj} + \delta_5(v^s)\left(1 + \frac{[(M')^s]^2}{v^s}\right)\mathrm{tr}\ A^s v^s,\end{aligned} \tag{2.44}$$

where $\delta(\rho) \to 0$ as $\rho \to \infty$.

From (2.5), (2.39) and (2.44) we obtain

$$\begin{aligned}\frac{1}{2}A^s_{ij}v^s_{ij} \geqslant \frac{1}{2}\frac{\partial B^s}{\partial p^s_l}v^s_l + \sqrt{v^s}\delta^s B^s - c\ \mathrm{tr}\ A^s\frac{(v^s_x)^2}{v^s}\\ - \left\{1 + \frac{[(M')^s]^2}{v^s}\right\}\delta_5(v^s)\ \mathrm{tr}\ A^s v^s.\end{aligned} \tag{2.45}$$

In (2.45) we substitute

$$v^s = z(x)\bar{v}^s, \quad z(x) > 0. \tag{2.46}$$

Then, noting that

$$
\begin{aligned}
v_i^s &= z_i \bar{v}^s + z \bar{v}_i^s, \\
v_{ij}^s &= z_{ij} \bar{v}^s + z_i \bar{v}_j^s + z_j \bar{v}_i^s + z \bar{v}_{ij}^s, \\
A_{ij}^s v_{ij}^s &= z A_{ij}^s \bar{v}_{ij}^s + 2 A_{ij}^s \bar{v}_i^s z_j + A_{ij}^s z_{ij} \bar{v}^s, \\
\frac{(v_x^s)^2}{v^s} &= \frac{\Sigma z_i^2}{z} \bar{v}^s + 2 z_i \bar{v}_i^s + z \frac{(\bar{v}_x^s)^2}{\bar{v}^s},
\end{aligned}
\tag{2.47}
$$

we obtain from (2.45) the inequality

$$
\begin{aligned}
\frac{1}{2} z A_{ij}^s \bar{v}_{ij}^s + \Sigma b_k \bar{v}_k^s \geqslant \Big\{ & -\frac{1}{2} A_{ij}^s z_{ij} + \frac{1}{2} \frac{\partial B^s}{\partial p_l^s} z_l + z \frac{\delta^s B^s}{\sqrt{v^s}} \\
& - c \operatorname{tr} A^s \frac{\Sigma z_i^2}{z} - \Big\{ 1 + \frac{[(M')^s]^2}{v^s} \Big\} \delta_5 (v^s) \operatorname{tr} A^s z \Big\} \bar{v}^s.
\end{aligned}
\tag{2.48}
$$

Recalling conditions (2.37), we obtain from (2.48)

$$
\begin{aligned}
\frac{1}{2} z A_{ij}^s \bar{v}_{ij}^s + \sum b_k \bar{v}_k^s \geqslant \Big\{ & -\frac{1}{2} A_{ij}^s z_{ij} - \delta_6 (v^s) \sum_{l=1}^n | z_l | \operatorname{tr} A^s \\
& - \frac{c \Sigma z_i^2}{z} \operatorname{tr} A^s - \Big\{ 1 + \frac{(M')^s}{\sqrt{v^s}} + \frac{[(M')^s]^2}{v^s} \Big\} \delta_7 (v^s) z \operatorname{tr} A^s \Big\} \bar{v}^s,
\end{aligned}
\tag{2.49}
$$

where the form of b_k is of no importance for what follows. Suppose we have fixed a number s such that

$$
\max_{\bar{\Omega}} | u_x^s | = M \equiv \max_{\substack{\bar{\Omega} \\ t=1, \dots, N}} | u_x^t |.
\tag{2.50}
$$

Clearly $(M')^s \leq M$ and $\max_{\Omega} v^s = M^2$.

We now show that the ratio $[(M')^s]^2/v^s$ is bounded at the point x_0 at which $\bar{v}^s(x)$ has its maximum. In fact,

$$
\frac{[(M')^s]^2}{v^s} \leqslant \frac{M^2}{\bar{v}^s (x_0)} = \frac{M^2}{\bar{M}^2},
\tag{2.51}
$$

where $\bar{M}^2 = \max_{\bar{\Omega}} \bar{v}^s$. But $M^2/\bar{M}^2 \leq \max_{\bar{\Omega}} z(x)$, since

$$
M^2 = \max_{\bar{\Omega}} v^s \leqslant \max_{\bar{\Omega}} z(x) \max_{\bar{\Omega}} \bar{v}^s = \max_{\bar{\Omega}} z(x) \bar{M}^2.
\tag{2.52}
$$

Thus at the point x_0 we have

$$
\begin{aligned}
\frac{1}{2} z A_{ij}^s \bar{v}_{ij}^s + \sum b_k \bar{v}_k^s \geqslant \Big\{ & -\frac{1}{2} A_{ij} z_{ij} - \delta_6 (v^s) \sum_{l=1}^n | z_l | \operatorname{tr} A^s \\
& - c \frac{\sum_{i=1}^n z_i^2}{z} \operatorname{tr} A^s - \delta_8 (v^s) z \operatorname{tr} A^s \Big\} \bar{v}^s.
\end{aligned}
\tag{2.53}
$$

We set

$$
z = \alpha + d^2 - |x|^2,
\tag{2.54}
$$

where d is the diameter of the domain Ω and $\alpha > 1$ is a sufficiently large number. Then

$$
z_i = -2x_i, \quad z_{ij} = -2\delta_{ij} \qquad (\delta_{ij} \text{ is Kronecker delta}).
$$

We now obtain at the point x_0, instead of (2.53),

$$\begin{aligned}\frac{1}{2} zA^s_{ij}\bar{v}^s_{ij} + \sum b_k \bar{v}^s_k \geqslant \Big\{ \mathbf{tr}\, A^s - \Big[2nd\delta_6(v^s) \\ + \frac{4cnd^2}{\alpha} + \delta_8(v^s)(\alpha + d^2) \Big] \mathbf{tr}\, A^s \Big\} \bar{v}^s. \end{aligned} \tag{2.55}$$

On taking a sufficiently large α and assuming that $v^s(x_0)$ is sufficiently large, say $v^s(x_0) \geqq L$, where L is a known number, determined by the constants n, d, α and the functions δ_6 and δ_8, we shall have

$$\frac{1}{2} zA^s_{ij}\bar{v}^s_{ij} + \sum b_k \bar{v}^s_k \Big|_{x=x_0} > 0. \tag{2.56}$$

But this contradicts the fact that the opposite inequality

$$\frac{1}{2} zA^s_{ij}\bar{v}^s_{ij} + \sum b_k \bar{v}^s_k \Big|_{x=x_0} \leqslant 0 \tag{2.57}$$

must hold at the point where $\bar{v}^s$ has its maximum. The function $\bar{v}^s$ is thus unable to attain its maximum at points $x \in \Omega$ at which $v^s(x) \geqq L$. Consequently

$$\begin{aligned}\max_{\bar{\Omega}} \bar{v}^s &\leqslant \max \Big\{ \max_{\partial\Omega} \bar{v}^s, \max_{\Omega,\, v^s \leqslant L} \bar{v}^s \Big\} \\ &= \max \Big\{ \max_{\partial\Omega} \frac{v^s}{z}, \max_{\Omega,\, v^s \leqslant L} \frac{v^s}{z} \Big\} \leqslant \max \Big\{ \max_{\partial\Omega} \frac{v^s}{z}, \max_{\bar{\Omega}} \frac{L}{z} \Big\}. \end{aligned} \tag{2.58}$$

Since $1 \leqq z \leqq \alpha + d^2$, we have

$$M^2 = \max_{\bar{\Omega}} v^s \leqslant \max_{\bar{\Omega}} z(x) \max_{\bar{\Omega}} \bar{v}^s \leqslant (\alpha + d^2) \max \Big\{ \max_{\partial\Omega} v^s, L \Big\}. \tag{2.59}$$

This proves Theorem 2.2.

NOTE. If $A^s_{ij}(x, u^t, p^s)$ satisfies the auxiliary condition

$$\sum_{i,j=1}^{n} A^s_{ij} \geqslant \varepsilon\, \mathbf{tr}\, A^s, \quad s = 1, \ldots, N, \ \varepsilon > 0, \tag{2.60}$$

the result of Theorem 2.2 will be retained provided that the conditions $|B^s_{p^s}| \leqq \delta_2 \mathscr{E}^s_2/|p^s|$ in (2.37) are replaced by the following wider conditions as $|p^s| \to \infty$:

$$|B^s_{p^s}| \leqslant \mu_1 \frac{\mathscr{E}^s_2}{|p^s|} \quad (s = 1, \ldots, N). \tag{2.61}$$

The proof of this statement is exactly similar to the proof of the assertion made in [2], Russian pp. 88-89, English pp. 48-49.

§3. Existence theorems

It may be mentioned first that, by using the result due to Ladyženskaja and Ural′ceva regarding a bound for the Hölder constant of the first derivatives of the solutions of quasilinear uniformly elliptic equations, we can prove, as in the case of a single quasilinear elliptic equation (see [5], p. 417), the following general existence theorem for the elliptic system

$$\begin{aligned} A^s_{ij}(x, u^t, u^s_x) u^s_{ij} &= B^s(x, u^t, u^s_x) \ \text{in}\ \Omega, \\ u^s &= \varphi^s \ \text{on}\ \partial\Omega, \\ s &= 1, \ldots, N. \end{aligned} \tag{3.1}$$

Theorem 3.1. *Let the boundary $\partial\Omega$ of the domain Ω and the boundary functions φ^s $(s=1,\cdots,N)$ belong to class C^3. Suppose that for any system of functions $\{v^t\}_1^N$ satisfying the conditions*

$$
\begin{gathered}
v^t \in C^2(\bar{\Omega}),\\
A^s_{ij}(x, v^t, v^s)\,v^s_{ij} = \tau B^s(x, v^t, v^s_x) \text{ in } \Omega,\\
v^s = \tau\varphi^s \text{ on } \partial\Omega,\\
s=1,\dots,N,\quad 0\leqslant\tau\leqslant 1,
\end{gathered}
\tag{3.2}
$$

we have proved the a priori estimate

$$
\max_{\substack{\bar{\Omega}\\ s=1,\dots,N}} (|v^s| + |v^s_x|) \leqslant M
\tag{3.3}
$$

with a constant M independent of v^s and τ. Then the Dirichlet problem (3.1) *has at least one solution $u(x)=\{u^s(x)\}_1^N$ belonging to the case $C^2(\Omega)$, whatever the boundary functions φ^s.*

When proving Theorem 3.1, we have to use (as in [**5**]) the following familiar lemma on the fixed points of a completely continuous operator.

Lemma (**Leray-Schauder, Schaefer**). *Let T be a completely continuous operator mapping a Banach space V into itself. The operator T will then have at least one fixed point, provided that there exists a constant c, independent of τ, such that, for any $v\in V$ satisfying the equation*

$$
v=\tau Tv, \qquad \tau\in[0,1],
$$

we have

$$
\|v\| \leqq c,
$$

where $\|\cdot\|$ is the norm in V.

Theorem 3.1, in conjunction with Theorems 1.1, 1.2, 2.1 and 2.2, enables us to prove the following existence theorems.

Theorem 3.2. *Let Ω be a bounded convex domain in E_n with a boundary $\partial\Omega$ belonging to the class C^2. Let the coefficients $A^s_{ij}(x,u^t,p^s)$ and $B^s(x,u^t,p^s)$ of the system* (3.1) *be continuously differentiable with respect to all their arguments and satisfy the following conditions: as $|p^s|\to\infty$*

$$
\begin{gathered}
|(\sqrt{(A^s)^\tau})_x| + |p^s|\,|(\sqrt{(A^s)^\tau})_u| \leqslant \delta_0(|p^s|)\sqrt{\frac{\mathscr{E}^s_1}{\omega^s}},\\
|B^s| \leqslant \psi(|p^s|)\,|p^s|\,\mathscr{E}^s_1 + \delta_1(|p^s|)\,\mathscr{E}^s_2,\\
|B^s - B^s_{p^s}p^s| \leqslant \mu\mathscr{E}^s_1,\quad |B^s_x| + |p^s|\,|B^s_u| \leqslant \delta_2(|p^s|)\frac{\mathscr{E}^s_1}{\omega^s},
\end{gathered}
\tag{3.4}
$$

where

$$
\omega^s = \omega^s(|p^s|) = \sup_{x\in\bar{\Omega},\ |u^t|\leqslant m,\ \sigma^s=\frac{p^s}{|p^s|},\ |\tau|=1} \frac{|(\sqrt{(A^s)^\tau})_{p^s}p^s|^2\,|p^s|^2}{\mathscr{E}^s_1},
\tag{3.5}
$$

$\mathscr{E}^s_1 = \sum_{i,j=1}^n A^s_{ij}p^s_ip^s_j$, $\mathscr{E}^s_2 = \operatorname{tr} A^s|p^s|$, $\psi(\rho)>0$ is a decreasing function satisfying the condition $\int^\infty (1/\rho^2\psi(\rho))\,d\rho=\infty$, and $\delta_i(\rho)$ are positive decreasing functions such that $\delta_i(\rho)\to 0$ as $\rho\to\infty$. Then, if the a priori estimate

$$\max_{\substack{\overline{\Omega} \\ s=1,\ldots,N}} |v^s| \leqslant m$$

(with a constant m independent of u^s and τ) holds for any solution of the problem (3.2), *the problem* (3.1) *will be solvable no matter what the boundary functions φ^s belonging to the class C^2.*

Theorem 3.3. *If the coefficients of the system* (3.1) *satisfy, in addition to conditions* (3.4), *the auxiliary conditions*

$$\mathcal{E}_1^s \geqslant \frac{\operatorname{tr} A^s}{\psi(|p^s|)}, \quad \int^{\infty} \frac{d\rho}{\rho^2\psi(\rho)} = \infty, \quad s = 1, \ldots, N, \tag{3.6}$$

the result of Theorem 3.2 *will also hold when the domain Ω (with a boundary belonging to class C^2) is not convex.*

Theorem 3.2 is proved by means of Theorems 1.1, 2.1 and 3.1. We have to assume to start with that the boundary $\partial\Omega$ and functions φ^s belong to the class C^3, and the coefficients $A^s_{ij}(x,u^t,p^s)$ and $B^s(x,u^t,p^s)$ to the class C^2. Under these assumptions, we find by conditions (3.4) that for a solution $v(x) \in C^2(\overline{\Omega})$ of the problem (3.2) we have

$$\max_{\substack{\overline{\Omega} \\ s=1,\ldots,N}} |v^s_x| \leqslant M.$$

In fact, the bound for $\max_{\partial\Omega; s=1,\cdots,N}|v^s_x|$ is an immediate consequence of Theorem 1.1 and the condition $|B^s| \leq \psi(|p^s|)|p^s|\mathcal{E}^s_1 + \delta_1(|p^s|)\mathcal{E}^s_2$. In order to obtain the bound for $\max_{\overline{\Omega}; s=1,\cdots,N}|v^s_x|$ by means of Theorem 2.1, we first have to regularize the solution $v^s(x)$, since Theorem 2.1 is concerned with solutions of the class $C^3(\Omega) \cap C^1(\overline{\Omega})$. However, the coefficients $A^s_{ij}(x,v^t,v^s_x)$ and $B^s(x,v^t,v^s_x)$ (with arbitrary $v^s(x) \in C^2(\overline{\Omega})$), considered as functions of x, belong to $C^1(\overline{\Omega})$ for any $s = 1,\cdots,N$; furthermore, the boundary data belong to the class C^3. In view of these facts, we find by Schauder's theorem that $v^s \in C^{2+\lambda}(\overline{\Omega})$ with arbitrary $\lambda < 1$. The coefficients A^s_{ij} and B^s, regarded as functions of x, thus belong to $C^{1+\lambda}(\overline{\Omega})$. On again applying Schauder's theorem, we now find that $v^s \in C^{3+\lambda}(\overline{\Omega})$. All in all, the required bound for $\max_{\overline{\Omega}; s=1,\cdots,N}|v^s_x|$ is obtained in accordance with Theorem 2.1. In conjunction with the postulated bound for $\max_{\overline{\Omega}; s=1,\cdots,N}|v^s|$, and using Theorem 3.1, this proves the existence of at least one solution of the problem (3.1). Superfluous conditions regarding the smoothness of the boundary, of the boundary functions φ^s, and of the coefficients A^s_{ij} and B^s may be eliminated by the standard method, using approximation and considerations of compactness (in this connection we again have to apply the important Ladyženskaja-Ural′ceva estimate for the norm of $C^{1+\lambda}$). The considerations indicated above are described, for the case of a single elliptic equation, in [**5**], pp. 452-453, and other places.

It may also be noted that the various sufficient conditions guaranteeing the a priori estimate $\max_{\overline{\Omega}; s=1,\cdots,N}|v^s| \leq m$ are outlined, for example, in [**1**, **5**].

Theorem 3.3 is proved in exactly the same way, provided that account is also taken of Theorem 1.2.

Theorem 3.4. *Let Ω be a bounded convex domain in E_n, with a boundary $\partial\Omega$ belonging to the class C^2. Let the coefficients $A^s_{ij}(x,u^t,p^s)$ and $B^s(x,u^t,p^s)$ of the system* (3.1) *belong to the class C^1 and satisfy the following conditions: as $|p^s|\to\infty$*

$$|(\sqrt{(A^s)^\tau})_p|\leqslant\mu\frac{\sqrt{\operatorname{tr}A^s}}{|p^s|},\tag{3.7}$$

$$|(A^s)^\tau_x|+|p^s||(A^s)^\tau_x|\leqslant\delta_0(|p^s|)(A^s)^\tau,$$

$$\begin{gathered}|B^s+|B^s_x|+|p^s||B^s_u|\leqslant\delta_1(|p^s|)\mathscr{E}^s_2,\\ |B^s_{p^s}|\leqslant\delta_2(|p^s|)\frac{\mathscr{E}^s_2}{|p^s|},\end{gathered}\tag{3.8}$$

where $\mathscr{E}^s_2=\operatorname{tr}A^s|p^s|$ and $\delta_i(\rho)$ are positive decreasing functions, such that $\delta_i(\rho)\to 0$ as $\rho\to\infty$.[1] *The problem* (3.1) *is then solvable with any φ^s belonging to the class C^2.*

Theorem 3.4 is proved by means of Theorems 1.1, 2.2 and 3.1, using arguments similar to those used in the proof of Theorem 3.1. The unconditional form of Theorem 3.4 is due to the fact that the a priori estimate $\max_{\bar\Omega;s=1,\dots,N}|u^s|\leq m$ follows from the condition $|B^s|\leq\delta_1(|p^s|)\mathscr{E}^s_2$. For, consider the sth equation of the system (3.1) and substitute in it

$$u^s=z(x)\bar u^s,\tag{3.9}$$

where $z(x)>0$. Then

$$zA^s_{ij}\bar u^s_{ij}+2A^s_{ij}z_i\bar u^s_j=B^s-A^s_{ij}z_{ij}\bar u^s.\tag{3.10}$$

Assuming that $\max_{\bar\Omega}|u^s|=\max_{\bar\Omega}u^s>0$ (if not, we should consider the equation $A^s_{ij}(-u^s)_{ij}=-B^s$), we set $z(x)=1+d^2-|x|^2$, where d is the diameter of the domain Ω, which contains the origin as an interior point. Then at the point $x_0\in\Omega$ at which $\bar u^s(x)$ attains its maximum (where $\bar u^s(x_0)>0$, since $\max_{\bar\Omega}u^s(x)>0$ and $z(x)>0$) we have

$$zA^s_{ij}\bar u^s_{ij}=B^s+2\operatorname{tr}A^s\bar u^s(x_0).$$

Noting now that at x_0

$$\begin{aligned}B^s(x,u^t,(z\bar u^s)_x)|&\leqslant\delta_4(|(z\bar u^s)_x|)|(z\bar u^s)_x|\operatorname{tr}A^s\\&=\delta_4(|z_x|\bar u^s(x_0))|z_x|\bar u^s(x_0)\operatorname{tr}A^s\leqslant\delta_4(2d\bar u^s(x_0))2d\bar u^s(x_0)\operatorname{tr}A^s,\end{aligned}$$

we obtain the inequality

$$zA^s_{ij}\bar u^s_{ij}|_{x=x_0}\geqslant[2-2d\delta_4(2d\bar u^s(x_0))]\operatorname{tr}A^s\bar u^s(x_0).\tag{3.11}$$

On now specifying L by the condition

$$2-2d\delta_4(2dL)=1,\tag{3.12}$$

we conclude that either $\bar u^s(x_0)\leq L$, or at x_0

$$zA^s_{ij}\bar u^s_{ij}|_{x=x_0}>0,$$

---

[1] We assume that $\rho\delta(\rho)$ is increasing as $\rho\to\infty$.

which contradicts our assumption that $\bar{u}^s(x)$ attains its maximum at x_0. From this we in fact obtain the required a priori estimate

$$\max_{\bar{\Omega}} u^s(x) \leqslant \max\left\{\max_{\partial\Omega} u^s(x),\ L\right\} = \max\left\{\max_{\partial\Omega} |\varphi^s|,\ L\right\}.$$

The following result, for a single elliptic equation, is a particular consequence of Theorem 3.4.

THEOREM 3.4′. *Let Ω be a bounded convex domain in E_n, $n \geqq 2$, with boundary belonging to the class C^2. Let the coefficients $A_{ij}(x,u,p)$ and $B(x,u,p)$ of the equation*

$$A_{ij}(x,u,u_x)u_{ij} = B(x,u,u_x) \tag{3.13}$$

belong to the class C^1 and satisfy the following conditions: as $|p| \to \infty$

$$\begin{gathered} |(\sqrt{A^\tau})_p| \leqslant \mu \frac{\sqrt{\operatorname{tr} A}}{|p|}, \\ |A^\tau_x| + |p|\,|A^\tau_u| \leqslant \delta_0(|p|)\,A^\tau, \\ |B| + |B_x| + |p|\,|B_u| + |p|\,|B_p| < \delta_1(|p|)\,\mathscr{E}_2, \end{gathered} \tag{3.14}$$

where $\mathscr{E}_2 = \operatorname{tr} A\,|p|$, $A = \|A_{ij}\|$, and $\delta_i(\rho)$ are positive decreasing functions such that $\delta_i(\rho) \to 0$ as $\rho \to \infty$. The Dirichlet problem

$$A_{ij}u_{ij} = B \quad \text{in } \Omega, \qquad u = \varphi \text{ on } \partial\Omega \tag{3.15}$$

is then solvable with any φ belonging to the class C^2.

Note that the conditions on the growth of the coefficient B

$$\begin{gathered} |B| + |B_x| + |p|\,|B_u| + |p|\,|B_p| \leqslant \delta(|p|)\,\mathscr{E}_2, \\ \delta(\rho) \to 0 \ \text{ for } \rho \to \infty, \end{gathered} \tag{3.16}$$

are in a sense unimprovable (from the point of view of preserving the solvability of the problem (3.15) with any boundary function $\varphi \in C^\infty$). For in fact the Dirichlet problem

$$\Delta u - B = 0 \text{ in } \Omega, \qquad u = \varphi \text{ on } \partial\Omega, \tag{3.17}$$

where $B = \lambda u + f(x)$, and λ is an eigenvalue of the Laplace operator, is not in general solvable (with arbitrary $f(x) \in C^\infty$). For the equation (3.17) we have $\mathscr{E}_2 = |p|$ and $B_u = \lambda$, so that $|p|B_u = \lambda\mathscr{E}_2$.

II. SYSTEMS WITH IDENTICAL MATRICES $\|A^s_{ij}\| = \|A_{ij}\|$

Now we shall consider elliptic systems of the form

$$A_{ij}(x,u^t,u^t_x)u^s_{ij} = B^s(x,u^t,u^t_x), \qquad s = 1,\cdots,N. \tag{0.1′}$$

§4. Estimation of $\max_{\partial\Omega;\,s=1,\cdots,N}|u^s_x|$

THEOREM 4.1. *Let the coefficients $A_{ij}(x,u^t,p^t)$ and $B^s(x,u^t,p^t)$ of the system (0.1′) depend continuously on their arguments and satisfy the conditions*

$$|B^s(x,u^t,p^t)| \leqslant \delta_1(|p|)\mathscr{E}_1 + \delta_2(|p|)\,\mathscr{E}_2, \quad s = 1,\ldots,N, \tag{4.1}$$

where $\mathscr{E}_1 = \sum_{i,j,s} A_{ij}p^s_i p^s_j$, $\mathscr{E}_2 = \operatorname{tr} A\,|p|$, $A = \|A_{ij}\|$, $|p| = \sum_t |p^t|$, and $\delta_i(\rho)$

are positive decreasing functions such that $\delta_i(\rho)\to 0$ *as* $\rho\to\infty$. *Let* $u(x)=\{u^t(x)\}_1^N$ *be a solution of the Dirichlet problem*

$$A_{ij}(x, u^t, u^t_x)\,u^s_{ij}=B^s(x, u^t, u^t_x) \text{ in } \Omega, \quad u^s=\varphi^s \text{ on } \partial\Omega, \quad s=1,\dots,N, \tag{4.2}$$

such that $u^t\in C^2(\Omega)\cap C^1(\overline{\Omega})$, $t=1,\cdots,N$, *while*

$$\max_{\substack{\overline{\Omega}\\ t=1,\dots,N}} |u^t_x|\leqslant m. \tag{4.3}$$

Finally, let the domain Ω *be strictly convex, and let its boundary* $\partial\Omega$ *and the functions* φ^s *(continued into* $\overline{\Omega}$*) belong to the class* C^3. *For* $M_0\equiv\max_{\partial\Omega}|u_x|$ *we then have the inequality*

$$M_0\leqq c+c_1\delta_2(M)\,M, \tag{4.4}$$

where $M=\max_{\overline{\Omega}}|u_x|$, *and the quantities* c_0 *and* c_1 *depend only on* m, *the functions* $\delta_i(\rho)$ *of condition* (4.1), *the norms* $\|\varphi^t\|_{C^2(\overline{\Omega})}$, *and* k^{-1} *and* K, *where* k *and* K *are the same as in Theorem* 1.1.

PROOF. We take a number s and a point $y_0\in\partial\Omega$ such that

$$\left|\frac{\partial u^s(y_0)}{\partial n}\right|=\max_{\substack{\partial\overline{\Omega}\\ t=1,\dots,N}}\left|\frac{\partial u^t}{\partial n}\right|, \tag{4.5}$$

where n is the direction of the inward normal to $\partial\Omega$. It can be assumed without loss of generality that

$$u^t(y_0)=0,\quad t=1,\cdots,N. \tag{4.6}$$

For clarity, suppose that $\partial u^s(y_0)/\partial n>0$. Then, following Ladyženskaja and Ural′ceva (see [1], Chapter VIII, §3), we introduce the function

$$v=u^s+\sum_{t=1}^N (u^t)^2. \tag{4.7}$$

Obviously

$$\frac{\partial v(y_0)}{\partial n}=\frac{\partial u^s(y_0)}{\partial n}=\max_{\substack{\partial\Omega\\ t=1,\dots,N}}\left|\frac{\partial u^t}{\partial n}\right|. \tag{4.8}$$

It follows from the system (0.1′) that

$$(A_{ij}u^s_{ij}-B^s)+2\sum_{t=1}^N u^t(A_{ij}u^t_{ij}-B^t)=0.$$

This last equation can be written as

$$A_{ij}v_{ij}-2\sum_{t=1}^N A_{ij}u^t_iu^t_j-B=0, \tag{4.9}$$

where $B=B^s+2\sum_1^N u^tB^t$. We now consider (4.9) as an equation of the form

$$a_{ij}(x)\,v_{ij}-b(x)=0,$$

where $b(x)=2\sum_1^N A_{ij}u^t_iu^t_j+B$. We introduce the function

$$\omega(x) = \varphi^s(x) + \sum_{t=1}^{N} [\varphi^t(x)]^2 + h(d) \equiv \varphi(x) + h(d), \tag{4.10}$$

where $d = d(x)$ in the domain $N \subset \Omega$, and N and $d(x)$ are the same as in §1. By Lemma 1.2,

$$a_{ij}(x)\omega_{ij} - b(x) \leqslant F\frac{h'' + Kh'}{h'^2} - k \operatorname{tr} \|a_{ij}\| h' + a_{ij}\varphi_{ij} - b, \tag{4.11}$$

where $F = a_{ij}(p - p_0)_i(p - p_0)_j$, $p = p_0 + \nu h'$, $p_0 = \varphi_x$, ν is the unit inward normal to $\partial\Omega$ at the point $y \in \partial\Omega$, $a_{ij} = a_{ij}(x)$ and $b = b(x)$. It may be assumed below without loss of generality that $\operatorname{tr}\|a_{ij}\| = \operatorname{tr}\|A_{ij}\| = 1$. Then

$$|a_{ij}\varphi_{ij}| \leqq c_\varphi, \tag{4.12}$$

where c_φ depends only on $\|\varphi\|_{C^2(\overline{\Omega})}$.

Further, recalling the condition (4.1), we have

$$\begin{aligned} -b &= -2\sum_{t=1}^{N} t_{ij}u_i^t u_j^t - B^s - 2\sum_{t=1}^{N} u^t B^t \\ &\leqslant -2\sum_{t=1}^{N} A_{ij}u_i^t u_j^t - (1 + 2Nm)\left[\delta_1(|u_x|)\sum_{t=1}^{N} A_{ij}u_i^t u_j^t + \delta_2(|u_x|)|u_x|\right]. \end{aligned} \tag{4.13}$$

We shall assume that $\max_{\overline{\Omega}}|u_x| \geqq L$, where L is defined by the condition

$$(1 + 2Nm)\delta_1(L) = 1 \tag{4.14}$$

(if this is not the case, a bound for $\max_{\overline{\Omega}}|u_x|$ is already obtained).

It then follows from (4.11) and (4.13) that

$$a_{ij}(x)\omega_{ij} - b \leqslant F\frac{h'' + Kh'}{h'^2} + h'\left(-k + \frac{c_\varphi}{h'} + \frac{\delta_2(M)M}{h'}\right), \tag{4.15}$$

where $M = \max_{\overline{\Omega}}|u_x|$. Let $h' \geqq \alpha$, where

$$\alpha = \frac{c_\varphi + \delta_2(M)M}{k}. \tag{4.16}$$

We then obtain from (4.15)

$$a_{ij}\omega_{ij} - b \leqslant F\frac{h'' + Kh'}{h'^2}. \tag{4.17}$$

We now choose a function $h(d)$ such that

$$\left.\begin{aligned} &h'' + Kh' = 0 \qquad \text{for } 0 \leqslant d \leqslant \delta, \\ &h(0) = 0, \quad h(\delta) = q, \quad h'(d) \leqslant \alpha, \end{aligned}\right\} \tag{4.18}$$

where $q \geqq \max_{\overline{\Omega}}|\varphi| + m$. Obviously (see [5], p. 436) all these conditions are satisfied by the function

$$h(d) = q\frac{1 - e^{-KD}}{1 - e^{-K\delta}}, \tag{4.19}$$

where $h'(d) \leqq \alpha + qK$. Then on the boundary of N, i.e. at the points where $d(x) = 0$ or $d(x) = \delta$, we have the inequality

$$v(x) \leqq \omega(x).$$

Applying Lemma 1, we then find that for all $x \in N$

(4.20) $$v(x) \leqq \omega(x).$$

Since

(4.21) $$v(y) = \omega(y), \qquad y \in \partial\Omega,$$

we have

(4.22) $$\frac{\partial v(y)}{\partial n} \leqslant \frac{\partial \omega(y)}{\partial n}, \quad y \in \partial\Omega.$$

This last inequality holds for all boundary points, including y_0. Hence, recalling (4.6) and (4.1),

(4.23) $$M_0' = \max_{\substack{\partial\Omega \\ t=1,\dots,N}} \left|\frac{\partial u^t}{\partial n}\right| = \frac{\partial v(y_0)}{\partial n} \leqslant \frac{\partial \varphi(y_0)}{\partial n} + h'(0) \leqslant c_\varphi + qK + \alpha.$$

But then, noting that $|\partial u^t/\partial\tau|$ (where τ is any tangential direction) is known (we have a bound for $|\partial u^t/\partial\tau|$ in terms of c_φ), denoting $M_0 = \max_{\partial\Omega}|u_x| = \max_{\partial\Omega; t=1,\dots,N}|u_x^t|$, and using (4.16), we obtain the final inequality

(4.24) $$M_0 \leqslant 2c_\varphi + qK + c_\varphi k^{-1} + \delta_2(M) M k^{-1}.$$

Theorem 4.1 is proved.

We shall pay special attention to the case of a uniformly elliptic system, for which

(4.25) $$A_{ij}\xi_i\xi_j \geqq c_0 \operatorname{tr} A |\xi|^2.$$

Then, obviously, with $|p| \neq 0$

(4.26) $$\mathscr{E}_2 \equiv \operatorname{tr} A|p| \leqslant \frac{A_{ij}p_ip_j}{c_0|p|} \equiv \frac{\mathscr{E}_1}{c_0|p|}.$$

It can be assumed in this case that (4.1) is satisfied with $\delta_2 = 0$. Under these conditions the restriction that Ω be convex can be lifted.

THEOREM 4.2. *Let the coefficients $A_{ij}(x, u^t, p^t)$ and $B^s(x, u^t, p^t)$ of the system (0.1′) depend continuously on their arguments and satisfy the conditions*

(4.27) $$|B^s(x, u^t, p^t)| \leqslant \delta_1(|p|)\mathscr{E}_1, \quad s = 1, \dots, N,$$

together with the condition (4.25). Let the boundary $\partial\Omega$ and functions φ^s (continued into $\overline{\Omega}$) belong to the class C^3. Then a bound can be obtained for $\max_{\overline{\Omega}; s=1,\dots,N}|u_x^s|$, which depends only on $m = \max_{\overline{\Omega}; t=1,\dots,N}|u_x^t|$, $\delta_1(\rho)$, the norm $\|\varphi^s\|_{C^2(\overline{\Omega})}$, and K, where K is the same as in Theorem 1.1.

PROOF. As in the previous case, we introduce a function $v = u^s + \sum_1^N (u^t)^2$ under the conditions (4.5) and (4.6). We can regard (4.9) for this function as a linear equation

(4.28) $$a_{ij}(x) v_{ij} - b(x) = 0,$$

where $a_{ij} = A_{ij}$, $b(x) = 2\sum_1^N A_{ij} u_i^t u_j^t + B$ and $B = B^s + 2\sum_1^N u^t B^t$. We also introduce a function $\omega(x)$ in accordance with (4.10). In view of Lemma 1.2, we shall use on this occasion the inequality

(4.29) $$a_{ij}\omega_{ij} - b \leqslant F\frac{h''}{h'^2} + K \operatorname{tr} A h' + a_{ij}\varphi_{ij} - b.$$

It may be assumed without loss of generality that $\operatorname{tr} A = 1$. Then

$$|a_{ij}\varphi_{ij}| \leqq c_\varphi, \tag{4.30}$$

where c_φ depends on the same quantities as in (4.12). Further, recalling (4.27), we have

$$\begin{aligned} -b &= -2\sum_{t=1}^{N} A_{ij}u_i^t u_j^t - B^s - 2\sum_{t=1}^{N} u^t B^t \\ &\leqslant -2\sum_{t=1}^{N} A_{ij}u_i^t u_j^t - (1 + 2Nm)\,\delta_1(|u_x|)\sum_{t=1}^{N} A_{ij}u_i^t u_j^t. \end{aligned} \tag{4.31}$$

Assuming that $\max_{\bar\Omega}|u_x| \geqq L$, where L is defined by the condition

$$(1 + 2Nm)\delta_1(L) = 1 \tag{4.32}$$

(if the reverse is the case, a bound for $\max_{\bar\Omega}|u_x|$, and along with this a bound for $\max_{\partial\Omega}|u_x|$, is already obtained), we find that

$$a_{ij}\omega_{ij} - b \leqslant F\frac{h''}{h'^2} + h'(K + c_\varphi), \tag{4.33}$$

if $h' \geqq 1$. Noting that $F = A_{ij}\nu_i\nu_j h'^2$ (where ν is the unit inward normal), and that

$$(A_{ij}\nu_i\nu_j)^{-1} \leqq c_0^{-1}, \tag{4.34}$$

which follows from (4.25), we have

$$a_{ij}\omega_{ij} - b \leqslant F\frac{h'' + ch'}{h'^2}, \tag{4.35}$$

where $c = (K + c_\varphi)/c_0$. We now choose a function $h(d)$ such that

$$\left.\begin{aligned} &h'' + ch' = 0 \quad \textbf{for} \quad 0 \leqslant d \leqslant \delta, \\ &h(0) = 0, \ h(\delta) = q, \ h'(d) \geqslant 1, \end{aligned}\right\} \tag{4.36}$$

where $q = \max_{\bar\Omega}|\varphi| + m$. All these conditions are satisfied by the function

$$h(d) = q\frac{1 - e^{-cd}}{1 - e^{-c\delta}}, \tag{4.37}$$

where $h'(d) \leqq 1 + qc$. We can then obtain, in exactly the same way as in the previous theorem, the inequality

$$\frac{\partial v(y_0)}{\partial n} \leqslant \frac{\partial \omega(y_0)}{\partial n} \leqslant \left|\frac{\partial \varphi(y_0)}{\partial n}\right| + 1 + qc, \tag{4.38}$$

whence follows the required bound for $\max_{\partial\Omega;\,t=1,\dots,N}|\partial u^t/\partial n|$, and along with this, the bound for $\max_{\partial\Omega;\,t=1,\dots,N}|u_x^t|$. Q.E.D.

§5. An estimate for $\max_{\bar\Omega;\,s=1,\dots,N}|u_x^s|$

Here we consider systems of the type

$$A_{ij}(x, u^t, |u_x|)\,u_{ij}^s = B^s(x, u^t, u_x^t), \qquad s = 1, \dots, N. \tag{5.1}$$

As in §2, we shall use the notation

$$A^\tau = A_{ij}\tau_i\tau_j, \tag{5.2}$$

where $|\tau| = 1$, and

$$\delta^s D = \delta^s D(x, u^t, u_x^t) = \frac{\partial D}{\partial u^t} u_k^t \frac{u_k^s}{|u_x^s|} + \frac{\partial D}{\partial x_k} \frac{u_k^s}{|u_x^s|}, \tag{5.3}$$

where summation is assumed over all possible indices t and k.

To the sth equation of the system (5.1) we apply the operator $u_k^s \partial/\partial x_k$. Then, using the notation (5.3) and introducing the functions

$$v^s = \sum_{k=1}^{n} (u_k^s)^2, \qquad s = 1, \ldots, N, \tag{5.4}$$

we obtain the identity

$$\frac{1}{2} A_{ij} v_{ij}^s = A_{ij} u_{ki}^s u_{kj}^s + \frac{\partial B^s}{\partial p_l^t} u_{lk}^t u_k^s - \frac{1}{2} \sum_{t=1}^{N} \frac{\partial A_{ij}}{\partial |p|} v_k^t \frac{u_k^s}{|u_x|} u_{ij}^s + \sqrt{v^s} \delta^s (B^s - A_{ij} u_{ij}^s). \tag{5.5}$$

Summing the identity (5.5) over $s = 1, \cdots, N$ and introducing the function $v = \sum_1^N v^s$, we get

$$\begin{aligned} \frac{1}{2} A_{ij} v_{ij} = \sum_{s=1}^{N} A_{ij} u_{ki}^s u_{kj}^s - \frac{1}{2} \left(\sum_{s=1}^{N} \frac{\partial A_{ij}}{\partial |p|} \frac{u_k^s}{|u_x|} u_{ij}^s \right) v_k \\ + \sum_{s=1}^{N} \frac{\partial B^s}{\partial p_l^t} u_{lk}^t u_k^s + \sum_{s=1}^{N} \sqrt{v^s} \delta^s (B^s - A_{ij} u_{ij}^t). \end{aligned} \tag{5.6}$$

Theorem 5.1. *Let the coefficients $A_{ij}(x, u^t, |p|)$ and $B^s(x, u^t, p^t)$ of the system (5.1) (where $|p| = (\sum_{t,l} (p_l^t)^2)^{1/2}$) belong to the class C^1 and satisfy the following conditions: as $|p| \to \infty$*

$$\begin{gathered} |(\sqrt{A^\tau})_x| + |p| \, |(\sqrt{A^\tau})_u| \leqslant \delta_0(|p|) \sqrt{\frac{\mathscr{E}_1}{\omega}}, \\ |B_p^s| \leqslant \delta_1(|p|) \sqrt{\frac{\mathscr{E}_1 \lambda}{\omega}}, \quad |B_x^s| + |p| \, |B_u^s| \leqslant \delta_2(|p|) \frac{\mathscr{E}_1 |p|}{\omega}, \\ |B^s| \leqslant \delta_3(|p|) \mathscr{E}_1, \\ \omega = \omega(|p|) = \sup_{x \in \overline{\Omega}, \, |u^t| \leqslant m, \, \sigma = \frac{p}{|p|}, \, |\tau| = 1} \frac{(\sqrt{A^\tau})_p^2 |p|^2}{\lambda}, \end{gathered} \tag{5.7}$$

where $\mathscr{E}_1 = \sum_{i,j,s} A_{ij} p_i^s p_j^s$, $\delta_i(\rho) \to 0$ as $\rho \to \infty$, $A^\tau = A_{ij} \tau_i \tau_j$ with $|\tau| = 1$, and λ is the least eigenvalue of the matrix $A = \| A_{ij} \|$.

Then, for any solution $u(x) = \{u^s(x)\}_1^N$ of the system (5.1) such that $u^s(x) \in C^3(\Omega) \cap C^1(\overline{\Omega})$, $s = 1, \cdots, N$, where

$$\begin{gathered} \max_{\overline{\Omega}, \; s=1, \ldots, N} |u^s| \leqslant m, \\ \max_{\partial\Omega, \; s=1, \ldots, N} |u_x^s| \leqslant M_0, \end{gathered} \tag{5.8}$$

the quantity $M = \max_{\overline{\Omega}; s=1, \cdots, N} |u_x^s|$ will depend only on m, M_0, the functions $\delta_i(\rho)$, and the constant m of conditions (5.7).

Proof. We multiply (5.6) by the function $f(v)$ (where $f(\rho) > 0$ for $\rho > 0$) and introduce the function

(5.9) $$w = \int_0^v f(\tau)\, d\tau.$$

It then follows from (5.6) that

(5.10)
$$\frac{1}{2} A_{ij} w_{ij} = \frac{1}{2} \frac{f'}{f^2} A_{ij} w_i w_j + f \sum_{s=1}^{N} A_{ij} u^s_{ki} u^s_{kj}$$
$$- \frac{1}{2} \sum_{s=1}^{N} \frac{\partial A_{ij}}{\partial |p|} u^s_{ij} w_k \frac{u^s_k}{|u_x|} + f \sum_{s=1}^{N} \frac{\partial B^s}{\partial p^t_l} u^t_{ek} u^s_k + f \sum_{s=1}^{N} \sqrt{v^s} \delta^s (B^s - A_{ij} u^s_{ij}).$$

We now introduce a function $\bar{w}$ such that

(5.11) $$w = \sum_{s=1}^{N} z(u^s)\, \bar{w}, \qquad z(\rho) > 0;$$

here

$$w_i = \sum_{s=1}^{N} [z'(u^s) u^s_i \bar{w} + z(u^s) \bar{w}_i],$$

(5.12)
$$w_{ij} = \sum_{s=1}^{N} [z''(u^s) u^s_i u^s_j \bar{w} + z'(u^s) u^s_{ij} \bar{w} + z'(u^s) u^s_i \bar{w}_j$$
$$+ z'(u^s) u^s_j \bar{w}_i + z(u^s) \bar{w}_{ij}].$$

We then obtain from (5.10) the identity

(5.13)
$$\frac{1}{2} \sum_{s=1}^{N} z(u^s) A_{ij} \bar{w}_{ij} + \sum_k b_k \bar{w}_k = -\frac{1}{2} \sum_{s=1}^{N} z''(u^s)\, \mathscr{E}^s_1 \bar{w}$$
$$+ \frac{1}{2} \frac{f'}{f^2} A_{ij} \left(\sum_{s=1}^{N} z'(u^s) u^s_i \right) \left(\sum_{t=1}^{N} z'(u^t) u^t_j \right) + f \sum_{s=1}^{N} A_{ij} u^s_{ki} u^s_{kj}$$
$$- \frac{1}{2} \sum_{s=1}^{N} \frac{\partial A_{ij}}{\partial |p|} u^s_{ij} \frac{u^s_k}{|u_x|} \left(\sum_{t=1}^{N} z'(u^t) u^t_i \right) \bar{w} - \frac{1}{2} \sum_{s=1}^{N} z'(u^s) B^s \bar{w}$$
$$+ f \sum_{s=1}^{N} \frac{\partial B^s}{\partial p^t_e} u^t_{ek} u^s_k + f \sum_{s=1}^{N} \sqrt{v^s} (B^s - A_{ij} u^s_{ij}),$$

where the form of b_k is of no consequence for what follows, $\mathscr{E}^s_1 = A_{ij} p^s_i p^s_j$, and allowance is made for the fact that $A_{ij} u^s_{ij} = B^s$. Just as in the case of a single elliptic equation, if we introduce the matrices $\|\tilde{u}^s_{ij}\|$ and $\|\tilde{A}_{ij}\|$ such that $\|\tilde{u}^s_{ij}\| = T^* \|u^s_{ij}\| T$, where $\|\tilde{u}_{ij}\|$ is diagonal and $\|\tilde{A}_{ij}\| = T^* \|A_{ij}\| T$, and use the condition

(5.14) $$(\sqrt{A^\tau})^2_p \leqslant \frac{c\omega\lambda}{v},$$

which follows from the definition of $\omega = \omega(v)$, we obtain the inequality

(5.15)
$$(A_{ij})_{|p|} u^s_{ij} \frac{u^s_k}{|u_x|} \left(\sum_{t=1}^{N} z'(u^t) u^t_i \right) \bar{w} \leqslant \frac{1}{2} f \tilde{A}_{ii} (\tilde{u}^s_{ii})^2$$
$$+ c_0 \frac{\omega\lambda}{f} \left(\sum_{t=1}^{N} z'(u^t) u^t_i \right) \frac{\bar{w}^2}{v} = \frac{1}{2} f A_{ij} u^s_{ki} u^s_{kj} + c_0 \frac{\omega\lambda}{f} \left(\sum_{t=1}^{N} z'(u^t) u^t_i \right) \frac{\bar{w}^2}{v}.$$

We have taken into account the fact that $A_{ii} = A^{\tau^i} = (A\tau^i, \tau^i)$, where the (unit) vector τ^i is the ith column of the matrix T.

We shall now assume that the function $f(\rho)$ is so chosen that it satisfies the conditions

$$\frac{f'}{f} = 2c_0 \frac{\omega}{v}, \quad f(v) > 0 \quad \text{for} \quad v > 0. \tag{5.16}$$

Then

$$c_0 \frac{1}{f} \frac{\omega\lambda}{v} \left(\sum_{t=1}^{N} z'(u^t) u_i^t \right)^2 \overline{w}^2 \leqslant \frac{1}{2} \frac{f'}{f^2} A_{ij} \left(\sum_{t=1}^{N} z'(u^t) u_i^t \right) \left(\sum_{s=1}^{N} z'(u^s) u_j^s \right). \tag{5.17}$$

It then follows from (5.13), (5.15) and (5.17) that

$$\begin{aligned} \sum_{s=1}^{N} z(u^s) A_{ij} \overline{w}_{ij} + \sum b_k \overline{w}_k \geqslant & \sum_{s=1}^{N} z''(u^s) \mathcal{E}_1^s \overline{w} + f \sum_{s=1}^{N} A_{ij} u_{ki}^s u_{kj}^s \\ & - \sum_{s=1}^{N} z(u^s) B^s \overline{w} + 2f \sum_{s=1}^{N} \frac{\partial B^s}{\partial p_t^s} u_{lk}^t u_k^s + 2f \sum_{s=1}^{N} \sqrt{v^s}\, \delta^s (B^s - A_{ij} u_{ij}^s). \end{aligned} \tag{5.18}$$

We now consider the expression $2f \sum_{s=1}^{N} \sqrt{v^s}\, \delta^s (A_{ij} u_{ij}^s)$. We have

$$\begin{aligned} |2f \sqrt{v^s}\, \delta^s (A_{ij} u_{ij}^s)| = & |\sqrt{f} \sqrt{\tilde{A}_{ii}}\, \tilde{u}_{ii}^s| \left| 2\sqrt{f} \sqrt{v^s} \frac{\delta^s (\tilde{A}_{ii})}{\sqrt{\tilde{A}_{ii}}} \right| \\ & \leqslant \frac{f}{2} \tilde{A}_{ii} (\tilde{u}_{ii}^s)^2 + cfv^s |\delta^s (\sqrt{\tilde{A}_{ii}^s})|. \end{aligned} \tag{5.19}$$

By definition of δ^s and the conditions (5.7) it now follows that

$$\left| \sum_{s=1}^{N} 2f \sqrt{v^s}\, \delta^s (A_{ij} u_{ij}^s) \right| \leqslant \frac{f}{2} \sum_{s=1}^{N} A_{ij} u_{ki}^s u_{kj}^s + cfv\delta_0(v) \frac{\mathcal{E}_1}{\omega}; \tag{5.20}$$

from this and (5.18) we obtain

$$\begin{aligned} \sum_{s=1}^{N} z(u^s) A_{ij} \overline{w}_{ij} + \sum_{k} b_k \overline{w}_k \geqslant & \sum_{s=1}^{N} z''(u^s) \mathcal{E}_1 \overline{w} + \frac{1}{2} f \sum_{s=1}^{N} A_{ij} u_{ki}^s u_{kj}^s \\ & - \sum_{s=1}^{N} z'(u^s) B^s \overline{w} + 2f \sum_{s=1}^{N} \frac{\partial B^s}{\partial p_t^s} u_{lk}^t u_k^s - c_1 fv\delta_3(v) \frac{\mathcal{E}_1}{\omega} + 2f \sum_{s=1}^{N} \sqrt{v^s}\, \delta^s B^s. \end{aligned} \tag{5.21}$$

Using condition (5.7), we obtain further

$$\begin{aligned} 2f \sum_{s=1}^{N} \frac{\partial B^s}{\partial p_t^s} u_{lk}^t u_k^s \leqslant & \frac{1}{2} f \sum_{s=1}^{N} A_{ij} u_{ki}^s u_{kj}^s + \frac{c_2}{\lambda} f \left| \frac{\partial B^s}{\partial p_t^s} \right|^2 v \\ & \leqslant \frac{1}{2} f \sum_{s=1}^{N} A_{ij} u_{ki}^s u_{kj}^s + c_2 fv\delta_1(v) \frac{\mathcal{E}_1}{\omega}; \end{aligned} \tag{5.22}$$

$$\begin{gathered} \left| 2f \sum_{s=1}^{N} \sqrt{v^s}\, \delta^s B^s \right| \leqslant c_3 fv\delta_2(v) \frac{\mathcal{E}_1}{\omega}; \\ \left| \sum_{s=1}^{N} z'(u^s) B^s \overline{w} \right| \leqslant \delta_3(v) \mathcal{E}_1 \sum_{s=1}^{N} z'(u^s). \end{gathered} \tag{5.23}$$

It was shown in [4], formula (2.17), that

$$fv \leqq 4c_0\omega w. \tag{5.24}$$

It then follows from (5.21) – (5.24) that

$$\begin{aligned}\sum_{s=1}^{N} z(u^s) A_{ij}\bar{w}_{ij} + \sum_{k=1}^{n} b_k \bar{w}_k \geqslant \Big\{\sum_{s=1}^{N} - z''(u^s)\mathcal{E}_1^s \\ - \delta_3(v)\sum_{s=1}^{N} |z'(u^s)|\,\mathcal{E}_1 - c_5\delta_4(v)\sum_{s=1}^{N}\mathcal{E}_1 z(u^s)\Big\}\,\bar{w}.\end{aligned} \tag{5.25}$$

We set

$$z(\rho) = 1 + e^m - e^\rho. \tag{5.26}$$

Since

$$z'(\rho) = -e^\rho, \qquad z''(\rho) = -e^\rho, \tag{5.27}$$

we have

$$-z''(u^s) \geqslant e^{-m}, \quad |z'(u^s)| \leqslant e^m, \quad |z(u^s)| \leqslant 1 + e^m. \tag{5.28}$$

It then follows from (5.25) that

$$\sum_{s=1}^{N} - z''(u^s)\mathcal{E}_1^s - \delta_3(v)\sum_{s=1}^{N}\mathcal{E}_1 z(u^s) \geqslant e^{-m} - \delta_3(v) N e^m - c_5\delta_4(v) N(1 + e^m). \tag{5.29}$$

We assume that $\max_{\bar\Omega}\bar{w}$ is attained at some point $x_0 \in \Omega$ at which $v(x_0) \geqq L$, where L is given by the condition

$$e^{-m} - \delta_3(L) N e^m - c_5\delta_4(L) N(1 + e^m) = \frac{1}{2}. \tag{5.30}$$

By (5.25) and (5.29) we must have at this point

$$\sum_{s=1}^{N} z(u^s) A_{ij}\bar{w}_{ij} + \sum_{k=1}^{n} b_k \bar{w}_k > 0, \tag{5.31}$$

which contradicts our assumption that $\bar{w}$ attains its maximum at x_0. Consequently

$$\max_{\bar\Omega} \bar{w} \leqslant \max\Big\{\max_{\partial\Omega}\bar{w},\ \max_{\Omega,\, v\leqslant L}\bar{w}\Big\} \leqslant \max\Big\{\max_{\partial\Omega}\frac{w}{N},\ \frac{L}{N}\Big\} \tag{5.32}$$

and

$$\begin{aligned}\max_{\bar\Omega} w \leqslant \max_{\bar\Omega}\sum_{s=1}^{N} z(u^s)\max_{\bar\Omega}\bar{w} \leqslant N(1+e^m)\max\left\{\frac{\int_0^{M_0} f(\tau)\,d\tau}{N},\ \frac{\int_0^{L} f(\tau)\,d\tau}{N}\right\} \\ = (1+e^m)\max\left\{\int_0^{M_0} f(\tau)\,d\tau,\ \int_0^{L} f(\tau)\,d\tau\right\},\end{aligned} \tag{5.33}$$

from which the required bound for $M = \max_{\bar\Omega} v$ obviously follows. This proves Theorem 5.1.

Assume that the "unattached terms" B^s of the system (5.1) have the form

$$B^s(x, u^t, p_l^t) = B^s(x, u^t, |p|), \quad |p| = \Big(\sum_{t,l}(p_l^t)^2\Big)^{1/2}. \tag{5.34}$$

In this case we have

THEOREM 5.1′. *Let the coefficients* $A_{ij}(x,u^t,|p|)$ *and* $B^s(x,u^t,|p|)$ *of the system* (5.1) *belong to the class* C^1 *and satisfy the following conditions: as* $|p|\to\infty$

$$
\begin{gathered}
|(\sqrt{A^\tau})_x|+|p||(\sqrt{A^\tau})_u|\leqslant\delta_0(|p|)\sqrt{\frac{\mathscr{E}_1}{\omega}},\\
|B^s|+|p||B^s_{|p|}|\leqslant\delta_1(|p|)\mathscr{E}_1,\quad |B^s_x|+|p||B^s_u|\leqslant\delta_2(|p|)\frac{\mathscr{E}_1|p|}{\omega},\qquad(5.35)\\
\omega=\omega(|p|)=\sup_{\substack{x\in\bar\Omega,\,|u^s|\leqslant m\\ \sigma=\frac{p}{|p|},\,|\tau|=1}}\frac{(\sqrt{A^\tau})^2_{|p|}|p|^2}{\lambda},
\end{gathered}
$$

where $\mathscr{E}_1$, $\delta_i(\rho)$, A^τ *and* λ *are the same as in Theorem* 5.1. *Then, for any solution* $u(x)=\{u^s(x)\}_1^N$ *of the system* (5.1) *belonging to the class* $C^3(\Omega)\cap C^1(\bar\Omega)$ *and satisfying the conditions* (5.6), *the quantity* $M=\max_{\bar\Omega;s=1,\dots,N}|u^s_x|$ *depends only on* m, M_0, *and the functions* $\delta_i(\rho)$ *and constant* μ *of conditions* (5.35).

PROOF. We consider the inequality (5.21), allowing for the fact that the term $2f\sum_{s=1}^N(\partial B^s/\partial p^t_l)u^t_{lk}u^s_k$ has the form

$$
\begin{aligned}
2f\sum_{s=1}^N\frac{\partial B^s}{\partial p^t_l}u^t_{lk}u^s_k&=2f\sum_{s=1}^N\frac{\partial B^s}{\partial|p|}\frac{u^t_lu^t_{lk}u^s_k}{|u_x|}=f\sum_{s=1}^N\frac{\partial B^s}{\partial|p|}v_k\frac{u^s_k}{|u_x|}=\sum_{s=1}^N\frac{\partial B^s}{\partial|p|}w_k\frac{u^s_k}{|u_x|}\\
&=\sum_{s=1}^N\frac{\partial B^s}{\partial|p|}\left(\sum_{t=1}^N z'(u^t)u^t_k\right)\frac{u^s_k}{|u_x|}\bar w+\sum_{k=1}^n b_k\bar w_k.
\end{aligned}\qquad(5.36)
$$

It follows from conditions (5.35) that

$$
\left|\sum_{s=1}^N\frac{\partial B^s}{\partial|p|}|p|\left(\sum_{t=1}^N z'(u^t)\frac{u^t_k}{|u_x|}\frac{u^s_k}{|u_x|}\right)\bar w\right|\leqslant c\sum_{s=1}^N|z'(u^s)|\mathscr{E}_1\bar w.\qquad(5.37)
$$

Then, on also recalling the inequalities (5.22)–(5.24), we obtain an inequality of the type (5.25), by means of which we obtain, in the same way as in the proof of Theorem 5.1, the required bound for $\max_{\bar\Omega;s=1,\dots,N}|u^s_x|$. Q.E.D.

We shall now prove

THEOREM 5.2. *Let the coefficients* $A_{ij}(x,u^t,|p|)$ *and* $B(x,u^t,|p^t|)$ *belong to the class* C^1 *and satisfy the following conditions: as* $|p|\to\infty$

$$
\left.\begin{gathered}
|(\sqrt{A^\tau})_{|p|}|\leqslant\mu\sqrt{\frac{\operatorname{tr}A}{|p|^2}},\\
|A^\tau_x|+|p||A^\tau_u|\leqslant\delta_1(|p|)A^\tau,
\end{gathered}\right\}\qquad(5.38)
$$

$$
|B^s_p|\leqslant\delta_2(|p|)\sqrt{\frac{\mathscr{E}_2\lambda}{|p|}},\quad |B^s_x|+|p||B^s_u|\leqslant\delta_3(|p|)\mathscr{E}_2,
$$

where $A^\tau=A_{ij}\tau_i\tau_j$ *with* $|\tau|=1$, *and* $\delta_i(\rho)\to0$ *as* $\rho\to\infty$. *Then, for every solution* $u(x)=\{u^s(x)\}_1^N$ *of the system* (5.1) (*such that* $u^s(x)\in C^3(\Omega)\cap C^1(\bar\Omega)$, $s=1,\dots,N$) *satisfying the conditions*

$$
\max_{\substack{\bar\Omega\\ s=1,\dots,N}}|u^s|\leqslant m,\qquad \max_{\substack{\partial\Omega\\ s=1,\dots,N}}|u^s_x|\leqslant M_0,\qquad(5.39)
$$

the quantity $M=\max_{\bar\Omega;s=1,\dots,N}|u^s_x|$ *depends only on* m, M_0, *and the functions* $\delta_i(\rho)$ *and constant* μ *of conditions* (5.38), *and also on the diameter* d *of* Ω.

Proof. Consider the identity (5.6). As in the case of a single elliptic equation (see [**5**]), we first obtain the bound

$$\left|\sum_{s=1}^{N}\frac{\partial A_{ij}}{\partial|p|}\frac{u_k^s}{|u_x|}u_{ij}^s\right|\leqslant\frac{1}{2}\sum_{s=1}^{N}A_{ij}u_{ki}^su_{kj}^s+c\,\mathrm{tr}\,A\frac{v_x^2}{v}. \tag{5.40}$$

In fact, on introducing as above the matrices $\|\tilde{u}_{ij}^s\|$ and $\|\widetilde{A}_{ij}\|$, and recalling (5.30), we have

$$\begin{aligned}\left|\sum_{s=1}^{N}\frac{\partial A_{ij}}{\partial|p|}\frac{u_k^s}{|u_x|}u_{ij}^sv_k\right|&=\left|\sum_{s=1}^{N}\frac{\partial\tilde{A}_{ii}}{\partial|p|}\frac{u_k^s}{|u_x|}\tilde{u}_{ii}^sv_k\right|\leqslant\frac{1}{4}\sum_{s=1}^{N}\tilde{A}_{ii}(\tilde{u}_{ii}^s)^2\\&+c\frac{\mathrm{tr}\,\tilde{A}}{v}v_x^2=\frac{1}{4}\sum_{s=1}^{N}A_{ij}u_{ki}^su_{kj}^s+c\frac{\mathrm{tr}\,A}{v}v_x^2,\end{aligned} \tag{5.41}$$

where $v_x^2=\sum_1^n v_k^2$. Further, from the conditions on the growth of A_x^τ and A_u^τ we have

$$|\delta^s(A^\tau)|\leqq\delta_4(v)A^\tau. \tag{5.42}$$

Then

$$\begin{aligned}\left|\sum_{s=1}^{N}\sqrt{v^s}\,\delta^s(A_{ij}u_{ij}^s)\right|&\leqslant\frac{1}{4}\sum_{s=1}^{N}\tilde{A}_{ii}(\tilde{u}_{ii}^s)^2+c\delta_4(v)\,\mathrm{tr}\,\tilde{A}v\\&=\frac{1}{4}\sum_{s=1}^{N}A_{ij}u_{ki}^su_{kj}^s+c\delta_4(v)\,\mathrm{tr}\,Av.\end{aligned} \tag{5.43}$$

Recalling (5.41) and using (5.43), we obtain from (5.6) the inequality

$$\begin{aligned}\frac{1}{2}A_{ij}v_{ij}\geqslant\frac{1}{2}\sum_{s=1}^{N}A_{ij}u_{ki}^su_{kj}^s-c\frac{v_x^2}{v}\,\mathrm{tr}\,A-c\delta_4(v)\,v\,\mathrm{tr}\,A\\+\sum_{s=1}^{N}\frac{\partial B^s}{\partial p_l^t}u_{lk}^tu_k^s+\sum_{s=1}^{N}\sqrt{v^s}\,\delta^sB^s.\end{aligned} \tag{5.44}$$

From the conditions on the growth of B_p^s, B_u^s and B_x^s we obtain

$$\begin{aligned}\left|\sum_{s=1}^{N}\frac{\partial B^s}{\partial p_l^t}u_{lk}^tu_k^s\right|&\leqslant\frac{1}{2}\sum_{s=1}^{N}A_{ij}u_{ki}^su_{kj}^s+\delta_s(v)\,\mathrm{tr}\,A\,v,\\\left|\sum_{s=1}^{N}\sqrt{v^s}\,\delta^sB^s\right|&\leqslant\delta_6(v)\,\mathrm{tr}\,Av.\end{aligned} \tag{5.45}$$

It then follows from (5.44) that

$$\frac{1}{2}A_{ij}v_{ij}\geqslant\left\{-c_1\frac{v_x^2}{v}-c_2\delta_7(v)\right\}\mathrm{tr}\,A\,v. \tag{5.46}$$

We now make the substitution

$$v=z(x)\bar{v},\quad z(x)>0, \tag{5.47}$$

where

$$z(x)=\alpha+d^2-|x|^2, \tag{5.48}$$

where d is the diameter of Ω, and α is a sufficiently large positive number (which will be fixed later). Since

$$v_i = -2x_i\bar{v} + z\bar{v}_i,\quad v_{ij} = -2\delta_{ij}\bar{v} - 2x_i\bar{v}_j - 2x_j\bar{v}_i + z\bar{v}_{ij}, \tag{5.49}$$

it follows from (5.46) that

$$\frac{1}{2} zA_{ij}\bar{v}_{ij} + \sum b_k\bar{v}_k \geqslant \left[2 - \frac{4c_1d^2}{z} - c_2z\delta_7(v)\right] \mathbf{tr}\, A\bar{v}, \tag{5.50}$$

where the form of b_k is of no importance for what follows. On now choosing α in such a way that

$$\frac{4c_1d^2}{z} \leqslant \frac{4c_1d^2}{\alpha} = \frac{1}{2}, \tag{5.51}$$

and assuming that $M = \max_{\bar{\Omega}} v(x) \geqq L$, where L is determined by the condition

$$c_2(\alpha + d^2)\,\delta_7(L) = \frac{1}{2}, \tag{5.52}$$

we obtain the inequality

$$\frac{1}{2} zA_{ij}\bar{v}_{ij} + \sum_{k=1}^{n} b_k\bar{v}_k > 0, \tag{5.53}$$

whence the bound for $M = \max_{\bar{\Omega};\, s=1,\ldots,N} |u_x^s|^2$ may be obtained as in the previous cases. Q.E.D.

Note that the article [**6**] was also concerned with estimating $\max_{\bar{\Omega}}|u_x|$ for the solutions of a system of the type (5.1). Our Theorem 5.1 reveals, however, that the conditions on the growth of the coefficients imposed in [**6**] are not an essential feature of the topic.

Under the auxiliary condition (5.34) we have

Theorem 5.2′. *Let the coefficients* $A_{ij}(x, u^t, |p|)$ *and* $B^s(x, u^t, |p^t|)$ *of the system* (5.1) *belong to the class* C^1 *and satisfy the following conditions: as* $|p| \to \infty$

$$\begin{gathered} |(\sqrt{A^\tau})_{|p|}| \leqslant \frac{\mu\sqrt{\mathbf{tr}\ A}}{|p|},\quad |A_x^\tau| + |p||A_u^\tau| \leqslant \delta_1(|p|)\,A^\tau, \\ |B_p^s| \leqslant \delta_2(|p|)\frac{\mathscr{E}_2}{|p|},\quad |B_x^s| + |p||B_u^s| \leqslant \delta_3(|p|)\,\mathscr{E}_2. \end{gathered} \tag{5.54}$$

Then, for any solution $u(x) = \{u^s(x)\}_1^N$ *of the system* (5.1) *belonging to the class* $C^3(\Omega) \cap C^1(\bar{\Omega})$ *and satisfying the conditions* (5.39), *the quantity* $M = \max_{\bar{\Omega};\, s=1,\ldots,N}|u_x^s|$ *depends only on* m, M_0, *the functions* $\delta_i(\rho)$ *and constant* μ *of conditions* (5.54), *and the diameter* d *of* Ω.

Proof. Consider the inequality (5.44). In view of the condition (5.34) and using (5.54), we have

$$\begin{aligned} \left|\sum_{s=1}^{N} \frac{\partial B^s}{\partial p_l^t} u_{lk}^t u_k^s\right| &= \left|\sum_{s=1}^{N} \frac{\partial B^s}{\partial |p|}\frac{u_l^t u_{lk}^t u_k^s}{|u_x|}\right| \leqslant \left|\delta_4(v)\ \mathbf{tr}\ A \sum_{k=1}^{n} v_k\right| \\ &= \left|\delta_4(v)\ \mathbf{tr}\ A \sum_{k=1}^{n} (z\bar{v}_k - 2x_k\bar{v})\right|. \end{aligned} \tag{5.55}$$

It follows from this that the inequality (5.50) is again obtained for the function $\bar{v}$ defined by (5.47); and in view of this we can find the required bound for $\max_{\bar{\Omega};\, s=1,\ldots,N}|u_x^s|$ by the same method as above. Q.E.D.

§6. Existence theorems

The following can be proved in the same way as Theorem 3.1.

THEOREM 6.1. *Let the boundary* $\partial\Omega$ *of the domain* Ω *and the boundary functions* φ^s $(s = 1, \cdots, N)$ *belong to the class* C^3. *Suppose that for any system of functions* $\{v^s(x)\}_1^N$ *satisfying the conditions*

$$(6.1)\qquad \begin{gathered} v^s \in C^2(\bar{\Omega}), \\ A_{ij}(x, v^t, |v_x|)\, v^s_{ij} = \tau B^s(x, v^t, v^t_x) \text{ in } \Omega, \\ v^s = \varphi^s \text{ on } \partial\Omega, \\ s = 1, \ldots, N, \quad \tau \in [0, 1], \end{gathered}$$

we have proved the a priori estimate

$$(6.2)\qquad \max_{\substack{\bar{\Omega} \\ s=1, \ldots, N}} (|v^s| + |v^s_x|) \leqslant M$$

with a constant M *which is independent of* v^s *and* τ. *Then the Dirichlet problem*

$$(6.3)\qquad \begin{gathered} A_{ij}(x, u^t, |u_x|)\, u^s_{ij} = B^s(x, u^t, u^t_x) \text{ in } \Omega, \\ u^s = \varphi^s \text{ on } \partial\Omega, \end{gathered}$$

has at least one solution $u(x) = \{u^s(x)\}_1^N$ *of the class* $C^2(\bar{\Omega})$, *whatever the boundary functions* φ^s.

THEOREM 6.2. *Let* Ω *be a strictly convex domain bounded in* E_n *with a boundary* $\partial\Omega$ *belonging to the class* C^2, *and let the coefficients* A_{ij} *and* B^s *of the system* (5.1) *belong to the class* C^1 *and satisfy the conditions* (4.1) *and* (5.7) *or* (4.1) *and* (5.35). *Then, if any solution of the problem*

$$(6.4)\qquad \begin{gathered} A_{ij} u^s_{ij} = \tau B^s \text{ in } \Omega, \quad u^s = \tau\varphi^s \text{ on } \partial\Omega, \\ s = 1, \ldots, N, \quad 0 \leqslant \tau \leqslant 1 \end{gathered}$$

is subject to the a priori estimate

$$(6.5)\qquad \max_{\substack{\bar{\Omega} \\ s=1, \ldots, N}} |u^s| \leqslant m,$$

the Dirichlet problem (6.3) *will have at least one solution, whatever the boundary functions* φ^s *belonging to* C^2. *If the coefficients* A_{ij} *also satisfy the auxiliary condition* (4.25), *the problem* (6.3) *must then be solvable in the case of an arbitrary nonconvex domain* Ω *with boundary* $\partial\Omega \in C^2$.

The proof of this last theorem is exactly similar to the proof of Theorem 3.2 and is based on Theorem 6.1 and the a priori estimates for $\max_{\partial\Omega}|u^s_x|$ and $\max_{\bar{\Omega}}|u^s_x|$ obtained in Theorems 4.1 (4.1′) and 5.1 or 5.1′.

Now let the coefficients of the system (5.1) satisfy the conditions (4.1) and (5.38), or (4.1) and (5.54). In this case we have the following unconditional form of the existence theorem.

THEOREM 6.3. *Let* Ω *be a strictly convex bounded domain in* E_n *with a boundary* $\partial\Omega$ *belonging to the class* C^2, *and let the coefficients* A_{ij} *and* B^s *of the system* (5.1) *belong to the class* C^1 *and satisfy the conditions* (4.1) *and* (5.38), *or* (4.1) *and* (5.54). *The Dirichlet problem* (6.3) *will then have at least one solution, no matter what the boundary functions* $\varphi^s \in C^2$.

PROOF. We take the sth equation of the system (5.1) and substitute in it

(6.6) $$u^s = z(x)\bar{u}^s,$$

where $z(x) > 0$. Then

(6.7) $$zA_{ij}\bar{u}^s_{ij} + 2A_{ij}z_i\bar{u}^s_j = B^s - A_{ij}z_{ij}\bar{u}^s.$$

We assume that $\max_{\bar{\Omega}}|u^s| = \max_{\bar{\Omega}} u^s > 0$ (if not, we consider the equation $A_{ij}(-u^s_{ij}) = -B^s$). We then set $z(x) = 1 + d^2 - |x|^2$, where d is the diameter of the domain Ω, which contains the origin. Then, at the point $x_0 \in \Omega$ at which the function $\bar{u}^s$ attains its maximum (this maximum is obviously positive, since $\max_{\bar{\Omega}} u^s(x) > 0$ and $z(x) > 0$), we have

(6.8) $$zA_{ij}\bar{u}_{ij} = B^s + 2\,\mathrm{tr}\,A\bar{u}^s(x_0).$$

Noting the condition $|B^s| \leq \delta(|p|)\,\mathrm{tr}\,A$ and the fact that $|u^s_x| = |z_x(x_0)|\bar{u}^s(x_0)$ at the point x_0, we obtain from (6.8)

(6.9) $$zA_{ij}\bar{u}_{ij}\big|_{x=x_0} \geqslant (2 - 2\delta d)\,\bar{u}^s(x_0)\ \mathrm{tr}\ A\big|_{x=x_0},$$

where $\delta = \delta(2d\bar{u}^s(x_0))$.[2] Let the number L be such that

(6.10) $$2 - 2d\delta(2dL) = 1.$$

Then either $\bar{u}^s(x_0) \leq L$, or else at x_0 we have

(6.11) $$zA_{ij}\bar{u}^s_{ij}\big|_{x=x_0} > 0,$$

which contradicts our hypothesis that $\bar{u}^s$ attains its maximum at x_0. Hence

(6.12) $$\max_{\bar{\Omega}} u^s(x) \leqslant \max\{\max_{\partial\Omega} u^s,\ L\} = \max\{\max_{\partial\Omega}|\varphi^s|,\ L\}.$$

Having obtained an a priori estimate of the type (6.5), we can prove our existence Theorem 6.3 in exactly the same way as the previous theorem, by using Theorem 6.1 and the a priori estimates for $\max_{\partial\Omega;\,s=1,\dots,N}|u^s_x|$ and $\max_{\bar{\Omega};\,s=1,\dots,N}|u^s_x|$ established in Theorems 4.1 (4.1′) and 5.2 or 5.2′.

§7. Notes

Let us mention in conclusion that the results of Theorems 1.1, 1.1′, 2.1, 2.2, 3.3, 3.4, 3.4′, 4.1, 4.2, 5.1, 5.1′, 5.2, 5.2′ and 6.2 naturally remain in force if the inequalities (1.6), (1.26), (2.6), (2.37), (3.4), (3.7), (3.8), (3.14), (4.1), (4.27), (5.7), (5.35), (5.38) and (5.54) are satisfied only for all the $u = (u^1, \cdots, u^N)$ which satisfy the inequality

$$\max_{\substack{\Omega \\ t=1,\dots,N}} |u^t| \leqslant m = \mathrm{const}.$$

Note also that, in the statements of some of the above-mentioned theorems, it has not been indicated explicitly that the relevant inequalities may be satisfied only as $|p^s| \to \infty$.

[2] We assume that the function $\rho\delta(\rho)$ is increasing.

Bibliography

[1] O. A. Ladyženskaja and N. N. Ural'ceva, *Linear and quasilinear equations of elliptic type*, "Nauka", Moscow, 1964; English transl., Academic Press, New York, 1968. MR **35** # 1955; **39** # 5941.

[2] A. V. Ivanov, *The Dirichlet problem for quasilinear nonuniformly elliptic equations of second order*, Zap. Naučn. Sem. Leningrad. Otdel. Mat. Inst. Steklov. (LOMI) **19** (1970), 79-94 = Sem. Math. V. A. Steklov Math. Inst. Leningrad **19** (1970), 43-52. MR **44** # 7126.

[3] ———, *On the Dirichlet problem for quasilinear nonuniformly elliptic equations of second order*, Trudy Mat. Inst. Steklov. **116** (1971), 34-54 = Proc. Steklov Inst. Math. **116** (1971), 29-51.

[4] ———, *On the structure of a quasilinear second-order elliptic equation*, Zap. Naučn. Sem. Leningrad. Otdel. Mat. Inst. Steklov. (LOMI) **21** (1971), 5-33 = J. Soviet Math. **1** (1973), 393-415.

[5] J. Serrin, *The problem of Dirichlet for quasilinear elliptic differential equations with many independent variables*, Philos. Trans. Roy. Soc. London Ser. A **264** (1969), 413-496. MR **43** # 7772.

[6] N. M. Ivočkina and A. P. Oskolkov, *Nonlocal estimates for the first derivatives of solutions of the first boundary problem for certain classes of nonuniformly elliptic and nonuniformly parabolic equations and systems*, Trudy Mat. Inst. Steklov. **110** (1970), 65-101 = Proc. Steklov Inst. Math. **110** (1970), 72-115. MR **45** # 5564.

Translated by
D. E. BROWN

Trudy Mat. Inst. Steklov.
125 (1973)

Proc. Steklov Inst. Math.
125 (1973)

ON THE QUESTION OF AN ADMISSIBLE LIMITING GROWTH OF THE RIGHT SIDE OF A QUASILINEAR ELLIPTIC EQUATION

A. V. IVANOV

Abstract. A class of quasilinear elliptic equations $A_{ij}(x,u,u_x)u_{ij} = B(x,u,u_x)$ having a nonempty intersection with the class of uniformly elliptic equations is distinguished for which the solvability of the Dirichlet problem is established for any right sides $B(x,u,p)$ having an order of growth $o(\mathscr{E}_1 \ln|p|)$, where $\mathscr{E}_1 = A_{ij}(x,u,p)p_i p_j$, for $|p| \to \infty$.

Bibliography: 5 items.

We consider in a domain $\Omega \subset E_n$ the quasilinear elliptic equation

$$A_{ij}(x,u,u_x)u_{ij} = B(x,u,u_x), \tag{1}$$

where $u_x = \operatorname{grad} u$ and $u_{ij} = u_{x_i x_j}$. It will always be assumed that the coefficients $A_{ij}(x,u,p)$ and right side $B(x,u,p)$ are of class C^1.

Of importance for the theory of solvability of the multidimensional ($n > 2$) Dirichlet problem

$$A_{ij}u_{ij} = B \text{ in } \Omega, \quad u = \varphi \text{ on } \partial\Omega \qquad (\partial\Omega \text{ is the boundary of } \Omega) \tag{2}$$

were the works of O. A. Ladyženskaja and N. N. Ural′ceva [1], which were primarily devoted to uniformly elliptic equations characterized by the condition $\Lambda \leqq \text{const}\,\lambda$, where $\lambda = \lambda(x,u,p)$ and $\Lambda = \Lambda(x,u,p)$ are respectively the least and greatest eigenvalues of the matrix $A = \| A_{ij}(x,u,p) \|$.

It is now known (see [2]) that an admissible (for the solvability of problem (2)) limiting growth of $B(x,u,p)$ for $|p| \to \infty$ is determined by two majorants: $\mathscr{E}_1 = A_{ij}p_ip_j$ and $\mathscr{E}_2 = \operatorname{tr} A\,|p|$. The present paper discusses only conditions on an admissible growth of B that can be formulated in terms of the majorant $\mathscr{E}_1$. It was shown in [3] that a growth of B for $|p| \to \infty$ even more rapid than $B = O(\mathscr{E}_1)$, viz. $B = o(\mathscr{E}_1 \ln|p|)$, is admissible for the solvability of the Dirichlet problem if an a priori estimate of the norm $\|u\|_{C^\gamma(\bar\Omega)}$ with $\gamma \in (0,1]$ is valid for the equations of form (1) with such right sides. More precisely, the results of [3] (concerning the estimation of $\max_{\bar\Omega}|u_x|$ in terms of $\max_{\partial\bar\Omega}|u_x|$ and $\|u\|_{C^\gamma(\bar\Omega)}$), the results of J. Serrin [4] on the estimation of $\max_{\partial\Omega}|u_x|$ and the results of Ladyženskaja and Ural′ceva [1] on the estimation of $\|u_x\|_{C^\gamma(\bar\Omega)}$ for the solutions of uniformly elliptic equations imply the following theorem.

THEOREM 1. *Suppose the coefficients $A_{ij}(x,u,p)$, the right side $B(x,u,p)$ (for all sufficiently large $|p|$), the boundary $\partial\Omega$ and the boundary function φ satisfy the conditions*

$$|A^\tau_p p| \leqslant c_0\sqrt{\frac{A^\tau \mathscr{E}_1}{|p|^2}}, \quad |A^\tau_p|\,|p| \leqslant c_1\sqrt{\frac{A^\tau \mathscr{E}_1}{|p|^{2\varepsilon}}}, \quad \|A\| \leqslant c_2\frac{\mathscr{E}_1}{|p|^{2\varepsilon}},$$
$$|\delta A^\tau| \leqslant c_3\sqrt{A^\tau \mathscr{E}_1}, \quad \mathscr{E}_1 \geqslant c_4\frac{\operatorname{tr} A\,|p|}{\ln|p|}; \tag{3}$$

AMS (MOS) subject classifications (1970). Primary 35J25, 35J60.

$$|B|\leqslant c_5\mathscr{E}_1 \ln|p|,\ |B_p p - B|\leqslant \delta_1(|p|)\,\mathscr{E}_1 \ln|p|,$$

(4) $$|B_p|\leqslant c_6\frac{\mathscr{E}_1}{|p|^{2\varepsilon}},\quad \frac{|\delta B|}{|p|}\leqslant c_7\mathscr{E}_1\ln|p|;$$

(5) $$\partial\Omega\in C^2,\quad \varphi\in C^2;$$

where

$$A^\tau = A_{ij}\tau_i\tau_j,\ |\tau|=1,\ \mathscr{E}_1 = A_{ij}p_ip_j,$$

$$\|A\| = \Big(\sum_{ij}A_{ij}^2\Big)^{1/2},\ \delta = \frac{p_k}{|p|}\frac{\partial}{\partial x_k} + |p|\frac{\partial}{\partial u},\qquad \epsilon > 0,$$

the c_i are positive constants and $\delta_1(\rho)$ is a decreasing continuous function such that $\delta_1(\rho)\to 0$ as $\rho\to\infty$.

If an arbitrary solution $v(x)\in C^2(\overline{\Omega})$ of the problem

(6) $$A_{ij}(x,v,v_x)v_{ij} = \tau B(x,v,v_x)\ \text{in}\ \Omega,\quad v=\tau\varphi\ \text{on}\ \partial\Omega,\quad \tau\in[0,1]$$

satisfies the estimate

(7) $$\|v\|_{C^\gamma(\overline{\Omega})} = \max_{\overline{\Omega}}|v| + \max_{x,\,x'\in\overline{\Omega}}\frac{|v(x)-v(x)'|}{|x-x'|^\gamma}\leqslant m,$$

where $\gamma\in[0,1]$ and m does not depend on either $\tau\in[0,1]$ or v, then problem (2) has at least one solution $u(x)\in C^2(\overline{\Omega})$.

REMARK 1. If $\mathscr{E}_1\ln|p| = o(\operatorname{tr}A|p|)$, the result of Theorem 1 is valid under the additional assumption that the domain $\overline{\Omega}$ is convex. But in this case, as was shown in [**5**], an even more rapid growth of B for $|p|\to\infty$ than $B=o(\mathscr{E}_1\ln|p|)$ is possible.

The result of Theorem 1 on the admissibility of the growth

(8) $$B = o(\mathscr{E}_1\ln|p|)\quad\text{for}\quad |p|\to\infty$$

is, however, conditional inasmuch as an estimate of $\|v\|_{C^\gamma(\overline{\Omega})}$ is postulated in the theorem for the solutions of the one-parameter family of problems (6). In the present paper we distinguish a class of elliptic equations of form (1) for which the result (8) is unconditional.

THEOREM 2. *Suppose the coefficients $A_{ij}(x,u,p)$, the right side $B(x,u,p)$ (for all sufficiently large $|p|$), the boundary $\partial\Omega$ and the boundary function φ satisfy the conditions*

$$n\Big(\frac{1}{4}+\varepsilon\Big)\frac{|A_p^\tau p|^2|p|^2}{A^\tau} + (\mathscr{E}_1)_p p\leqslant \mu\mathscr{E}_1,\quad \mu<1,\quad \forall\tau\perp\frac{p}{|p|};$$

(9) $$|\delta A^\tau|\leqslant c_1\sqrt{A^\tau\mathscr{E}_1\ln|p|}\qquad \forall\tau\perp\frac{p}{|p|};$$

$$|\delta\mathscr{E}_1|\leqslant c_2\mathscr{E}_1|p|;\quad \mathscr{E}_1\ln|p|\geqslant c_3\operatorname{tr}A|p|;$$

(10) $$|B|\leqslant c_4\mathscr{E}_1\ln|p|,\quad |B-B_pp|\leqslant\delta_1(|p|)\,\mathscr{E}_1\ln|p|,\quad \frac{|\delta B|}{|p|}\leqslant c_5\mathscr{E}_1\ln|p|;$$

(11) $$\partial\Omega\in C^2,\quad \varphi\in C^2,$$

where

$$A^\tau = A_{ij}\tau_i\tau_j,\ |\tau|=1,\ \mathscr{E}_1=A_{ij}p_ip_j,\ \delta=\frac{p_k}{|p|}\frac{\partial}{\partial x_k}+|p|\frac{\partial}{\partial u},\ \varepsilon>0,.$$

the c_i are positive constants and $\delta_1(\rho)$ is a decreasing continuous function such that $\delta_1(\rho) \to 0$ as $\rho \to \infty$.

Suppose further that the coefficients $A_{ij}(x,u,p)$ and right side $B(x,u,p)$ satisfy any of the known sufficient conditions (see [**1,4**]*) guaranteeing the a priori estimate*

$$\max_{\Omega} |v| \leqslant m \tag{12}$$

for any solution $v(x) \in C^2(\overline{\Omega})$ of problem (6) *with the constant m not depending on either $\tau \in [0,1]$ or v. Then problem* (2) *has at least one solution $u(x) \in C^2(\overline{\Omega})$.*

Remark 2. If $\mathscr{E}_1 \ln|p| = o(\operatorname{tr} A|p|)$ for $|p| \to \infty$, the result of Theorem 2 is valid under the additional assumption that the domain Ω is convex. But in this case, as was shown in [**5**], an even more rapid growth of B for $|p| \to \infty$ than $B = o(\mathscr{E}_1 \ln|p|)$ is possible.

The classes of elliptic matrices $\|A_{ij}\|$ defined by conditions (3) and (9) have a nonempty intersection. Consider, for example, the matrix $A = \|A_{ij}\|$, where

$$A_{ij} = |p|^{-2}\left(\delta_{ij} + k\frac{p_i}{|p|}\frac{p_j}{|p|}\right), \quad k = \text{const} > 0. \tag{13}$$

This matrix is the matrix of a uniformly elliptic equation, since for all τ with $|\tau| = 1$

$$A^\tau = A_{ij}\tau_i\tau_j = |p|^{-2}[1 + k(\sigma, \tau)^2], \quad \sigma = \frac{p}{|p|} \tag{14}$$

and

$$|p|^{-2} \leqslant A_{ij}\tau_i\tau_j \leqslant (k+1)|p|^{-2}. \tag{15}$$

(Thus $\lambda = |p|^{-2}$ and $\Lambda = (k+1)|p|^{-2}$.) Inasmuch as

$$\mathscr{E}_1 = A_{ij}p_ip_j = 1 + k, \tag{16}$$

it is easily seen that conditions (3) are valid for $\epsilon = 1$ and suitable constants c_0, c_1, c_2, c_3. On the other hand, by taking into account the fact that for $\tau \perp \sigma$

$$A_p^\tau p = -2|p|^{-2}, \quad A^\tau = |p|^{-2}, \quad \frac{|A_p^\tau p|^2 |p|^2}{A^\tau} = 4, \quad (\mathscr{E}_1)_p\, p = 0, \tag{17}$$

we see that the inequality

$$n\left(\frac{1}{4} + \varepsilon\right)\frac{|A_p^\tau p|^2|p|^2}{A^\tau} + (\mathscr{E}_1)_p\, p \leqslant \mu\mathscr{E}_1 \text{ (where } \mu < 1 \text{ and } \tau \perp \sigma) \tag{18}$$

is valid if

$$n\left(\frac{1}{4} + \varepsilon\right)4 \leqslant (k+1). \tag{19}$$

Thus for any $k > n - 1$ it is always possible to find numbers $\mu < 1$ and $\epsilon > 0$ such that (19) and hence (18) are valid. The rest of conditions (9) are obviously satisfied. Therefore a matrix A defined by (13) (for $k > n - 1$) belongs to each of the classes discussed in Theorems 1 and 2.

Proof. Theorem 2 is proved by establishing an a priori estimate of $\max_{\overline{\Omega}}|u_x|$ (in terms of $\max_{\partial\Omega}|u_x|$ and $\max_{\overline{\Omega}}|u|$). This result is a generalization of Theorem 3.1 in the author's paper [**2**]. We rewrite conditions (9) and (10) in the form

$$n\left(\frac{1}{4}+\varepsilon\right)\frac{|A^{\tau}_{p}p|^2|p|^2}{A^{\tau}}+(\mathcal{E}_1)_p\,p\leqslant\mu\mathcal{E}_1\quad(\mu<1,\ \tau\perp\sigma),$$

$$(20)\qquad |\delta A^{\tau}|\leqslant\bar{c}_1\sqrt{A^{\tau}\mathcal{E}_1}\quad(\tau\perp\sigma),\quad |\delta\mathcal{E}_1|\leqslant c_2\mathcal{E}_1|p|,\quad \mathcal{E}_1\ln|p|\geqslant c_3\,\mathrm{tr}\,A|p|,$$

$$|B-B_p p|\leqslant\alpha\mathcal{E}_1,\quad \frac{|\delta B|}{|p|}\leqslant\bar{c}_4\mathcal{E}_1,$$

where

$$(21)\qquad \bar{c}_1=c_1\sqrt{\ln M},\quad \alpha=\nu\ln M,\quad \bar{c}_4=c_4\ln M,\quad M=\max_{\bar{\Omega}}|u_x|.$$

By considering conditions (20) only for all p satisfying the inequality $|p|>L$, where $\delta_1(L)=\nu$, we can assume that the quantity ν is sufficiently small (the number ν will be fixed below), since the function $\delta_1(\rho)$ in condition (10) is decreasing. We note that conditions (20) coincide in form with the conditions of Theorem 3.1 in [**2**], but that some of the constants in (20) depend on M.

As in [**2**], we introduce the functions

$$(22)\qquad u=\ln\bar{u}/K,\qquad K=\mathrm{const},$$

and

$$(23)\qquad \bar{v}=\bar{u}_x^2=\sum_{k=1}^{n}\bar{u}_k^2.$$

In exactly the same way as in [**2**], we obtain for the function $\bar{v}$ the identity (see [**2**], formula (3.10))

$$(24)\qquad \frac{1}{2}A_{ij}\bar{v}_{ij}=A_{ij}\bar{u}_{ki}\bar{u}_{kj}+\frac{1}{2}[\bar{B}_{\bar{p}_l}-(A_{ij})_{\bar{p}_l}\bar{u}_{ij}]v_l+\sqrt{\bar{v}}(\bar{\delta}\bar{B}-\bar{\delta}A_{ij}\bar{u}_{ij}),$$

where

$$\bar{B}=K\bar{u}(B+K\mathcal{E}_1)\quad\text{and}\quad \bar{\delta}=\frac{\bar{p}_k}{|\bar{p}|}\frac{\partial}{\partial x_k}+|\bar{p}|\frac{\partial}{\partial\bar{u}}.$$

We note that if $D=D(x,u,p)=D(x,\ln\bar{u}/K,\bar{p}/K\bar{u})$, then

$$(25)\qquad \bar{\delta}D=\delta D-K|p|D_p p,\quad \delta D=\frac{p_k}{|p|}D_{x_k}+|p|D_u;$$

while if $D=D(\bar{u})$,

$$(26)\qquad \bar{\delta}D=K\bar{u}|p|D_{\bar{u}}.$$

Using (25), we have

$$(27)\qquad \sqrt{\bar{v}}\,\bar{\delta}A_{ij}\bar{u}_{ij}=(\delta A_{ij}\bar{u}_{ij}-K|p|(A_{ij})_p\,p\bar{u}_{ij})\sqrt{\bar{v}},$$

where $p=u_x$. We represent the matrix $\|\bar{u}_{ij}\|$ in the form $\|\bar{u}_{ij}\|=T\|\tilde{u}_{ij}\|T^*$, where T is an orthogonal matrix transforming the matrix $\|\bar{u}_{ij}\|$ into the diagonal matrix

$$\|\tilde{u}_{ij}\|=\begin{Vmatrix}\tilde{u}_{11}\dots 0\\ \vdots\quad\ \vdots\\ 0\ \dots\ \tilde{u}_{nn}\end{Vmatrix},$$

and let $\widetilde{A}=T^*AT$. We then have the relation

(28) $$\sqrt{\bar v}\,\delta A_{ij}\bar u_{ij}=\sqrt{\bar v}\,\delta\tilde A_{ii}\tilde u_{ii}-\sqrt{\bar v}\,K\,|p|\,(\tilde A_{ii})_p\,p\tilde u_{ii},$$

in which $\widetilde{A}_{ii}=A^{\tau^i}=(A\tau^i,\tau^i)$, where τ^i is the ith column of the orthogonal matrix T, so that $|\tau^i|=1$.

Consider the identity (24) as well as the equalities (27) and (28) at a point $x_0\in\Omega$ at which the function $\bar v(x)$ achieves its greatest value. We will assume in this connection that $|u_x(x_0)|>L$. Otherwise we would already have an estimate of $\max_{\bar\Omega}|u_x|$ since, if we let $v=|u_x|^2=\sum_1^n u_k^2$, we obtain the relations

(29) $$\bar v=K^2\bar u^2 v,\qquad v=\bar v/K^2\bar u^2$$

and

(30) $$\max_{\bar\Omega} v(x)\leqslant\frac{\max_{\bar\Omega}\bar v}{K^2\min\bar u^2}=\frac{\bar v(x_0)}{K^2\min\bar u^2}\leqslant\frac{K^2\max\bar u^2}{K^2\min\bar u^2}L^2\leqslant e^{4mK}L^2,$$

where $m=\max_{\bar\Omega}|u|$. Thus suppose $|u_x(x_0)|\geq L$. Now in order to make use of the first two of the conditions (20), we need to check that the vector $\tau^i\perp u_x(x_0)$. In fact, the definition of τ^i implies that $\|\bar u_{ij}\|\tau^i=\tilde u_{ii}\tau^i$, whence $(\|\bar u_{ij}\|\tau^i,\bar u_x)=\tilde u_{ii}(\tau^i,\bar u_x)$. But inasmuch as at x_0 the derivatives $\bar v_l(x_0)=0$ (the necessary condition for an extremum), we see that

(31) $$(\|\bar u_{ij}\|\tau^i,\ \bar u_x)=\bar u_{kl}\tau^i_l\bar u_k=\frac{1}{2}\bar v_l\tau^i_l=0.$$

Thus at x_0 either $\tilde u_{ii}=0$ or $\tau^i\perp\bar u_x(x_0)$ and hence $\tau^i\perp u_x(x_0)$ (inasmuch as $u_x=\bar u_x/K\bar u$ and $\bar u>0$). If $\tilde u_{ii}=0$, the summand with the corresponding index in (28) simply vanishes; while if $\tau^i\perp u_x$, we can use conditions (20) in estimating (28). Applying these conditions, we get

(32) $$\begin{aligned}\left|\sqrt{\bar v}\,K\,|p|\,(\tilde A_{ii})_p\,p\tilde u_{ii}\right|&=\left|\sqrt{\tilde A_{ii}}\,\tilde u_{ii}\right|\cdot\left|\frac{(\tilde A_{ii})_p\,p}{\sqrt{\tilde A_{ii}}}\sqrt{\bar v}\,K\,|p|\right|\\&\leqslant\frac{1}{1+4\varepsilon_1}\tilde A_{ii}\tilde u_{ii}^2+\left(\frac{1}{4}+\varepsilon_1\right)n\,\frac{|(A^{\tau^i})_p\,p|^2}{A^{\tau^i}}K^2|p|^2\\&\leqslant(1-\varepsilon_2)A_{ij}u_{ki}u_{kj}+[\mu\mathcal{E}_1-(\mathcal{E}_1)_p\,p]K^2\bar v,\end{aligned}$$

where $\epsilon_2=1-(1+4\epsilon_1)^{-1}$, and

(33) $$|\delta\tilde A_{ii}\tilde u_{ii}\sqrt{\bar v}\,|\leqslant\varepsilon_2\tilde A_{ii}\tilde u_{ii}^2+\frac{1}{\varepsilon_2}\frac{|\delta\tilde A_{ii}|^2}{\tilde A_{ii}}\bar v\leqslant\varepsilon_2 A_{ij}u_{ki}u_{kj}+c\mathcal{E}_1\bar v,$$

where $c=n\bar c_1^2/\epsilon_2$. It follows from (24), (28), (32) and (33) that the inequality

(34) $$A_{ij}\bar v_{ij}\geqslant\sqrt{\bar v}\,\delta\bar B-K^2[\mu\mathcal{E}_1-(\mathcal{E}_1)_p\,p]\,\bar v-c\mathcal{E}_1\bar v$$

is valid at x_0.

We now transform the term $\sqrt{\bar v}\,\delta\bar B$ with the use of (25), (26) and (29):

(35) $$\begin{aligned}\sqrt{\bar v}\,\delta\bar B&=\bar v\frac{\delta\bar B}{K\bar u\,|p|}=\frac{\bar v}{K\bar u\,|p|}[K\bar u\delta(B+K\mathcal{E}_1)+K^2\bar u\,|p|\,(B+K\mathcal{E}_1)]\\&=\bar v\left[\frac{\delta B}{|p|}-KB_p p+K\frac{\delta\mathcal{E}_1}{|p|}-K^2(\mathcal{E}_1)_p\,p+KB+K^2\mathcal{E}_1\right]\\&=\bar v\left[K(B-B_p p)+\frac{\delta B}{|p|}+K^2(\mathcal{E}_1-(\mathcal{E}_1)_p\,p)+K\frac{\delta\mathcal{E}_1}{|p|}\right].\end{aligned}$$

Using conditions (20), we conclude from (34) and (35) that the following inequality holds at x_0:

$$A_{ij}\bar{v}_{ij} \geqslant [(1-\mu)K^2 - c^{(1)}K - c^{(2)}]\,\mathscr{E}_1\bar{v}, \tag{36}$$

where $c^{(1)} = \alpha + c_2$ and $c^{(2)} = n\bar{c}_1^2/\epsilon_2 + \bar{c}_4$. We put

$$K = \frac{2c^{(1)}}{1-\mu} = \frac{2\alpha}{1-\mu} + \frac{2c_2}{1-\mu} = \frac{2\nu}{1-\mu}\ln M + \frac{2c_2}{1-\mu}. \tag{37}$$

Then

$$\begin{aligned}(1-\mu)K^2 - c^{(1)}K - c^{(2)} &= \frac{1}{2}K^2 - C^{(2)} \\ &= \frac{1}{2}\left(\frac{2\nu}{1-\mu}\ln M + \frac{2c_2}{1-\mu}\right)^2 - \left(\frac{nc_1 \ln M}{\varepsilon_2} + c_4 \ln M\right).\end{aligned} \tag{38}$$

We first choose ν so that $8m\nu/(1-\mu) < 2$; suppose, for example,

$$\frac{8m\nu}{1-\mu} = 1, \quad \nu = \frac{1-\mu}{8m}. \tag{39}$$

We will now assume that the quantity M is so large that the expression (38) is positive. Otherwise the inequality

$$\frac{1}{2}\left(\frac{2\nu}{1-\mu}\ln M + \frac{2c_2}{1-\mu}\right)^2 - \left(\frac{nc_1 \ln M}{\varepsilon_2} + c_4 \ln M\right) \leqslant 0 \tag{40}$$

would at once imply the required estimate

$$M \leq e^q, \quad q = q(\nu, \mu, c_1, c_2, c_4, \epsilon_1). \tag{41}$$

Then at $x_0 \in \Omega$ we obtain the inequality

$$A_{ij}\bar{v}_{ij} > 0, \tag{42}$$

which is incompatible with the assumption that the point $x_0 \in \Omega$ is a point at which the function $\bar{v}(x)$ has an absolute maximum. Consequently $\bar{v}(x)$ must take its greatest value on the boundary $\partial\Omega$. Then, using (29), we obtain the estimate

$$\max_{\bar{\Omega}} v \leqslant \frac{\max_{\bar{\Omega}} \bar{v}}{K^2 \min \bar{u}^2} \leqslant \frac{\max_{\partial\Omega} \bar{v}}{K^2 \min \bar{u}^2} \leqslant \frac{K^2 \max \bar{u}^2}{K^2 \min \bar{u}^2} \max_{\partial\Omega} v \leqslant e^{4mK} \max_{\partial\Omega} v. \tag{43}$$

Combining (30) and (43), we have

$$M^2 = \max_{\bar{\Omega}} v \leqslant e^{4mK} \max\{\max_{\partial\Omega} |u_x|^2,\ L^2\}, \tag{44}$$

where

$$m = \max_{\bar{\Omega}} |u| \qquad \text{and} \qquad K = \frac{2\nu}{1-\mu}\ln M + \frac{2c_2}{1-\mu}.$$

Substituting the latter value in (44), we get

$$M^2 \leqslant M^{\frac{8m\nu}{1-\mu}} e^{\frac{8mc_2}{1-\mu}} \max\{\max_{\partial\Omega} |u_x|^2,\ L^2\}. \tag{45}$$

Using (39), we get

$$M = \max_{\bar{\Omega}} |u_x| \leqslant e^{\frac{8mc_2}{1-\mu}} \max \{\max_{\partial\Omega} |u_x|^2, L^2\}. \tag{46}$$

Combining (41) and (46), we finally get

$$\max_{\bar{\Omega}} |u_x| \leqslant e^{\frac{8mc_2}{1-\mu}} \max \left\{\max_{\partial\Omega} |u_x|^2, L^2, e^{\left(q - \frac{8mc_2}{1-\mu}\right)}\right\}, \tag{47}$$

where $q = q(\nu, \mu, c_1, c_2, c_4, \epsilon_1)$.

Bibliography

[1] O. A. Ladyženskaja and N. N. Ural'ceva, *Linear and quasilinear equations of elliptic type,* "Nauka", Moscow, 1964; English transl., Academic Press, New York, 1968. MR **35** # 1955; **39** # 5941.

[2] A. V. Ivanov, *Structure of quasilinear second-order elliptic equations,* Zap. Naučn. Sem. Leningrad. Otdel. Mat. Inst. Steklov. (LOMI) **21** (1971), 5-33 = J. Soviet Math. **1** (1973), 393-415.

[3] ———, *The limit growth of the coefficients of quasilinear elliptic equations,* Zap. Naučn. Sem. Leningrad. Otdel. Mat. Inst. Steklov. (LOMI) **27** (1972), 73-90 = J. Soviet Math. (to appear). MR **47** # 5441.

[4] J. Serrin, *The problem of Dirichlet for quasilinear elliptic differential equations with many independent variables,* Philos. Trans. Roy. Soc. London Ser. A **264** (1969), 413-496. MR **43** # 7772.

[5] A. V. Ivanov, *The Dirichlet problem for quasilinear nonuniformly elliptic equations of second order,* Zap. Naučn. Sem. Leningrad. Otdel. Mat. Inst. Steklov. (LOMI) **19** (1970), 79-94 = Sem. Math. V. A. Steklov Math. Inst. Leningrad **19** (1970), 43-52. MR **44** # 7126.

Translated by
S. SMITH

Trudy Mat. Inst. Steklov.
125 (1973)

Proc. Steklov. Inst. Math.
125 (1973)

A PRIORI ESTIMATES FOR SOLUTIONS OF THE DIRICHLET PROBLEM FOR MULTIDIMENSIONAL QUASILINEAR EQUATIONS OF ELLIPTIC TYPE

N. M. IVOČKINA

Abstract. A class of nonuniformly elliptic quasilinear equations is indicated, whose coefficients $a_{ij}(x, u, u_x)$ depend on their arguments in a special way; and for this class a priori estimates of $\max_\Omega |u_x|$ and interior estimates of the Hölder norm of the derivatives of the solutions of the Dirichlet problem are constructed. The estimates are independent of the smoothness properties of the coefficients of the equation.
Bibliography: 4 items.

We shall consider the problem of the smoothness of the solutions of the first boundary value problem for multidimensional quasilinear elliptic equations of the type

$$a_{ij}(x, u, u_x) u_{ij} = f(x, u, u_x), \tag{1}$$

$$u|_{\partial\Omega} = \varphi(x), \quad x = (x^1, \cdots, x^n), \tag{2}$$

where $u_{ij} = \partial^2 u/\partial x^i \partial x^j$, summation is performed over repeated indices, and $\partial\Omega$ is the boundary of a bounded domain Ω of E^n.

The present study is closely linked with the theory of quasilinear elliptic equations and systems developed by O. A. Ladyženskaja and N. N. Ural′ceva in [**1**], and represents a continuation of the articles [**2, 3**]. Here we shall construct nonlocal estimates for the gradient of the solutions of a particular class of problems (1), (2). Our attention is mainly concerned with obtaining an estimate for the first derivatives along a fixed direction, in which no part is played by the smoothness properties of the coefficients of (1). The eigenvalues of the quadratic form

$$a_{ij}(x, u, p)\xi^i\xi^j > 0, \quad i, j = 1, \ldots, n, \tag{3}$$

can have arbitrary growth order with respect to p. As a corollary of these results we construct an interior estimate for the Hölder norm of the first derivatives of the solutions of problem (1), (2).

Our method of obtaining the estimates follows in essence the method developed in the book [**1**].

§1. Estimation of $\int_\Omega |u_l|^p dx$

Before constructing any a priori estimates for the solutions of the problem (1), (2), we shall state the conditions under which it is possible to estimate the first derivatives of the solution of this problem on the boundary $\partial\Omega$. In fact we have the following lemma, proved in [**3**]:

LEMMA 1. *Let $u(x)$ be a solution of the Dirichlet problem* (1), (2) *belonging to $C^1(\Omega) \cap W_2^2(\Omega)$.*[1] *Assume that the ellipticity condition*

AMS (MOS) subject classifications (1970). Primary 35B45, 35J60, 35J25.

[1] The notation for the spaces and norms is the same here as in [**1**].

$$\nu\xi^{i2} \leqslant a_{ij}(x, u, p)\xi^i\xi^j$$

holds for the equation (1), *while*

$$|f(x, u, p)| \leqslant \mu(1+|p|)^2, \quad p=(p_1, \ldots, p_n)$$

and $\max_\Omega |u| = M_0 < \infty$. *Then, if* Ω *is a strictly convex domain of Euclidean space* E^n *with boundary* $\partial\Omega \in C^2$, *we have*

$$\max_{\partial\Omega} |u_x| < A,$$

where the constant A *depends on* M_0, ν, $|\varphi|_{C^2(\partial\Omega)}$, μ, *and the boundary* $\partial\Omega \in C^2$.

We shall assume henceforth that the problem (1), (2) is such that the assertion of Lemma 1 holds for it.

To obtain our estimate for $\int_\Omega |u_l|^p dx$, we shall require a further simple proposition, which is a particular case of a lemma proved in [**3**].

Lemma 2. *Let the function* $u(x)$, $x=(x^1, \cdots, x^n)$, *continuous in the closure of the bounded domain* Ω *of* E^n, *have a finite integral* $\int_\Omega (1+u_l^{2s})u_{ll}^2 dx$ *for some* l *and any finite* $s>0$. *Assume also that* $\max_\Omega|u| = M_0 < \infty$ *and* $\max_{\partial\Omega}|u_l| = M_1 < \infty$. *Then*

$$\int_\Omega (1+u_l^{2(s+2)})\,dx \leqslant c_1(s, M_0)\int_\Omega (1+u_l^{2s})u_{ll}^2 dx + c_2(M_1^{2s+3}, M_0, \text{meas}\,\partial\Omega, \text{meas}\,\Omega). \tag{4}$$

Since the proof of (4) is particularly simple, we shall quote it.

Proof. We integrate by parts in the integral

$$\begin{aligned}\int_\Omega (1+u_l^{2(s+2)})\,dx &= \int_\Omega u_l^{2s+3}u_l dx + \text{meas}\,\Omega \\ &= -(2s+1)\int_\Omega uu_l^{2s+2}u_{ll}dx + \int_{\partial\Omega} uu_l^{2s+3}dx + \text{meas}\,\Omega.\end{aligned} \tag{5}$$

We apply Cauchy's inequality to the product

$$(2s+3)uu_l^{2s+2}u_{ll} = u_l^{s+2}(2s+3)uu_l^s u_{ll} \leqslant \frac{1}{2}u_l^{2(s+2)} + \frac{(2s+3)^2}{2}M_0^2 u_l^{2s}u_{ll}^2. \tag{6}$$

Substituting (6) in (5), we obtain

$$\int_\Omega (1+u_l^{2(s+2)})\,dx \leqslant \frac{1}{2}\int_\Omega (1+u_l^{2(s+2)})\,dx + \frac{c_1(s, M_0)}{2}\int_\Omega u_l^{2s}u_{ll}^2 dx + \frac{c_2}{2}(M_0, M_1^{2s+3}, \text{meas}\,\partial\Omega, \text{meas}\,\Omega).$$

The inequality (4) follows from this. Q.E.D.

In this section we shall use an estimate obtained in [**4**], namely,

Lemma 3. *Let there be defined in a bounded domain* $\Omega \in E^2$ *with class* C^2 *boundary* $\partial\Omega$ *a function* $u(x)$, $u|_{\partial\Omega} = \psi(x)$, *for which the integral* $\int_\Omega u_{ij}^2 dx$ *is finite and* $|u_x| < \infty$. *Then for any continuous function* $g(p)$, $p=(p_1, p_2)$, *such that* $G(p) \geqq \alpha > 0$, α *a constant, we have*

$$\left| \int_\Omega G(u_x)(u_{11}u_{22} - u_{12}^2)\,dx \right| \leqslant c, \tag{7}$$

with a constant c which depends only on

$$M_0 = \max_\Omega |u|, \qquad \max_{|p| \leqslant \max_{\partial\Omega} |u_x|} |G(p)|,$$

the boundary $\partial\Omega$, *and* $\|\psi\|_{C^2(\partial\Omega)}$.

NOTE. Lemma 3 remains in force if, instead of requiring that $G(p)$ be continuous, we require that it be bounded for all finite values of p.

Let us now proceed to obtain an estimate for $\int_\Omega |u_l|^p dx$. We assume that the ellipticity condition for (1) is satisfied in the form

$$a_{ij}(x, u, p)\xi^i\xi^j \geqslant \nu \sum_{i=1}^n \xi^{i2}, \quad i, j = 1, \ldots, n, \tag{8}$$

where $x = (x^1, \cdots, x^n)$, $|u| \leqq M_0 < \infty$, $p = (p_1, \cdots, p_n)$ and $|p| < \infty$, while ν is a constant. For $i, j \neq l$, let the coefficients $a_{ij}(x,u,p)$ of (1) have the form

$$a_{ij}(x, u, p) = \tilde{a}_i(x^i)\, b_{jl}(p_j p_l)\, \delta_i^j, \tag{9}$$

$$a_{ij}^{\cdot}(x, u, p) = \tilde{a}_i(x^i)\, b_j(p_j)\, \delta_i^j, \tag{9'}$$

where δ_i^j is Kronecker delta, the functions $b_j(p_j)$ and $b_{jl}(p_j, p_l)$ are continuous, and the $\tilde{a}_i(x^i)$ satisfy the inequalities

$$\nu_i \leqslant \tilde{a}_i(x^i) \leqslant \mu_i, \quad i = k_1, \ldots, k_{n-1}; \tag{10}$$

ν_i and μ_i are constants.

We shall assume also that the $a_{li}(x,u,p)$ are bounded functions on the set $\mathfrak{M}\{x \in \Omega, |u| \leqq M_0, |p| < \infty\}$. We have

THEOREM 1. *Let the conditions listed above be satisfied for the problem* (1), (2). *Assume further that, for the solution* $u(x)$, *the following quantities are finite*:

$$\max_\Omega |u| = M_0 < \infty, \tag{11}$$

$$\max_\Omega |u_x| < \infty,$$

$$\int_\Omega |u_l|^{2s} u_{ll}^2 dx < \infty, \quad 0 < s < \infty \tag{12}$$

and that for $f(x,u,p)$

$$|f(x, u, p)| \leqslant \mu_0(\pi(p_l^2)\, p_l^2 + 1), \tag{13}$$

where $\pi(p_l^2) \geqq 0$ *and* $\pi(p_l^2) \to 0$ *as* $p_l \to \infty$, *while* μ_0 *is a constant. Then*

$$\int_\Omega |u_l|^{2(s+2)} dx \leqslant c_3(\nu_i, \mu_i, \nu, \mu, \mu_0, s, |u_l|_{\partial\Omega} |\varphi_{xx}|_{\partial\Omega}, \partial\Omega, \Omega). \tag{14}$$

PROOF. We multiply (1) by $(1 + u_l^{2s}) u_{ll}$:

$$(1 + u_l^{2s})\, a_{ij} u_{ij} u_{ll} = f(1 + u_l^{2s})\, u_{ll}. \tag{15}$$

We transform (15) as follows in the light of condition (9):

$$a_{ij}u_{li}u_{lj}(1+u_l^{2s}) = f(1+u_l^{2s})u_{ll} - a_{ii}(1+u_l^{2s})(u_{ii}u_{ll} - u_{il}^2). \tag{16}$$

We divide both sides of (16) by $\prod_{1;k_i\neq l}^{n-1}\tilde{a}_{k_i}$ (see (9′)):

$$\frac{a_{ij}u_{li}u_{lj}(1+u_l^{2s})}{\prod\limits_{k_i\neq l}\tilde{a}_{k_i}} = \frac{f(1+u_l^{2s})u_{ll}}{\prod\limits_{k_i\neq l}\tilde{a}_{k_i}} - \frac{(1+u_l^{2s})b_{il}(u_{ii}u_{ll}-u_{il}^2)}{\prod\limits_{k_i\neq i,l}\tilde{a}_{k_i}}. \tag{17}$$

We apply Cauchy's inequality to the first term on the right side of (17):

$$\left|\frac{f(1+u_l^{2s})u_{ll}}{\prod\limits_{k_i\neq l}\tilde{a}_{k_i}}\right| \leqslant \frac{\nu}{2}(1+u_l^{2s})u_{ll}^2 + \frac{1}{2\nu}\frac{f^2(1+u_l^{2s})}{\prod\limits_{k_i\neq l}\tilde{a}_{ki}^2}. \tag{18}$$

Using the ellipticity condition (8) and the condition (10), we find from (17) in the light of (18) that

$$\sum_{j=1}^{n}(1+u_l^{2s})u_{lj}^2 \leqslant c_4(\nu_i,\ \nu,\ \mu_i)f^2(1+u_l^{2s}) - \frac{2}{\nu}\frac{(1+u_l^{2s})b_{il}(u_{ii}u_{ll}-u_{il}^2)}{\prod\limits_{k_i\neq i,\,l}\tilde{a}_{k_i}}. \tag{19}$$

We integrate (19) over the domain Ω:

$$\begin{aligned}\int\limits_{\Omega}\sum_{j=1}^{n}(1+u_l^{2s})u_{lj}^2dx &\leqslant c_4\int\limits_{\Omega}f^2(1+u_l^{2s})\,dx \\ &-\frac{2}{\nu}\int\limits_{\Omega}\prod_{\substack{1\\ k_i\neq i,\,l}}^{n-2}\tilde{a}_{k_i}^{-1}(1+u_l^{2s})b_{il}(u_{ii}u_{ll}-u_{il}^2)\,dx = \mathscr{I}_1+\mathscr{I}_2.\end{aligned} \tag{20}$$

Consider the term $\mathscr{I}_2$ of (20):

$$\begin{aligned}\mathscr{I}_2 &= -\frac{2}{\nu}\int\limits_{\Omega}\prod_{\substack{1\\ k_i=i,\,l}}^{n-2}\tilde{a}_{k_i}^{-1}(1+u_l^{2s})b_{il}(u_{ii}u_{ll}-u_{il}^2)\,dx \\ &= -\frac{2}{\nu}\int\limits_{\Omega}\prod_{\substack{k_i\neq i,\,l\\ k_1\ldots k_{n-2}}}\tilde{a}_{k_i}^{-1}d\Omega_{k_1\ldots k_{n-2}}\int\limits_{\Omega_{il}}(1+u_l^{2s})b_{il}(u_{ii}u_{ll}-u_{il}^2)\,d\Omega_{il}.\end{aligned}$$

By Lemma 3,

$$\frac{2}{\nu}\left|\int\limits_{\Omega_{il}}(1+u_l^{2s})b_{il}(u_{ii}u_{ll}-u_{il}^2)\,d\Omega_{il}\right| \leqslant c_5,$$

where c_5 depends on the constants mentioned in (13), whence the following inequality is obtained for $|\mathscr{I}_2|$:

$$\begin{aligned}|\mathscr{I}_2| &\leqslant \left|\frac{2}{\nu}\int\limits_{\Omega}\prod_{\substack{k_i\neq l,\,i\\ k_1\ldots k_{n-2}}}^{n-2}\tilde{a}_{k_i}^{-1}d\Omega_{k_1\ldots k_{n-2}}\right|\left|\int\limits_{\Omega_{il}}(1+u_l^{2s})b_{il}(u_{ii}u_{ll}-u_{il}^2)\,d\Omega_{il}\right| \\ &\leqslant c_6(\mu_i,\ \text{meas}\ \Omega_{k_1\ldots k_{n-2}},\ c_5).\end{aligned} \tag{21}$$

If we use (21) to coarsen the inequality (20), we obtain an analog of the second basic inequality for elliptic equations:

$$\int\limits_{\Omega}\sum_{j=1}^{n}(1+u_l^{2s})u_{lj}^2dx \leqslant c_4\int\limits_{\Omega}f^2(1+u_l^{2s})\,dx + c_5. \tag{22}$$

In conjunction with (13), the last inequality gives us

$$\int_\Omega \sum_{j=1}^n (1+u_l^{2s}) u_{lj}^2 dx \leqslant c_7 \int_\Omega \pi(u_l^2)(1+u_l^{2s+4})\, dx + c_5. \tag{23}$$

Let A_k be the set of points at which $|u_l| \geqq K$. We can then rewrite (23) as

$$\begin{aligned}\int_\Omega \sum_{j=1}^n (1+u_l^{2s}) u_{lj}^2 dx &\leqslant c_8 \int_{\Omega\setminus A_k} (1+u_l^{2s+4})\, dx + \pi(K^2)\int_{A_k} (1+u_l^{2s+4})\, dx \\ &+ c_5 \leqslant c_8 (1+K^{2s+4}) \operatorname{meas}\Omega + \pi(K^2) \int_\Omega (1+u_l^{2s+4})\, dx + c_5.\end{aligned} \tag{24}$$

In view of the properties of $\pi(K^2)$, it is possible to choose K so large that $\pi(K^2) \leqq \epsilon$.

On taking a sufficiently small ϵ, we find from the inequality (4) of Lemma 2 and (24) that

$$\int_\Omega |u_l|^{2s+4} dx \leqslant c_3, \quad s > 0,$$

which proves Theorem 1.

In Theorem 1, the estimate (14) is obtained for the derivative with respect to any one of the coordinate directions. Clearly, if the coefficients $a_{ij}(x,u,p)$ of (1) have the form (9) for all i, j from 1 to n, a corresponding estimate may also be obtained for $u_x = (u_1, \cdots, u_n)$. In fact,

THEOREM 2. *For the problem* (1), (2), (8), *let conditions* (9′) *and* (10) *be satisfied for all i, j from* 1 *to n, together with conditions* (11) *and* (12). *Assume that $f(x,u,p)$ is such that*

$$|f(x, u, p)| \leqslant \mu\pi(|p|^2)(1+p^2), \tag{25}$$

$$p = (p_1, \ldots, p_n).$$

Then

$$\int_\Omega |u_x|^{2s} dx \leqslant c_3' (\nu_i, \mu_i, \nu, \mu, s, |u_x|_{\partial\Omega}, |\varphi_{xx}|_{\partial\Omega}, \partial\Omega, \Omega). \tag{26}$$

To prove Theorem 2, we simply sum the inequality (22) of Theorem 1 over l from $l=1$ to n:

$$\int_\Omega \sum_{l=1}^n \sum_{j=1}^n (1+u_l^{2s}) u_{lj}^2 dx \leqslant c_7' \int_\Omega \sum_{l=1}^n (1+u_l^{2s}) f^2 dx + c_5'. \tag{27}$$

We then observe that, by Young's inequality,

$$u_l^{2s} u_k^2 \leqslant \frac{s}{s+2} u_l^{2(s+2)} + \frac{2}{s+2} u_k^{2(s+2)}. \tag{28}$$

In the light of (28) and our assumption (25), we obtain from (27)

$$\int_\Omega \sum_{l=1}^n \sum_{j=1}^n (1+u_l^{2s}) u_{lj}^2 dx \leqslant c_9 \int_\Omega \sum_{l=1}^n (1+u_l^{2s+4})\, \pi(u_x^2)\, dx + c_{10}. \tag{29}$$

The rest of the proof of (26) is exactly the same as in the relevant part of the proof of Theorem 1.

§2. Estimate for $|u_x|$

To construct an interior nonlocal estimate for the gradient of the solution of problem (1), (2), we utilize the criterion for the boundedness of a function proved in [1]. We shall state this criterion as

LEMMA 4. *Let a function $v(x)$, summable in Ω, satisfy for arbitrary values of $k \geqq k_0 > 0$ the inequalities*

$$\int_{A_k} (v-k)\,dx \leqslant \gamma k^{\alpha} \operatorname{meas}^{1+\varepsilon} A_k, \tag{30}$$

where A_k denotes the set of points $x \in \Omega$ at which $v(x) > k$, and γ, α and ϵ are constants such that $\epsilon > 0$, $\gamma > 0$ and $0 \leqq \alpha \leqq 1+\epsilon$. Then $\operatorname{ess\,sup} v(x)$ *will not exceed some constant which depends only on γ, α, ϵ, k_0 and $\|v\|_{L_1(A_{k_0})}$.*

We have

THEOREM 3. *Let the conditions of Theorem 1 hold for the problem (1), (2). Assume also that, for the l mentioned in Theorem 1,*

$$|u_l|_{\partial\Omega} \leqq k_0.$$

Then

$$\max_{\Omega} |u_l| \leqslant c_{11}(c_3,\ \nu,\ \nu_i,\ \mu,\ \mu_i,\ k_0). \tag{31}$$

The methods developed in [1] will be utilized for proving (31). We multiply (1) by the function $(u_l - k)u_{ll}$, where k is a constant greater than k_0, and we transform the result in the same way as in Theorem 1:

$$a_{ij}u_{ij}(u_l-k)u_{ll} = a_{lj}u_{li}(u_l-k)u_{lj} + a_{ii}(u_l-k)(u_{ii}u_{ll}-u_{il}^2) = f(u_l-k)u_{ll}$$

or

$$a_{ij}u_{lj}(u_l-k)u_{li} = f(u_l-k)u_{ll} - a_{ii}(u_l-k)(u_{ii}u_{ll}-u_{il}^2). \tag{32}$$

Applying condition (8) in (32) and applying Cauchy's inequality to $f(u_l-k)u_{ll}$:

$$f(u_l-k)u_{ll} \leqslant \frac{\nu}{2}(u_l-k)u_{ll}^2 + \frac{f^2(u_l-k)}{2\nu},$$

we arrive at the inequality

$$\sum_{i=1}^{n}(u_l-k)u_{li}^2 \leqslant c_{12}f^2(u_l-k) - \frac{2}{\nu}a_{ii}(u_l-k)(u_{ii}u_{ll}-u_{il}^2). \tag{33}$$

We multiply (33) by the product $\prod_{k_i \neq l}\tilde{a}_{k_i}^{-1}(x^{k_i})$, $i = 1, \cdots, n-1$, and integrate the result over the set $A_k\{u_l > k\}$, where $k \geqq k_0$:

$$\begin{aligned}\sum_{i=1}^{n}\int_{A_k}\prod_{k_i\neq l}\tilde{a}_{k_i}^{-1}(u_l-k)u_{li}^2\,dx &\leqslant c_{13}\int_{A_k} f^2(u_l-k)\,dx \\ &-\frac{2}{\nu}\int_{A_k}\prod_{k_i\neq l,\ i}\tilde{a}_{k_i}^{-1}b_{il}(u_l-k)(u_{ii}u_{ll}-u_{il}^2)\,dx = \mathscr{I}_1 + \mathscr{I}_2.\end{aligned} \tag{34}$$

We observe that the continuous positive function $b_{il}(u_i, u_l)$ can be written in the form

$$b_{il}(u_i, u_l) = \tilde{b}_{il}(u_i, u_l) + \frac{\partial \tilde{b}_{il}(u_i, u_l)}{\partial u_i} u_i,$$

where

$$\tilde{b}_{il} = \frac{1}{u_i} \int_0^{u_i} b_{il}(\tau, u_l)\, d\tau.$$

We integrate by parts in the integral

$$\begin{aligned}
&\int_{A_k} \prod_{k_i \neq i,\, l} \tilde{a}_{k_i}^{-1} \tilde{b}_{il}(u_i, u_l)(u_l - k)(u_{ii}u_{ll} - u_{il}^2)\, dx \\
&= -\int_{A_k} \prod_{k_i \neq i,\, k} \tilde{a}_{k_i}^{-1} \Big[\frac{\partial \tilde{b}_{il}}{\partial u_l} u_i (u_l - k) u_{il} u_{ll} + \frac{\partial \tilde{b}_{il}}{\partial u_i} u_i (u_l - k) u_{ii} u_{ll} + \tilde{b}_{il} u_i u_{li} u_{ll} \\
&\qquad + \tilde{b}_{il} u_i (u_l - k) u_{lli} - \frac{\partial \tilde{b}_{il}}{\partial u_l} u_i (u_l - k) u_{il} u_{ll} \\
&\qquad - \frac{\partial \tilde{b}_{il}}{\partial u_i} u_i (u_l - k) u_{il}^2 - \tilde{b}_{il} u u_i u_{ll} u_{li} - \tilde{b}_{il} u_i (u_l - k) u_{lli} \Big] dx \\
&= -\int_{A_k} \prod_{k_i \neq i,\, l} \tilde{a}_{k_i}^{-1} \frac{\partial \tilde{b}_{il}}{\partial u_i} u_i (u_l - k)(u_{ii}u_{ll} - u_{il}^2)\, dx.
\end{aligned} \tag{35}$$

In view of (35) we have the equation

$$\begin{aligned}
\mathscr{I}_2 &= -\frac{\nu}{2} \int_{A_k} \prod_{k_i \neq i,\, l} \tilde{a}_{k_i}^{-1} b_{il} (u_l - k)(u_{ii}u_{ll} - u_{il}^2)\, dx \\
&= -\frac{\nu}{2} \int_{A_k} \prod_{k_i \neq i,\, l} \tilde{a}_{k_i}^{-1} \Big[\tilde{b}_{il} + \frac{\partial \tilde{b}_{il}}{\partial u_i} u_i \Big] (u_l - k)(u_{ii}u_{ll} - u_{il}^2)\, dx = 0.
\end{aligned} \tag{35'}$$

While the chain of equations (35) is only possible for differentiable $\tilde{b}_{il}(u_i, u_l)$, it will clearly also hold for continuous $\tilde{b}_{il}(u_i, u_l)$, as may be shown by approximating the $\tilde{b}_{il}$ by differentiable functions. The boundary integrals in (35) vanish, since $(u_l - k)|_{\partial A_k} = 0$. Thus, applying (10) and taking account of (35′), we find from (34) that

$$\sum_{i=1}^{n} \int_{A_k} (u_l - k) u_{li}^2 dx \leqslant c_{14} \int_{A_k} f^2 (u_l - k)\, dx \tag{36}$$

for arbitrary $k \geq k_0 \geq 0$.

We find a bound for the integral on the right side of (36) by using (13) and Hölder's inequality:

$$\int_{A_k} f^2 (u_l - k)\, dx \leqslant c_{15} \int_{A_k} |u_l|^5 dx \leqslant (\operatorname{meas} A_k)^{\frac{n-2}{n} + 2\varepsilon} \left[\int_{A_k} |u_l|^{\frac{5n}{2(1-\varepsilon n)}} dx \right]^{\frac{2(1-\varepsilon n)}{n}} \tag{37}$$

where $0 < \epsilon < 1/n$ is a constant. Since, by Theorem 1, $\int_\Omega |u_l|^q dx$ is bounded by a known constant for any finite positive q, it follows from (37) and (36) that

$$\int_{A_k} (u_l - k) u_{li}^2 dx \leqslant c_{16} (\operatorname{meas} A_k)^{\frac{n-2}{n} + 2\varepsilon} \tag{38}$$

Consider $\int_{A_k} (u_l - k) dx$. We apply Hölder's inequality and obtain

(39) $$\int_{A_k} (u_l - k)\,dx \leqslant \text{meas}^{1/3} A_k \left[\int_{A_k} (u_l - k)^{3/2}\,dx \right]^{2/3}.$$

To obtain our further inequalities, we use the imbedding theorem, which holds for any function $v(x) \in \overset{\circ}{W}{}^1_m(\Omega)$ in a bounded domain Ω:

(40) $$\| v \|_{L_p} \leqslant c\,(\text{meas}\,\Omega)^{\frac{1}{n} - \frac{n-m}{nm}} \| v_x \|_{L_m(\Omega)},$$

where $m > 1$ and $p \leqq nm/(n-m)$ (see [1], Chapter II, §2). In (40) we put $p = 3/2$, $m = 2$, $v(x) = u_l - k$ and $\Omega = A_k$:

(41) $$\left[\int_{A_k} (u_l - k)^{3/2}\,dx \right]^{2/3} \leqslant \frac{3}{2}\,c\,(\text{meas}\,A_k)^{\frac{2}{s} - \frac{n-2}{2n}} \left[\int_{A_k} (u_l - k) \sum_{i=1}^{n} u_{li}^2\,dx \right]^{1/2}.$$

The inequalities (39), (38) and (41) give, for arbitrary $k \geqq k_0 > 0$,

(42) $$\int_{A_k} (u_l - k)\,dx \leqslant c_{17}\,\text{meas}^{1+\epsilon} A_k,$$

which is the same as the inequality (30) of Lemma 4 if we put $v(x) = u_l - k$, $\gamma = c_{17}$ and $\alpha = 0$. This shows that u_l is bounded from above. To find a lower bound for u_l, we carry out all the arguments of Theorem 3 for $-u_l$. Q.E.D.

If we now require that all the coefficients of (1) be expressible in the form (9′), and that (13) hold for f, we can obtain an estimate for the gradient of the solution of problem (1), (2). We have

Theorem 4. *Under the conditions of Theorem* 2, $\max_\Omega |u_x|$ *is bounded by a constant which depends on the known parameters of the problem* (1), (2).

The proof is a slight modification of the proof of Theorem 3, carried out for every $l = 1, \cdots, n$.

§3. Estimate for $|u_x|_\alpha$

The fact that the first derivatives of the solutions of problem (1), (2) belong to the Hölder class $C^\alpha(\Omega')$ will be proved in the same way as that used by Ladyženskaja and Ural′ceva in [1] for the case of quasilinear elliptic equations; as the criterion for functions to belong to the Hölder class we take Lemma 4, which is proved in Chapter II, §6 of [1].

Lemma 5. *For the functions* $w^{1,2} = \pm v(x) \in W^1_m(\Omega)$, $n \geqq m > 1$, *with* $\text{ess sup}_\Omega |v| = M_0$ *in any sphere* $K_\rho \subset \Omega$, *let the following series of inequalities be satisfied for arbitrary* $\rho \in (0,1)$ *and all* $k \geqq \max_{K_\rho} w^i(x) - \delta$:

(43) $$\int_{A_{k,\rho}} |w^i_x|^m \xi^m dx \leqslant \gamma \int_{A_{k,\rho}} (w^i - k)^m |\xi_x|^m\,dx + (\text{meas}\,A_{k,\rho})^{1-\frac{m}{q}}, \quad i = 1, 2,$$

where $q > n \geqq 2$, $A_{k,\rho} = K_\rho \cap A_k\{w^i \geqq k\}$, *and* $\xi(x)$ *is a nonnegative smooth function vanishing on* ∂K_ρ. *Then* $v(x)$ *will satisfy a Hölder condition inside* Ω *with an exponent* α *which depends on* γ, δ, m, M_0, Ω, *and the distance to the boundary* $\partial\Omega$.

Theorem 5. *Let* $u(x)$ *be a solution of the problem* (1), (2), (9′) *which belongs* *to* $W_2^2(\Omega)$, *with* $\max_\Omega |u| = M_0 < \infty$ *and* $\max_\Omega |u_l(x)| = M_1 < \infty$. *Then, under the conditions of Theorem* 1, $u_l(x)$ *will belong to the Hölder class* $C^\alpha(\Omega')$, *where* Ω' *is an interior subdomain of* Ω; *the value of the norm* $|u_l|_{\alpha,\Omega'}$ *depends on* ν, ν_i, μ, μ_i, M_0, M, *and the distance from* Ω' *to the boundary* $\partial\Omega$.

Proof. We multiply (1) by the function $u_{ll}\xi^2(x)$, where $\xi(x)$ is any nonnegative smooth function which vanishes in the closure of the complement of the sphere K_ρ, lying in the domain Ω':

$$a_{ij}u_{ij}u_{ll}\xi^2 = fu_{ll}\xi^2. \tag{44}$$

We transform (44) by the familiar method (see (17), (32)):

$$a_{ij}u_{ij}u_{ll}\xi^2 = a_{ij}u_{li}u_{lj}\xi^2 + \tilde{a}_i(x^i)b_i(u_i, u_l)(u_{ll}u_{ii} - u_{il}^2)\xi^2 = fu_{ll}\xi^2$$

or

$$a_{ij}u_{li}u_{lj}\xi^2 = fu_{ll}\xi^2 - \tilde{a}_i(x^i)b_i(u_i)(u_{ii}u_{ll} - u_{il}^2)\xi^2. \tag{45}$$

We multiply (45) by $\prod_{k_i \neq l}\tilde{a}_{k_i}^{-1}(x^{k_i})$, and then obtain estimates for the result by using the ellipticity condition (8), together with condition (10) and Cauchy's inequality:

$$\sum_{i=1}^n u_{li}^2\xi^2 \leqslant c_{18}f^2\xi^2 - c_{19}\prod_{k_i \neq l,\, i}\tilde{a}_{k_i}^{-1}(x^{k_i})b_i(u_i)(u_{ii}u_{ll} - u_{il}^2)\xi^2. \tag{46}$$

We integrate (46) over the set $A_{k,\rho} = A_k \cap K_\rho$, where $K_\rho \subset \Omega'$ and k is an arbitrary number:

$$\begin{aligned}\int_{A_{k,\rho}}\sum_{i=1}^n u_{li}^2\xi^2dx &\leqslant c_{18}\int_{A_{k,\rho}} f^2\xi^2dx \\ &\quad - c_{19}\int_{A_{k,\rho}}\prod_{k_i \neq i,\, l}\tilde{a}_{k_i}^{-1}b_i(u_{ii}u_{ll} - u_{il}^2)\xi^2dx = I_1 + I_2,\end{aligned} \tag{47}$$

since $|u_x| = M_1 < \infty$, and in view of the properties of the functions f and ξ, we have the bound for I_1:

$$I_1 \leq c_{20}(M_1, \xi, c_{18})\,\mathrm{meas}\,A_{k,\rho}. \tag{48}$$

Consider I_2. We integrate by parts:

$$\begin{aligned}I_2 &= -c_{19}\int_{A_{k,\rho}}\prod_{k_i \neq i,\, l}\tilde{a}_{k_i}^{-1}b_i(u_{ii}u_{ll} - u_{il}^2)\xi^2dx \\ &= c_{19}\int_{A_{k,\rho}}\prod_{k_i \neq i,\, l}\tilde{a}_{k_i}^{-1}b_i[(u_l - k)_l u_{ii} - (u_l - k)_i u_{li}]\xi^2dx \\ &= c_{19}\int_{A_{k,\rho}}\Big[\prod_{k_i \neq i,\, l}\tilde{a}_{k_i}^{-1}\frac{\partial b_i}{\partial u_i}(u_l - k)u_{il}u_{ii} + b_i(u_l - k)u_{iil} \\ &\quad + 2b_i(u_l - k)u_{ii}\xi\xi_l - \frac{\partial b_i}{\partial u_i}(u_l - k)u_{il}u_{ii} - b_i(u_l - k)u_{iil} \\ &\quad - 2b_i(u_l - k)u_{il}\xi\xi_i\Big]dx = 2c_{19}\int_{A_{k,\rho}}\prod_{k_i \neq i,\, l}\tilde{a}_{k_i}^{-1}[b_i(u_l - k)u_{ii}\xi\xi_l \\ &\qquad - b_i(u_l - k)u_{il}\xi\xi_i]\,dx.\end{aligned} \tag{49}$$

We express $\prod_{k_i \neq i,l} \tilde{a}_{k_i}^{-1} b_i u_{ii}$ by means of (1):

$$\prod_{k_i \neq i,\, l} \tilde{a}_{k_i}^{-1} b_i u_{ii} = - \prod_{k_i \neq l} \tilde{a}_{k_i}^{-1} [a_{il} u_{il} - f]$$

and substitute in (48):

$$I_2 = -2c_{19} \int_{A_{k,\rho}} \prod_{k_i \neq i,\, l} \tilde{a}_{k_i}^{-1} [\tilde{a}_i^{-1} (a_{il} u_{il} - f)(u_l - k) \xi\xi_l + b_i u_{li} (u_l - k) \xi\xi_i]\, dx. \tag{50}$$

We apply Cauchy's inequality to the integrand in (50):

$$\begin{aligned} \Big| \prod_{k_i \neq l,\, i} \tilde{a}_{k_i}^{-1} [(\tilde{a}_i^{-1} a_{il} u_{il} - f)(u_l - k) + b_i u_{li} (u_l - k) \xi\xi_i] \Big| \\ \leqslant 2c_{21} (u_l - k) \left[\sum_{i=1}^{n} |u_{li}| + 1 \right] \xi |\xi_x| \\ \leqslant c_{21} \left[\varepsilon \sum_{i=1}^{n} (u_{li}^2 + 1) \xi^2 + \frac{1}{\varepsilon} (u_l - k)^2 |\xi_x|^2 \right]. \end{aligned} \tag{51}$$

Here c_{21} depends on ν_i, M_0 and M_1.

Taking $\epsilon = 1/4c_{19}c_{21}$ in (51), we find the following inequality for I_2:

$$|I_2| \leqslant \frac{1}{2} \int_{A_{k,\rho}} \sum_{i=1}^{n} u_{li}^2 \xi^2 dx + c_{22} \left[\int_{A_{k,\rho}} (u_l - k)^2 |\xi_x|^2 dx + \text{meas}\, A_{k,\rho} \right]. \tag{52}$$

Using (48) and (52), we find from (47) that

$$\int_{A_{k,\rho}} |u_{lx}|^2 \xi^2 \leqslant c_{23} \left[\int_{A_{k,\rho}} (u_l - k)^2 \xi_x^2 dx + \text{meas}\, A_{k,\rho} \right]. \tag{53}$$

Note that (53) holds for arbitrary k. A series of inequalities of the type (53) will obviously also hold for $-u_l$, after which it follows from Lemma 4 with $v(x) = u_l$, $m = 2$, $q = \infty$ and $\gamma = c_{23}$ that $u_l(x)$ will satisfy at any point of Ω' a Hölder condition with exponent α which depends on the parameters of the problem. Q.E.D.

As a corollary to Theorem 5 we have

THEOREM 6. *Let $u(x)$ be a solution of problem* (1), (2), *belonging to* $W_2^2(\Omega)$, *with* $\max_\Omega |u(x)| = M_0 < \infty$ *and* $\max_\Omega |u_x(x)| = M_1 < \infty$. *Then, under the conditions of Theorem* 2, *$u_x(x)$ will belong to the Hölder class $C^\alpha(\Omega')$, where Ω' is an interior subdomain of Ω; the value of the norm $|u_x|_{\alpha,\Omega'}$ depends on ν, ν_i, μ, μ_i, M_0, M_1, and the distance from Ω' to the boundary $\partial\Omega$.*

Under the conditions of Theorem 6, equations of the type (1) have no distinctive direction l, so that Theorem 5 holds for all $u_i(x)$, $i = 1, \cdots, n$. Consequently $u_x(x) = (u_1, \cdots, u_n)$ will satisfy a Hölder condition with a known exponent α at any interior point of Ω.

Bibliography

[1] O. A. Ladyženskaja and N. N. Ural′ceva, *Linear and quasilinear equations of elliptic type*, "Nauka", Moscow, 1964; English transl., Academic Press, New York, 1968. MR **35** # 1955; **39** # 5941.

[2] N. M. Ivočkina and A. P. Oskolkov, *On estimates of first derivatives for nonuniformly elliptic nondivergent equations,* Problems of Appl. Math. and Geometric Simulation, Leningrad. Inž.-Stroĭt. Inst., Leningrad, 1968, pp. 5-7. (Russian)

[3] ———, *Nonlocal estimates of the first derivatives of solutions of the Dirichlet problem for nonuniformly elliptic quasilinear equations,* Zap. Naučn. Sem. Leningrad. Otdel. Mat. Inst. Steklov. (LOMI) **5** (1967), 37-109 = Sem. Math. V. A. Steklov Math. Inst. Leningrad **5** (1967), 12-34. MR **37** #4406.

[4] N. M. Ivočkina, *The Dirichlet problem for quasilinear two-dimensional second-order elliptic equations,* Problemy Mat. Anal., vyp. 2, Izdat. Leningrad. Univ., Leningrad, 1969, pp. 140-153 = Problems in Math. Anal., no. 2, Plenum Press, New York, 1971, pp. 115-125. MR **42** #674.

Translated by
D. E. BROWN

Trudy Mat. Inst. Steklov.
125 (1973)

Proc. Steklov Inst. Math.
125 (1973)

ON THE STRONG CONVERGENCE OF SOLUTIONS OF DIFFERENCE EQUATIONS TO GENERALIZED SOLUTIONS OF SECOND-ORDER LINEAR EQUATIONS OF THREE CLASSICAL TYPES

N. K. KORENEV

Abstract. Strong convergence in "energy" norms of solutions of certain very simple difference schemes to generalized solutions (belonging to "energy" classes) of boundary value and initial-boundary value problems is proved for second-order linear equations of elliptic, parabolic, and hyperbolic type, respectively (the first boundary condition is kept in mind). This is done for all of the given problems under the same minimal conditions on the data as those under which theorems of existence for these problems and weak convergence of the solutions of the difference schemes considered here to their solutions were proved earlier by O. A. Ladyženskaja.

Bibliography: 8 items.

A series of difference schemes was constructed by O. A. Ladyženskaja for second-order linear equations of elliptic, parabolic, and hyperbolic type, and under very mild assumptions on the smoothness of all data of the problem she proved their weak convergence in "energy" norms to functions which turned out to be generalized solutions of boundary value problems which belong to "energy" classes (cf., for example, [**1**–**3**, **5**]). At the same time existence theorems were proved for these equations. We formulated the problem (under the same assumptions on the data) of proving strong convergence in "energy" norms of solutions of the same difference schemes to generalized solutions (from "energy" classes) of differential equations by assuming that the latter are known.

Questions of strong convergence of solutions of a series of difference schemes for elliptic and parabolic equations were considered in [**6**–**8**]. Basically, smoother solutions were considered in them under stronger assumptions on the data of the problem than ours, but, on the other hand, the rate of convergence was investigated in terms of the steps of the lattice.

The following notation is used below:

E^n is n-dimensional Euclidean space; $x=(x_1,\cdots,x_n)$ is a point of this space.

To each $h>0$ there corresponds a decomposition of E^n by means of hyperplanes $x_i=k_ih$, $i=1,2,\cdots,n$, $k_i=0,\pm1,\pm2,\cdots$ into cubes $\omega^h(k)$ (k is a multi-index $k_1,\cdots,k_n$) defined thus:

$$\omega^h(k)=\{x\colon\ k_ih<x_i<(k_i+1)h,\ i=1,\ 2,\ \ldots,\ n\}.$$

E^{n+1} is the space of variables $(x,t)=(x_1,\cdots,x_n,t)$.

To each $\tau>0$ there corresponds a decomposition of E^{n+1} by means of hyperplanes $t_m=m\tau$, $m=0,\pm1,\pm2,\cdots$ so that, together with the decomposition of E^n into cubes, one has a decomposition of E^{n+1} into parallelepipeds $q^h(k,l)$ defined thus:

$$q^h(k,\ l)=\omega^h(k)\times(l\tau,\ (l+1)\tau).$$

AMS (*MOS*) *subject classifications* (1970). Primary 35A35, 35D99; Secondary 35J25, 35K15, 35K20, 35L15, 35L20.

Ω is a bounded region in E^n.

S is the boundary of Ω, and $\overline{\Omega} = \Omega + S$.

Q_{T_0} is a cylinder in E^{n+1}, and $\overline{Q}_{T_0} = \overline{\Omega} \times [0, T_0]$.

Γ_{T_0} is a lateral surface of $\overline{Q}_{T_0}$.

Ω_{t_0} is the cross-section of Q_{T_0} with the hyperplane $t = t_0$.

Ω_0 and Ω_{T_0} are its lower and upper bases. The closure of the union of the cubes $\omega^h(k)$, which is completely contained in Ω, is denoted by $\overline{\Omega}_h$.

S_h is the boundary of $\overline{\Omega}_h$.

$\overline{\Omega}_h \backslash S_h = \Omega_h$. In addition, the same symbols $\overline{\Omega}_h$, S_h and Ω_h will denote the following sets of vertices of the cubes $\omega^h(k)$: $\overline{\Omega}_h$ is the set of vertices of the cubes constituting $\overline{\Omega}_h$; S_h is the set of vertices belonging to S_h; $\Omega_h \backslash S_h$ is the set of "interior" vertices.

Q_T^h is the cylinder consisting of the parallelepipeds which are entirely contained in Q_{T_0}; the height of this cylinder is selected so that $T + \tau \leq T_0$, and it remains fixed in the sequel. With a change of the quantity τ the following condition will be assumed: $T/\tau = N$ is an integer.

Ω_{hi} is the cross-section of Q_T^h with the hyperplane $t_i = i\tau$.

In addition, the symbols $\overline{Q}_T^h$, Γ_{hT} and Q_T^h denote the following sets:

$\overline{Q}_T^h$ is the set of vertices of the parallelepipeds constituting $\overline{Q}_T^h$.

Γ_{hT} is the set of vertices belonging to the lateral surfaces $\Gamma_{hT} = S_h \times [0, T]$ of Q_T^h.

$Q_T^h = \overline{Q}_T^h \backslash \Gamma_{hT}$ is the set of "interior" vertices of Q_T^h.

Note that the sets of vertices Ω_{h0} and Ω_{hT} are contained in Q_T^h.

A network function defined on the vertices of the cubes which decompose E^n is denoted by $\alpha_h(k)$.

A function defined on the vertices of the parallelepipeds $q^h(k,l)$ which decompose E^{n+1} is denoted by $\alpha_h(k,l)$.

$$\tilde{\alpha}_h(x)\big|_{\omega^h(k)} = \alpha_h(k).$$

$$\tilde{\alpha}_h(x, t)\big|_{q^h(k, l)} = \alpha_h(k, l).$$

In the sequel we use the following notation for difference quotients: a "right" difference quotient with respect to the variable x_i:

$$\alpha_{hx_i}(k) = \frac{1}{h}[\alpha_h(\ldots, k_i + 1, \ldots) - \alpha_h(\ldots, k_i, \ldots)];$$

a "left" difference quotient with respect to x_i:

$$\alpha_{h\bar{x}_i}(k) = \frac{1}{h}[\alpha_h(\ldots, k_i, \ldots) - \alpha_h(\ldots, k_i - 1, \ldots)].$$

Higher order difference quotients are defined thus:

$$\alpha_{hx_ix_k} = (\alpha_{hx_i})_{x_k} = (\alpha_{hx_k})_{x_i}.$$

$\alpha_h'(x)$ is the multilinear extension of the function $\alpha_h(k)$:

$$\alpha_h'(x)\big|_{\bar{\omega}^h(k)} = \alpha_{hx_1\ldots x_n}(k)(x_1 - k_1h)(x_2 - k_2h)\ldots(x_n - k_nh) + \ldots + \alpha_h(k).$$

$\alpha_h'(x,t)$ denotes the multilinear extension of the function $\alpha_h(k,l)$ defined on $\bar{q}^h(k,l)$, where the extension is carried out with respect to the variables x and t.

The basic function spaces used below are the following:

$L_p(\Omega)$, $p \geqq 1$, is the space of functions which are integrable with power p over Ω, with the norm

$$\|\varphi\|_{p,\Omega} = \left\{ \int_\Omega |\varphi(x)|^p \, dx \right\}^{\frac{1}{p}}.$$

$L_{q,r}(Q_{T_0})$, $q, r \geqq 1$, is the space of functions which have the finite norm

$$\|\psi\|_{q,rQ_{T_0}} = \left\{ \int_0^{T_0} \left[\int_\Omega |\psi(x, t)|^q \, dx \right]^{\frac{r}{q}} dt \right\}^{\frac{1}{r}}.$$

$\overset{\circ}{W}{}_p^1(\Omega)$ is a Banach space; it is defined as the closure of the set of finite smooth functions in Ω with respect to the norm

$$\{\|\varphi_x\|_{p,\Omega}^p + \|\varphi\|_{p,\Omega}^p\}^{\frac{1}{p}} = \left\{ \int_\Omega \left(\sum_{i=1}^n \left| \frac{\partial\varphi}{\partial x_i} \right|^p + |\varphi|^p \right) dx \right\}^{\frac{1}{p}}.$$

$V_2^{1,0}(Q_{T_0})$ is a Banach space; it is defined as the closure of the set of smooth functions which are equal to zero near Γ_{T_0} with respect to the norm

$$|\psi|_{Q_{T_0}} = \max_{t \in [0, T_0]} \|\psi(x, t)\|_{2,\Omega} + \|\psi_x\|_{2, Q_{T_0}}.$$

Note that the elements of this space are continuous with respect to t in the norm of $L_2(\Omega)$. For details of these spaces cf. [1].

§1. Auxiliary propositions

Here we shall cite facts concerning network functions, and also network approximations of elements from certain function spaces (for the most part they are known), in the form in which they are used below.

Let there be defined a network function $\alpha_h(k)$ on $\overline{\Omega}_h$; $\alpha_h'(x)$ is its multilinear extension defined on the closed region $\overline{\Omega}_h$.

We have the inequality

$$\int_{\Omega_h} |\alpha_h'(x)|^p \, dx \leqslant 2^n h^n \sum_{k \in \overline{\Omega}_h} |\alpha_h(k)|^p. \tag{1.1}$$

Indeed, for the cube $\omega^h(k)$ it is obvious that

$$\int_{\omega^h(k)} |\alpha_h'(x)|^p \, dx \leqslant h^n \max_{k \in \overline{\omega}^h(k)} |\alpha_h(k)|^p \leqslant h^n \sum_{k \in \overline{\omega}^h(k)} |\alpha_h(k)|^p.$$

Summing this inequality over all $\omega^h(k) \subset \Omega_h$, we obtain

$$\int_{\Omega_h} |\alpha_h'(x)|^p \, dx \leqslant h^n \sum_{\omega^h(k) \subset \Omega_h} \sum_{k \in \overline{\omega}^h(k)} |\alpha_h(k)|^p;$$

by taking into account the fact that each vertex $k \in \overline{\Omega}_h$ belongs to perhaps only 2^n cubes, we arrive at (1.1). Moreover, we have the inequality

$$h^n \sum_{k \in \overline{\Omega}_h} |\alpha_h(k)|^p \leqslant 2^n (p+1)^n \int_{\Omega_h} |\alpha_h'(x)|^p \, dx. \tag{1.2}$$

Let us prove this. As a preliminary, note that it is sufficient to prove it for the cube $\omega^h(0)$.

Let us consider the case of one variable. At the points 0 and h let there be given a function $\alpha_h(i)$, $i=0,1$; $\alpha_h'(x)$ is its multilinear extension defined by the formula

$$\alpha_h'(x) = \alpha_{hx}(0)\, x + \alpha_h(0), \quad 0 \leqslant x \leqslant h.$$

Together with α_h let us consider a function β_h such that

$$|\alpha_h(i)| = |\beta_h(i)|, \quad i=0,\ 1$$

and $\beta_h(i)$ attains values of opposite sign at the points 0 and 1.

For the functions $\alpha_h(i)$ and $\beta_h(i)$ the equality

$$h\sum_{i=0}^{1} |\alpha_h(i)|^p = h\sum_{i=0}^{1} |\beta_h(i)|^p \tag{1.3}$$

is valid, and for their multilinear extensions $\alpha_h'(x)$ and $\beta_h'(x)$, by virtue of the fact that $|\beta_h'(x)| \leqq |\alpha_h'(x)|$, we have

$$\int_0^h |\beta_h'(x)|^p\, dx \leqslant \int_0^h |\alpha_h'(x)|^p\, dx. \tag{1.4}$$

We shall prove inequality (1.2) for the function $\beta_h'(x)$. Let us consider $\int_0^h |\beta_h'(x)|^p dx$.

By an immediate integration we obtain

$$\int_0^h |\beta_h'(x)|^p\, dx = \frac{h}{p+1}\, \frac{|\beta_h(0)|^{p+1} + |\beta_h(1)|^{p+1}}{|\beta_h(0)| + |\beta_h(1)|}.$$

Then the quotient of norms is equal to

$$\frac{\int_0^h |\beta_h'(x)|^p\, dx}{h\sum_{i=0}^{1} |\beta_h(i)|^p} = \frac{1}{(p+1)}\, \frac{|\beta_h(0)|^{p+1} + |\beta_h(1)|^{p+1}}{(|\beta_h(0)| + |\beta_h(1)|)\,(|\beta_h(0)|^p + |\beta_h(1)|^p)}.$$

Noting that

$$|\beta_h(1)|\,|\beta_h(0)|^p \leqslant \frac{|\beta_h(1)|^{p+1}}{p+1} + \frac{p\,(\beta_h(0)|^{p+1}}{p+1}$$

and

$$|\beta_h(0)|\ |\beta_h((1)|^p \leqq \frac{|\beta_h(0)|^{p+1}}{p+1} + \frac{p\,|\beta_h(1)|^{p+1}}{p+1},$$

we obtain

$$\frac{\int_0^h |\beta_h'(x)|^p\, dx}{h\sum_{i=0}^{1} |\beta_h(i)|^p} \geqslant \frac{1}{2\,(p+1)}.$$

Taking (1.3) and (1.4) into account, we obtain

$$h\sum_{i=0}^{1} |\alpha_h(i)|^p \leqslant 2\,(p+1) \int_0^h |\alpha_h'(x)|^p\, dx.$$

Moreover, by induction on the dimension it is easy to prove that

(1.5) $$h^n \sum_{k\in\bar{\omega}^h(k)} |\alpha_h(k)|^p \leqslant 2^n(p+1)^n \int_{\omega^h(k)} |\alpha_h'(x)|^p\,dx.$$

Summing (1.5) over the cubes $\omega^h(k) \subset \Omega_h$ and discarding the repeating terms on the left, we arrive at (1.2).

Thus we have

Lemma 1.1. *For the function $\alpha_h(k)$ and its multilinear extension $\alpha_h'(x)$, which are defined on $\bar{\Omega}_h$, one has the inequality*

(1.6) $$\frac{1}{2^n}\int_{\Omega_h} |\alpha_h'(x)|^p\,dx \leqslant h^n \sum_{k\in\bar{\Omega}_h} |\alpha_h(k)|^p \leqslant 2^n(p+1)^n \int_{\Omega_h} |\alpha_h'(x)|^p\,dx.$$

Remark. The constant $2^n(p+1)^n$ in (1.5) is precise. Indeed, for a function defined on the square $[0,h]\times[0,h]$ such that

$$\alpha_h(0,\ 0)=1;\quad \alpha_h(0,\ h)=-1;\quad \alpha_h(h,\ 0)=-1;\quad \alpha_h(h,\ h)=1,$$

we have that for $p=1$

$$h^2\sum_k |\alpha_h(k)| = 4h^2, \qquad \int_{[0,\,h]\times[0,\,h]} |\alpha_h'(x)|\,dx = \frac{h^2}{4}.$$

The quotient of norms turns out to be equal to 16, which coincides with the value of the constant calculated in (1.5).

Let the function $\alpha_h(k)$ be defined on $\bar{\Omega}_h$. Let us consider yet one more extension of this function, to be denoted by $\alpha_{h(i)}(x)$, which is defined in the following way: $\alpha_{h(i)}(x)$ is defined on the cube $\bar{\omega}^h(k)$, except the boundary k_i+1, as a multilinear function with respect to the variables $x_1,\cdots,x_{i-1}$, $x_{i+1},\cdots,x_n$ and as a constant with respect to x_i. Such a combined extension will be defined on a set that is wider than $\bar{\Omega}_h$, namely on $\bar{\Omega}_h+\mathbf{e}_ih$, where $+\mathbf{e}_ih$ denotes that to $\bar{\Omega}_h$ is added a layer of cubes in the direction of increase of the ith coordinate.

For such an extension, as is easy to see, the following will be true:

$$\frac{1}{2^{n-1}}\int_{\bar{\Omega}_h+\mathbf{e}_ih} |\alpha_{h(i)}(x)|^p\,dx \leqslant h^n \sum_{k\in\bar{\Omega}_h} |\alpha_h(k)|^p \leqslant 2^{n-1}(p+1)^{n-1} \int_{\bar{\Omega}_h+\mathbf{e}_ih} |\alpha_{h(i)}(x)|^p\,dx.$$
(1.7)

Let the function $\alpha_h(k)$ be defined on $\bar{\Omega}_h$ and $\alpha_h(k)=0$, $k\in S_h$. Let us extend the definition of $\alpha_h(k)$ by zero on the vertices of the cubes $\omega^h(k)$ which are situated outside $\bar{\Omega}_h$. Let $\alpha_h'(x)$ be the multilinear extension of $\alpha_h(k)$; by construction, it is equal to zero on S_h and outside $\bar{\Omega}_h$. Noting that $\partial\alpha_h'(x)/\partial x_i = (\alpha_{hx_i})_{(i)}(x)$, where the latter symbol denotes the combined extension of the difference quotient α_{hx_i}, and taking into account that $(\alpha_{hx_i})_{(i)}(x)$ vanishes outside $\bar{\Omega}_h$, we apply the last inequality to $\partial\alpha_h'(x)/\partial x_i$, and we obtain

(1.8) $$\frac{1}{2^{n-1}}\int_{\Omega_h}\left|\frac{\partial\alpha_h'(x)}{\partial x_i}\right|^p dx \leqslant h^n \sum_{k\in\bar{\Omega}_h} |\alpha_{hx_i}(k)|^p \leqslant 2^{n-1}(p+1)^{n-1}\int_{\Omega_h}\left|\frac{\partial\alpha_h'(x)}{\partial x_i}\right|^p dx.$$

To each function $\varphi(x)$ from $\overset{\circ}{W}{}_p^1(\Omega)$ or from $L_p(\Omega)$, and to each function $\psi(x,t)\in \overset{\circ}{V}{}_2^{1,0}(Q_{T_0})$, there correspond network approximations $\varphi_h(k)$ and $\psi_h(k,l)$, which are constructed in the following manner: $\varphi(x)$ and $\psi(x,t)$

are extended by zero for all $x \in E^n$ outside Ω. By the same token $\varphi(x)$ is defined on the entire space E^n, and $\psi(x,t)$ on the layer $E^n \times [0, T_0]$.

Put

$$\varphi_h(k) = \frac{1}{h^n} \int\limits_{\omega^h(k)} \varphi(x)\,dx, \tag{1.9}$$

$$\psi_h(k, l) = \frac{1}{\tau h^n} \int\limits_{q^h(k, l)} \psi(x, t)\,dx dt. \tag{1.10}$$

Thus $\varphi_h(k)$ is defined on the vertices of the cubes forming the decomposition of E^n, and $\psi_h(k,l)$ is defined on the vertices of the parallelepipeds $q^h(k,l)$ consisting of layers of height T in E^{n+1} which were selected above.

LEMMA 1.2. *Let $\varphi(x) \in L_p(\Omega)$, $p \geqq 1$, and $\psi(x,t) \in L_q(Q_{T_0})$, $q \geqq 1$. Then for their network approximations defined by* (1.7) *and* (1.8) *one has*

$$\left\{h^n \sum_k |\varphi_h(k)|^p\right\}^{\frac{1}{p}} \leqslant \|\varphi\|_{p, \Omega}, \tag{1.11}$$

$$\left\{\tau \sum_{l=0}^{N} h^n \sum_k |\psi_h(k, l)|^q\right\}^{\frac{1}{q}} \leqslant \|\psi\|_{q, Q_{T_0}}; \tag{1.12}$$

moreover,

$$\|\varphi - \tilde{\varphi}_h\|_{p, \Omega} \xrightarrow[h \to 0]{} 0, \tag{1.13}$$

$$\|\psi - \tilde{\psi}_h\|_{q, Q_T} \xrightarrow[h, \tau \to 0]{} 0. \tag{1.14}$$

In (1.11) and (1.12) the symbol $\sum_k$ denotes summation over all vertices in E^n.

LEMMA 1.3. *Let $\varphi(x) \in \overset{\circ}{W}{}_p^1(\Omega)$ and $\psi(x,t) \in \overset{\circ}{V}{}_2^{1,0}(Q_{T_0})$, and let $\varphi_h(k)$ and $\psi_h(k,l)$ be their network approximations defined by* (1.9) *and* (1.10). *Then*

$$\left\{h^n \sum_k |\varphi_{hx_i}(k)|^p\right\}^{\frac{1}{p}} \leqslant \left\|\frac{\partial \varphi}{\partial x_i}\right\|_{p, \Omega}, \tag{1.15}$$

$$\left\{\tau \sum_{l=0}^{N} h^n \sum_k |\psi_{hx_i}(k, l)|^q\right\}^{\frac{1}{q}} \leqslant \left\|\frac{\partial \psi}{\partial x_i}\right\|_{q, Q_{T_0}}, \tag{1.16}$$

$$\left\|\tilde{\varphi}_{hx_i} - \frac{\partial \varphi}{\partial x_i}\right\|_{p, \Omega} \xrightarrow[h \to 0]{} 0, \tag{1.17}$$

$$\left\|\tilde{\psi}_{hx_i} - \frac{\partial \psi}{\partial x_i}\right\|_{2, Q_T} \xrightarrow[h, \tau \to 0]{} 0, \tag{1.18}$$

$$\|\tilde{\psi}_h(x, t) - \psi(x, t)\|_{2, \Omega} \xrightarrow[h, \tau \to 0]{} 0, \tag{1.19}$$

where the convergence to zero in (1.19) *is uniform with respect to t.*

The proof of (1.11) – (1.18) nowhere differs from the proof of the analogous facts concerning Steklov averaging functions, and we shall omit it. The proof of (1.19) is easy to carry out if one considers the continuity of $\psi(x,t)$ with respect to t in the norm of $L_2(\Omega)$.

For functions $\varphi(x) \in \overset{\circ}{W}{}_p^1(\Omega)$ and $\psi(x,t) \in \overset{\circ}{V}{}_2^{1,0}(Q_{T_0})$ let us introduce their network approximations $\dot{\varphi}_h(k)$ and $\dot{\psi}_h(k,l)$ thus:

$$\dot{\varphi}_h(k) = \begin{cases} \varphi_h(k), & k \in \Omega_h, \\ 0, & k \in S_h \text{ and outside } \Omega_h, \end{cases} \tag{1.20}$$

$$\dot{\psi}_h(k, l) = \begin{cases} \psi_h(k, l), & k \in Q_T^h, \\ 0, & k \in \Gamma_{hT} \text{ and outside } Q_T^h, \end{cases} \tag{1.21}$$

where extension of the definition by zero in (1.21) applies only in the layer $E^n \times [0, T]$.

For such "truncated" approximations one has

Lemma 1.4. *For the network approximations $\dot{\varphi}_h(k)$, $\dot{\psi}_h(k,l)$ $\varphi_h(k)$ and $\psi_h(k,l)$ of the functions $\varphi(x) \in \overset{\circ}{W}{}_p^1(\Omega)$ and $\psi(x,t) \in \overset{\circ}{V}{}_2^{1,0}(Q_{T_0})$, under the condition that the boundary of Ω is such that there exists a direction along some one of the coordinates x_i so that the distance from S to S_h does not exceed Ch, one will have*

$$\left\{h^n \sum_k |\dot{\varphi}_{hx_i}(k) - \varphi_{hx_i}(k)|^p\right\}^{\frac{1}{p}} \xrightarrow[h \to 0]{} 0, \tag{1.22}$$

$$\left\{\tau \sum_{l=0}^{N} h^n \sum_k |\dot{\psi}_{hx_i} - \psi_{hx_i}|^2\right\}^{\frac{1}{2}} \xrightarrow[h, \tau \to 0]{} 0, \tag{1.23}$$

$$\|\dot{\tilde{\psi}}_h(x, t) - \tilde{\psi}_h(x, t)\|_{2, \Omega} \xrightarrow[h, \tau \to 0]{} 0, \tag{1.24}$$

where in (1.24) *the convergence to zero is uniform with respect to t. In addition,*

$$\|\dot{\psi}_h(x, t) - \psi(x, t)\|_{2, \Omega} \xrightarrow[h, \tau \to 0]{} 0 \quad \textit{uniformly in } t, \tag{1.25}$$

$$\|\dot{\tilde{\psi}}_h(x, t+\tau) - \psi(x, t)\|_{2, \Omega} \xrightarrow[h, \tau \to 0]{} 0 \quad \textit{uniformly in } t. \tag{1.26}$$

Because of the notation used below, we shall say that sums $h^n \sum_k \varphi_h(k)$ are frequently encountered; each term can also be written thus: $\int_{\omega^h(k)} \tilde{\varphi}(x)\,dx$. The symbol $\tilde{\varphi}_{hx_i}(x)$ denotes a piecewise constant extension of the difference quotient $\varphi_{hx_i}(k)$; and $\varphi_h'(x,t)$ is the multilinear extension with respect to x and t.

§2. The case of an elliptic equation

Let us consider the problem

$$-\frac{\partial}{\partial x_j}\left(a_{ij}\frac{\partial u}{\partial x_i} + a_j u\right) + b_i \frac{\partial u}{\partial x_i} + au = \frac{\partial F_i}{\partial x_i} - f, \tag{2.1}$$

$$u|_S = 0. \tag{2.2}$$

The Fredholm solvability in the space $\overset{\circ}{W}{}_2^1(\Omega)$ is proved for it in [2] under the assumption that the a_{ij} are bounded functions such that

$$\nu \sum_{i=1}^n \xi_i^2 \leqslant a_{ij}\xi_i\xi_j \leqslant \mu \sum_{i=1}^n \xi_i^2, \quad \nu, \mu = \text{const} > 0, \tag{2.3}$$

$$\left\|\sum_{i=1}^n a_i^2, \sum_{i=1}^n b_i^2\right\|_{q/2, \Omega}, \quad \|a\|_{q/2, \Omega} \leqslant \mu, \quad q > n, \tag{2.4}$$

(2.5)
$$\left\| \sum_{i=1}^{n} F_i^2 \right\|_{1,\Omega}, \quad \|f\|_{\frac{2\hat{n}}{\hat{n}+2},\Omega} < \infty,$$
$$\hat{n} = \begin{cases} n, & \dim \Omega = n > 2, \\ 2+\varepsilon, & \dim \Omega = 2. \end{cases}$$

We shall assume that the generalized solution $u(x) \in \overset{\circ}{W}{}_2^1(\Omega)$ is unique. Regarding possible sufficient conditions for uniqueness, see [**2**], Chapter III.

Let us consider the following difference scheme corresponding to the problem (2.1), (2.2):

$$-[a_{hij}v_{x_i} + a_{hj}v]_{\bar{x}_j} + b_{hi}v_{x_i} + a_h v = F_{hi\bar{x}_i} - f_h, \tag{2.6}$$

$$v|_{S_h} = 0. \tag{2.7}$$

The system (2.6) is considered at the points of Ω_h. The values of v must be defined at the points of $\bar{\Omega}_h$.

We shall assume that the system (2.6) – (2.7) is uniquely solvable for any values of the number h. This will be satisfied if we require the satisfaction of the following condition:

$$a_{ij}\xi_i\xi_j + (a_i + b_i)\xi_i\xi_0 + a\xi_0^2 \geqslant \nu_0 \sum_{i=1}^{n} \xi_i^2, \quad \nu_0 = \text{const} > 0. \tag{2.8}$$

For the reason for this see [**3**]. In [**3**] is proved the convergence of solutions of the difference scheme (2.6) – (2.7) to an exact solution of the problem (2.1), (2.2) in the following sense:

$$\int_\Omega \left(\tilde{v}_{x_i} - \frac{\partial u}{\partial x_i}\right) \Phi dx \xrightarrow[h\to 0]{} 0,$$
$$\int_\Omega (\tilde{v} - u)\Phi dx \xrightarrow[h\to 0]{} 0$$

for any $\Phi(x)$ from $L_2(\Omega)$. The proof in [**3**] is carried out for the case of smooth coefficients and a selfadjoint problem (2.1), (2.2); however, there is no difficulty in transferring it to the case being considered here. Let us note that the condition (2.8) is assumed to be satisfied for simplicity. Actually, it is sufficient to assume that the system (2.6) – (2.7) is uniquely solvable for all $h < h_0$.

A solution of (2.6) – (2.7) satisfies the following identity:

$$h^n \sum_{\Omega_h} [a_{hij}v_{x_i}\varphi_{hx_j} + a_{hj}v\varphi_{hx_j} + b_{hi}v_{x_i}\varphi_h + a_h v\varphi_h + F_{hi}\varphi_{hx_i} + f_h\varphi_h] = 0 \tag{2.9}$$

with any network function $\varphi_h(k)$ which is equal to zero on S_h and outside $\bar{\Omega}_h$. In this regard the function v must also be defined to be zero outside $\bar{\Omega}_h$.

We shall prove that under the conditions (2.3) – (2.5) and (2.8) we have

$$\int_\Omega \left(\left|\tilde{v}_{x_i} - \frac{\partial u}{\partial x_i}\right|^2 + |\tilde{v} - u|^2\right) dx \xrightarrow[h\to 0]{} 0. \tag{2.10}$$

For this it suffices to prove that

$$(2.11)\qquad \int_\Omega |\tilde{v}_{x_i} - \tilde{u}_{hx_i}|^2\,dx \xrightarrow[h\to 0]{} 0.$$

Indeed, by virtue of the triangle inequality

$$\left\{\int_\Omega \left|\tilde{v}_{x_i} - \frac{\partial u}{\partial x_i}\right|^2 dx\right\}^{1/2} \leqslant \left\{\int_\Omega |\tilde{v}_{x_i} - \tilde{u}_{hx_i}|^2\,dx\right\}^{1/2} + \left\{\int_\Omega \left|\tilde{u}_{hx_i} - \frac{\partial u}{\partial x_i}\right|^2 dx\right\}^{1/2}.$$

The convergence to zero of the last term is guaranteed by Lemma 1.4. It remains to prove the convergence to 0 of the expression (2.11).

As a preliminary let us remark that the expression

$$(2.12)\qquad h^n \sum_{\Omega_h} a_{hij}(v_{x_i} - \dot{u}_{hx_i})(v_{x_j} - \dot{u}_{hx_j})$$

is equivalent to (2.11) since the positive definiteness of a_{hij} follows from the positive definiteness of the matrix a_{ij}. Passing to the piecewise constant extensions of the network functions appearing in (2.9), we write it in integral form:

$$(2.13)\qquad \begin{aligned}\int_{\Omega_h} [\tilde{a}_{hij}\tilde{v}_{x_i}\tilde{\varphi}_{hx_j} + \tilde{a}_{hi}\tilde{v}\tilde{\varphi}_{hx_i} + \tilde{b}_{hi}\tilde{v}_{x_i}\tilde{\varphi}_h + \tilde{a}_h\tilde{v}\tilde{\varphi}_h \\ + \tilde{F}_{hi}\tilde{\varphi}_{hx_i} + \tilde{f}_h\tilde{\varphi}_h]\,dx = 0.\end{aligned}$$

By virtue of the fact that v and φ_h are equal to zero on S_h and outside $\overline{\Omega}_h$, their difference quotients and extensions $\tilde{v}_{x_i}$ and $\tilde{\varphi}_{hx_i}$ also turn out to be zero outside $\overline{\Omega}_h$. Therefore the integral in (2.13) can be written as an integral over all of Ω, i.e.

$$(2.14)\qquad \int_\Omega [\tilde{a}_{hij}\tilde{v}_{x_i}\tilde{\varphi}_{hx_j} + \tilde{a}_{hi}\tilde{v}\tilde{\varphi}_{hx_i} + \tilde{b}_{hi}\tilde{v}_{x_i}\tilde{\varphi}_h + \tilde{a}_h\tilde{v}\tilde{\varphi}_h + \tilde{F}_{hi}\tilde{\varphi}_{hx_i} + \tilde{f}_h\tilde{\varphi}_h]\,dx = 0.$$

Such a passage, like the passage from (2.9) to (2.14), will be repeatedly encountered below, and we shall not dwell too much on the details.

Let us multiply out the parentheses in (2.12) and replace the term $h^n \sum_{\overline{\Omega}_h} a_{hij} v_{x_i} v_{x_j}$ by its expression in terms of the "lowest" terms of the identity (2.9) with $\varphi_h = v$.

After the passage to the piecewise constant extension and continuation by zero of v and $\dot{u}_h$ to E^n, we obtain the following expression:

$$(2.15)\qquad \begin{aligned}\int_\Omega \tilde{a}_{hij}(\tilde{v}_{x_i} - \tilde{u}_{hx_i})(\tilde{v}_{x_j} - \tilde{u}_{hx_j})\,dx \\ = \int_\Omega [\tilde{a}_{hij}\tilde{u}_{hx_i}\tilde{u}_{hx_j} - 2\tilde{a}_{hij}\tilde{v}_{x_i}\tilde{u}_{hx_j} - \tilde{a}_{hj}\tilde{v}\tilde{v}_{x_j} - \tilde{b}_{hi}\tilde{v}_{x_i}\tilde{v} \\ - \tilde{a}_h\tilde{v}^2 - \tilde{F}_{hi}\tilde{v}_{x_i} - \tilde{f}_h\tilde{v}]\,dx.\end{aligned}$$

In (2.15) we pass to the limit as h tends to 0. Let us remark that $\tilde{a}_{hij}$ converges to a_{ij} almost everywhere; $\tilde{a}_{hj}$, $\tilde{b}_{hi}$, $\tilde{a}_h$, $\tilde{F}_{hi}$ and $\tilde{f}_h$ converge strongly to a_j, b_i, a, F_i and f in the norms of the spaces to which the latter belong; $\tilde{u}_{hx_i}$ converges strongly in $L_2(\Omega)$ to $\partial u/\partial x_i$ and $\tilde{v}_{x_i}$ converges weakly in $L_2(\Omega)$ to $\partial u/\partial x_i$. From the weak convergence of $\tilde{v}_{x_i}$ to $\partial u/\partial x_i$ also follows the weak convergence of $\partial v'/\partial x_i$ to $\partial u/\partial x_i$. Applying the theorem of imbedding of $W_2^1(\Omega)$ in

$L_{2n/(n-2)}(\Omega)$ to $v(x)$, we obtain that $v'(x)$ converges weakly to u in the norm of $L_{2n/(n-2)}(\Omega)$, and, by virtue of the equivalence of the norms of v' and $\tilde{v}$, $\tilde{v}$ converges weakly to u in $L_{2n/(n-2)}(\Omega)$. The complete continuity of the imbedding of $W_2^1(\Omega)$ into $L_{q'}(\Omega)$ with $q' < 2n/(n-2)$ permits one to assert the strong convergence of $\tilde{v}$ to u in $L_{q'}(\Omega)$, where $q' < 2n/(n-2)$.

This fact that a_j^2, b_i^2 and a^2 are elements of $L_{q/2}(\Omega)$ with $q > n$ and the strong convergence of $\tilde{v}$ to u in $L_q(\Omega)$ with $q' < 2n/(n-2)$ give

$$\lim_{h\to 0}\int_\Omega (\tilde{a}_{hj}\tilde{v}\tilde{v}_{x_j} + \tilde{b}_{hi}\tilde{v}_{x_i}\tilde{v} + \tilde{a}_h\tilde{v}^2)\,dx = \int_\Omega \left(a_j u\frac{\partial u}{\partial x_j} + b_i\frac{\partial u}{\partial x_i}u + au^2\right)dx.$$

We finally obtain

$$\lim_{h\to 0}\int_\Omega \tilde{a}_{hij}(\tilde{v}_{x_i} - \tilde{u}_{hx_i})(\tilde{v}_{xj} - \tilde{u}_{hx_j})\,dx$$
$$= -\int_\Omega \left[a_{ij}\frac{\partial u}{\partial x_i}\frac{\partial u}{\partial x_j} + a_j u\frac{\partial u}{\partial x_j} + b_i\frac{\partial u}{\partial x_i}u + au^2 + F_i\frac{\partial u}{\partial x_i} + fu\right]dx = 0.$$

We obtain equality to zero of the last expression by taking into account an identity which is satisfied by a generalized solution of (2.1), (2.2) from $\overset{\circ}{W}{}_2^1(\Omega)$.

Theorem 2.1. *A solution v of the difference scheme* (2.6) – (2.7) *corresponding to the problem* (2.1), (2.2) *under the conditions* (2.3) – (2.5) *and* (2.8) *possesses the following property*:

$$\int_\Omega \left(\left|\tilde{v}_{x_i} - \frac{\partial u}{\partial x_i}\right|^2 + |\tilde{v} - u|^2\right)dx \underset{h\to 0}{\to} 0.$$

§3. The case of a parabolic equation

We consider the problem

$$\frac{\partial u}{\partial t} - \frac{\partial}{\partial x_j}\left[a_{ij}(x,t)\frac{\partial u}{\partial x_i} + a_i(x,t)u\right] + b_i(x,t)\frac{\partial u}{\partial x_i} + a(x,t)u = \frac{\partial F_i}{\partial x_i} - f, \tag{3.1}$$

$$u|_{t=0} = u_0(x), \tag{3.2}$$

$$u|_{\Gamma_{T_0}} = 0, \tag{3.3}$$
$$(x,t) \in Q_{T_0} = \Omega \times [0, T_0].$$

A theorem of existence and uniqueness of a generalized solution from the space $\overset{\circ}{V}{}_2^{1,0}(Q_{T_0})$ (and even $\overset{\circ}{V}{}_2^{1,1/2}(Q_{T_0})$) is proved for it in [1] under the following hypotheses: the coefficients a_{ij} are bounded, in general discontinuous, functions which satisfy the inequalities

$$\nu\sum_{i=1}^n \xi_i^2 \leq a_{ij}\xi_i\xi_j \leq \mu\sum_{i=1}^n \xi_i^2; \quad \nu, \mu = \text{const} > 0. \tag{3.4}$$

The remaining coefficients are measurable functions which have the following bounded norms:

$$\left\|\sum_{i=1}^n a_i^2, \sum_{i=1}^n b_i^2, a\right\|_{q, Q_{T_0}} \leq \mu_1; \quad \frac{1}{q}\left(1 + \frac{n}{2}\right) < 1. \tag{3.5}$$

The free terms satisfy the following conditions:

$$\|\mathbf{F}\|_{2,Q_{T_0}}=\left\{\int\limits_{Q_{T_0}}\sum_{i=1}^{n}|F_i|^2\,dxdt\right\}^{1/2}\leqslant\mu_2,$$
(3.6)
$$\|f\|_{q_1,Q_{T_0}}\leqslant\mu_2,\quad \frac{1}{q_1}\left(1+\frac{n}{2}\right)<1+\frac{n}{4}.$$

The initial value

(3.7) $$u_0(x)\in L_2(\Omega).$$

Note that the conditions (3.5) and (3.6) are more stringent than in [1], where the case of various indices of symmetry with respect to x and t is considered.

For a solution $u(x,t)$ one has the inequality

(3.8)
$$\max_{t\in[0,T_0]}\|u(x,t)\|_{2,\Omega}+\|u_x\|_{2,Q_{T_0}}\equiv|u|_{Q_{T_0}}\leqslant C\left\{\|u_0\|_{2,\Omega}+\|\mathbf{F}\|_{2,Q_{T_0}}+\|f\|_{q_1,Q_{T_0}}\right\}.$$

Various difference schemes for the problem (3.1) – (3.3) are given in [2, 3], and the weak convergence in $\overset{\circ}{W}{}_2^{1,0}(Q_{T_0})$ of the corresponding solutions to a solution of (3.1) – (3.3) is proved.

We shall dwell on one of them, namely the implicit scheme

(3.9) $$v_{\bar t}-[a_{hij}v_{x_i}+a_{hj}v]_{\bar x_j}+b_{hi}v_{x_i}+a_hv=F_{ih\bar x_i}-f_h,$$

(3.10) $$v|_{t=0}=\dot u_{h0}(k),$$

(3.11) $$v|_{\Gamma_{hT}}=0.$$

The equations are assumed to be satisfied at the points $(k,l)\in Q_T^h$, $l\geqq 1$. It follows from [1, 3] that the system (3.9) – (3.11) is uniquely solvable for any ratio of the steps h and τ if τ is sufficiently small. The hypothesis on the boundedness of the coefficients is not essential, since the proof given in [1, 3] transfers to the case of the problem (3.1) – (3.3) under the conditions (3.4) – (3.7). The result is

$$\lim_{h,\tau\to 0}\left[\int\limits_{Q_T}(\tilde v-u)\Phi\,dxdt+\int\limits_{Q_T}\left(\tilde v_{x_i}-\frac{\partial u}{\partial x_i}\right)\Psi_i\,dxdt\right]=0$$

for any $\Phi,\Psi_i\in L_2(Q_T)$.

Let us cite some assertions which are valid for solutions of (3.9) – (3.11), and which will be used in the sequel.

One has the identity

(3.12)
$$h^n\sum_{\Omega_{hm}}v\varphi_n-h^n\sum_{\Omega_{h0}}\dot u_{h0}\varphi_h-\tau\sum_{l=0}^{m-1}h^n\sum_{\Omega_{hl}}v\varphi_{ht}$$
$$+\tau\sum_{l=1}^{m}h^n\sum_{\Omega_{hl}}(a_{hij}v_{x_i}\varphi_{hx_j}+a_{hi}v\varphi_{hx_i}+b_{hi}v_{x_i}\varphi_h+a_hv\varphi_h)$$
$$=-\tau\sum_{l=1}^{m}h^n\sum_{\Omega_{hl}}(F_{hi}\varphi_{hx_i}+f_h\varphi_h).$$

It is valid with any $\varphi_h(k,l)$ which is equal to zero on Γ_{hT} and at all of the vertices of the layer $E^n \times [0, T]$ and with $v(k,l)$ also defined by zero outside $\overline{Q}_T^h$ on the layer $E^n \times [0, T]$. Formula (3.12) is obtained as a result of multiplying equation (3.9) by $\varphi_h(k,l)$, summation over the cylinder $\overline{Q}^h_{m\tau} \subseteq \overline{Q}^h_T$, and transformation of the "principal" terms with the aid of the "summation by parts" formula.

Recall that the decomposition with respect to t is such that $T/\tau = N$ is an integer.

Putting $\varphi_h = v$, we obtain from (3.12) after simple transformations

$$\begin{aligned} h^n \sum_{\Omega_{hm}} |v|^2 + 2\tau \sum_{l=1}^{m} h^n \sum_{\Omega_{hl}} a_{h\,ij} v_{x_i} v_{x_j} &= -\tau^2 \sum_{l=1}^{m} h^n \sum_{\Omega_{hl}} |v_{\bar t}|^2 \\ &+ h^n \sum_{\Omega_{h0}} |\dot u_{h0}|^2 - 2\tau \sum_{l=1}^{m} h^n \sum_{\Omega_{hl}} a_{h_i} v v_{x_i} - 2\tau \sum_{l=1}^{m} h^n \sum_{\Omega_{hl}} b_{h_i} v_{x_i} v \\ &- 2\tau \sum_{l=1}^{m} h^n \sum_{\Omega_{hl}} (a_h |v|^2 + F_{h_i} v_{x_i} + f_h v). \end{aligned} \tag{3.13}$$

Note that (3.12) and (3.13) are valid for $m \leq N$, and $\dot u_{h0}$, in contrast with the averaging of coefficients which is carried out over x and t, is an average only over x since $u_0 = u_0(x)$.

From (3.13) one obtains an inequality which is analogous to (3.8) for an exact solution of (3.1) – (3.3):

$$\begin{aligned} |\tilde v|_{Q_T} \equiv \max_{0 \leqslant i \leqslant N} \|\tilde v\|_{2,\Omega_i} + \|\tilde v_x\|_{2,Q_T} \\ \leqslant C\left\{\|u_0\|_{2,\Omega} + \|\mathbf{F}\|_{2,Q_{T_0}} + \|f\|_{q_1,Q_{T_0}}\right\}. \end{aligned} \tag{3.14}$$

Let us note that for its deduction one has to use the theorem of imbedding of $\overset{\circ}{V}_2(Q_{T_0})$ in $L_q(Q_{T_0})$ with $(1+n/2)/q = n/4$, which is possible by virtue of the equivalence proved above of the norms of network functions and their multilinear extensions.

Let us also recall an identity which an exact solution of (3.1) – (3.3) satisfies:

$$\begin{aligned} &\int_{Q_T} u \frac{\partial \Phi}{\partial t}\, dxdt + \int_{\Omega_0} u_0 \Phi(x, 0)\, dx - \int_{\Omega_T} u(x, T)\, \Phi(x, T)\, dx \\ &- \int_{Q_T} \left[a_{ij} \frac{\partial u}{\partial x_i}\frac{\partial \Phi}{\partial x_j} + a_j u \frac{\partial \Phi}{\partial x_j} + b_i \frac{\partial u}{\partial x_i}\Phi + au\Phi + F_i \frac{\partial \Phi}{\partial x_i} + f\Phi \right] dxdt = 0. \end{aligned} \tag{3.15}$$

This identity is valid with any $\Phi(x,t) \in W_2^1(Q_{T_0})$ which is equal to zero on Γ_{T_0}.

We shall prove that with the satisfaction of the conditions (3.4) – (3.7) the following is true for solutions of the system (3.9) – (3.11):

$$\|\tilde v - u\|_{2,\Omega_t} + \left\|\tilde v_x - \frac{\partial u}{\partial x}\right\|_{2,Q_T} \xrightarrow[h,\tau \to 0]{} 0, \quad t \in [0, T]. \tag{A}$$

The scheme of the proof of this assertion is the following:

a) $\int_\Omega [v'(x,t) - u(x,t)]\varphi dx \to 0$ as $h, \tau \to 0$, where the convergence to zero is uniform with respect to t, and $\varphi(x)$ is any function from $L_2(\Omega)$.

b) The weak convergence in $L_2(\Omega)$ of v' to u uniformly in t gives the possibility of asserting the strong convergence in $L_2(Q_T)$ of v' to u, and by the same token also the strong convergence in $L_2(Q_T)$ of $\tilde v$ to u.

c) The strong convergence in $L_2(Q_T)$ of $\tilde{v}$ to u and the membership of $\tilde{v}$ in $L_q(Q_T)$ with $(1+n/2)/q=n/4$ gives the possibility of asserting that $\tilde{v}$ converges strongly in $L_q(Q_T)$ for any q such that $(1+n/2)/q>n/4$.

We assume that fact c) is known. Assertion b) is proved in exactly the same way as an analogous assertion for Galerkin approximations of the Hopf solution of the problem for the Navier-Stokes equation (cf. [4], Russian pp. 223-224) if one applies the reasoning cited in [4] to the difference $v'(x,t)-u(x,t)$ and uses the equivalence of the norms of multilinear, piecewise constant and combined extensions.

The assertions a) – c) are first proved only for a subsequence of decompositions. The assertion on the convergence of the entire sequence is obtained if one uses the uniqueness of the exact solution of the problem (3.1) – (3.3).

Let us prove a). Take φ to be a smooth function which does not depend on t and is finite on Ω. In this regard one can easily check that for $h<h_0$, where h_0 is some step, φ is equal to zero outside Ω_h. Let us assume that τ is sufficiently small. Let φ_h be the values of the function φ at the nodes of the lattice. We multiply the equation (3.9) by φ_h, and, taking φ_h and v to be extended by zero on the layer $E^n\times[0,T]$, we sum the resulting equality over $\overline{\Omega}_{hl}$ and over t_l from t_{m_1+1} to t_{m_2}. After summation by parts we obtain

$$
\begin{aligned}
h^n\sum_{\Omega_{hm_2}} v\varphi_h - h^n\sum_{\Omega_{hm_1}} v\varphi_h &= -\tau\sum_{l=m_1+1}^{m_2} h^n\sum_{\Omega_{hl}} a_{hij}v_{x_i}\varphi_{x_j} \\
&\quad -\tau\sum_{l=m_1+1}^{m_2} h^n\sum_{\Omega_{hl}}[a_{hi}v\varphi_{hx_i}+b_{hi}v_{x_i}\varphi_h+a_hv\varphi_h+F_{hi}\varphi_{hx_i}+f_h\varphi_n].
\end{aligned}
\tag{3.16}
$$

Passing to piecewise constant extensions in (3.16) and estimating the moduli of the terms, we obtain

$$
\begin{aligned}
\left|\int_{\Omega_{t''}}\tilde{v}\tilde{\varphi}_h dx-\int_{\Omega_{t'}}\tilde{v}\tilde{\varphi}_h dx\right| &\leqslant \left|\int_{Q_{t'+\tau,t''+\tau}}\tilde{a}_{hij}\bar{\tilde{v}}_{x_i}\tilde{\varphi}_{hx_j}dxdt\right| \\
&\quad+\left|\int_{Q_{t'+\tau,t''+\tau}}\tilde{a}_{hi}\tilde{v}\tilde{\varphi}_{hx_i}dxdt\right|+\left|\int_{Q_{t'+\tau,t''+\tau}}\tilde{b}_{hi}\tilde{v}_{x_i}\tilde{\varphi}_h dxdt\right| \\
&\quad+\left|\int_{Q_{t'+\tau,t''+\tau}}\tilde{a}_h\tilde{v}\tilde{\varphi}_h dxdt\right|+\left|\int_{Q_{t'+\tau,t''+\tau}}\tilde{F}_{hi}\tilde{\varphi}_{hx_i}dxdt\right| \\
&\quad+\left|\int_{Q_{t'+\tau,t''+\tau}}\tilde{f}_h\tilde{\varphi}_h dxdt\right|.
\end{aligned}
\tag{3.17}
$$

Here $t'=m_1\tau$ and $t''=m_2\tau$.

We estimate the terms appearing on the right by means of the Hölder inequality in the following manner:

$$
\begin{aligned}
\left|\int_{Q_{t'+\tau,t''+\tau}}\tilde{a}_{hij}\tilde{v}_{x_i}\tilde{\varphi}_{hx_j}dxdt\right| &\leqslant \int_{Q_{t'+\tau,t''+\tau}}|\tilde{a}_{hij}||\tilde{v}_{x_i}||\tilde{\varphi}_{hx_j}|dxdt \\
&\leqslant \mu\|\tilde{v}_x\|_{2,Q_{t'+\tau,t''+\tau}}\|\tilde{\varphi}_{hx}\|_{2,\Omega}|t''-t'|^{1/2};
\end{aligned}
$$

$$
\left|\int_{Q_{t'+\tau,t''+\tau}}\tilde{a}_{hi}\tilde{v}\tilde{\varphi}_{hx_i}dxdt\right|\leqslant\left\|\sum_{i=1}^n a_i^2\right\|^{1/2}_{q,Q_{t'+\tau,t''+\tau}}\|\tilde{v}\|_{q,Q_{t'+\tau,t''+\tau}}\|\tilde{\varphi}_{hx}\|_{2,\Omega}|t''-t'|^{1/2};
$$

$$\left|\int\limits_{Q_{t'+\tau,t''+\tau}} \tilde{b}_{hi}\tilde{v}_{x_i}\tilde{\varphi}dxdt\right| \leqslant \left\|\sum_{i=1}^{n} b_i^2\right\|_{q,\,Q_{t'+\tau,t''+\tau}} \|\tilde{v}_x\|_{2,\,Q_{t'+\tau,t''+\tau}}\|\tilde{\varphi}_h\|_{\bar{q},\,\Omega}|t''-t'|^{\frac{1}{q}};$$

$$\left|\int\limits_{Q_{t'+\tau,t''+\tau}} \tilde{a}_h\tilde{v}\tilde{\varphi}_h dxdt\right| \leqslant \|a\|_{q,\,Q_{t'+\tau,t''+\tau}}\|\tilde{v}\|_{q,\,Q_{t'+\tau,t''+\tau}}\|\tilde{\varphi}_h\|_{\bar{q},\,\Omega}|t''-t'|^{\frac{1}{q}};$$

$$\left|\int\limits_{Q_{t'+\tau,t''+\tau}} \tilde{F}_{hi}\tilde{\varphi}_{hx_i}dxdt\right| \leqslant \|\mathbf{F}\|_{2,\,Q_{t'+\tau,t''+\tau}}\|\tilde{\varphi}_{hx}\|_{2,\,\Omega}|t''-t'|^{1/2};$$

$$\left|\int\limits_{Q_{t'+\tau,t''+\tau}} \tilde{f}_h\tilde{\varphi}_h dxdt\right| \leqslant \|f\|_{q_1,\,Q_{t'+\tau,t''+\tau}}\|\varphi_h\|_{q_1',\Omega}|t''-t'|^{\frac{1}{q_1'}}.$$

This inequality shows us that for small τ

$$\left|\int\limits_{\Omega_{t''}} \tilde{v}\tilde{\varphi}_n dx - \int\limits_{\Omega_{t'}} \tilde{v}\tilde{\varphi}dx\right| \leqslant C|t''-t'|^{\delta}\|\varphi_{hx}\|_{2,\,\Omega}, \quad t'=m_1\tau; \quad t''=m_2\tau, \tag{3.18}$$

where C is determined by the data of the problem, and

$$\delta = \min\left[\frac{1}{2}, \frac{1}{\bar{q}}, \frac{1}{q_1'}\right];$$

for the notation cf. [1], p. 161; transl. p. 136.

Let us further assume in (3.12) that φ_h is the same as in (3.16), and let us subtract from both sides of (3.12) the expression

$$h^n\sum_{\Omega_{hm}} \dot{u}_h\varphi_h - h^n\sum_{\Omega_{h0}} \dot{u}_h\varphi_h$$
$$+\tau\sum_{l=1}^{m} h^n\sum_{\Omega_{hl}}[a_{hij}u_{hx_i}\varphi_{hx_j} + a_{hi}\dot{u}_h\varphi_{hx_i} + b_{hi}\dot{u}_{hx_i}\varphi_h + a_h\dot{u}_h\varphi_h].$$

After regrouping the terms and passing to piecewise constant extensions, we obtain the inequality

$$\begin{aligned}\left|\int\limits_{\Omega_t}(\tilde{v}-\tilde{u}_h)\tilde{\varphi}_h dx\right| \leqslant{}& \left|\int\limits_{\Omega_0}(\tilde{u}_0-\tilde{u}_h)\tilde{\varphi}_h dx\right| + \left|\int\limits_{Q_t}\tilde{a}_{hij}(\tilde{v}_{x_i}-\tilde{u}_{hx_i})\varphi_{hx_j}dxdt\right|\\ &+\left|\int\limits_{Q_t}\tilde{a}_{hi}(\tilde{v}-\tilde{u}_h)\tilde{\varphi}_{hx_i}dxdt\right| + \left|\int\limits_{Q_t}\tilde{b}_{hi}(\tilde{v}_{x_i}-\tilde{u}_{hx_i})\tilde{\varphi}_h dxdt\right|\\ &+\left|\int\limits_{Q_t}\tilde{a}_h(\tilde{v}-\tilde{u}_h)\tilde{\varphi}_h dxdt\right| + \left|\int\limits_{\Omega_0}\tilde{u}_h\tilde{\varphi}_h dx - \int\limits_{\Omega_t}\tilde{u}_h\tilde{\varphi}_h dx\right.\\ &-\int\limits_{Q_t}[\tilde{a}_{hij}\tilde{u}_{hx_i}\tilde{\varphi}_{hx_j} + \tilde{a}_{hi}\tilde{u}_h\tilde{\varphi}_{hx_i} + \tilde{b}_{hi}\tilde{u}_{hx_i}\tilde{\varphi}_h + \tilde{a}_h\tilde{u}_h\tilde{\varphi}_h\\ &\left.+\tilde{F}_{hi}\tilde{\varphi}_{hx_i} + \tilde{f}_h\tilde{\varphi}_h]dxdt\right|.\end{aligned} \tag{3.19}$$

Here $t = m\tau$; the term containing φ_{ht} transformed to 0 by virtue of the fact that φ does not depend on t.

Let us pass to the limit in (3.19) as $h, \tau \to 0$. Here we assume that the height of the cylinder Q_t is fixed with $t = m\tau$, where m is an integer. Taking into account the fact that $v'(x,t)$ converges weakly to $u(x,t)$ in $L_2(Q_T)$ and the membership of $v'(x,t)$ as well as $\tilde{v}$ in $L_q(Q_T)$ with $(1+n/2)/q < n/4$, we have weak convergence of $\tilde{v}$ to u in $L_q(Q_T)$ with $(1+n/2)/q < n/4$. Again using the weak convergence of $\tilde{v}_{x_i}$ to $\partial u/\partial x_i$ in $L_2(Q_T)$, we obtain that the first terms on the right in (3.19) give zero in the limit.

The last term also gives zero, since it passes to an identity which is valid for a generalized solution of (3.1) – (3.3) with $\varphi = \varphi(x)$.

Hence by (1.25) and (1.26) (cf. Lemma 1.4 of this paper) we easily obtain

$$\int_{\Omega_t} (v'(x, t) - u(x, t))\, \varphi(x)\, dx \xrightarrow[h,\, \tau \to 0]{} 0 \tag{3.20}$$

for any fixed $t \in [0, T]$. The weak equicontinuity of $v'(x,t)$ with respect to t in $L_2(\Omega)$ shows us that the convergence is uniform with respect to t. The fact that the smooth, finite (in Ω) functions are dense in $L_2(\Omega)$ guarantees the validity of (3.20) with any $\varphi(x) \in L_2(\Omega)$, if in this connection one considers that $|v'(x,t)|_{Q_T} \leqq \text{const}$.

If one also considers that the step functions are dense in the set of continuous functions, one obtains

$$\int_{\Omega_t} [v'(x, t) - u(x, t)]\, \varphi(x)\, dx \xrightarrow[h,\, \tau \to 0]{} 0; \quad t \in [0, T]. \tag{3.21}$$

We shall pass to the proof of assertion (A). For this we multiply out the parentheses in the expression

$$h^h \sum_{\bar{\Omega}_{hm}} |v - \dot{u}_h|^2 + 2\tau \sum_{l=1}^{m} h^n \sum_{\bar{\Omega}_{hl}} a_{hij} (v_{x_i} - \dot{u}_{hx_i})(v_{x_j} - \dot{u}_{hx_j})$$

and write the terms

$$h^n \sum_{\Omega_{hm}} v^2 + 2\tau \sum_{l=1}^{m} h^n \sum_{\Omega_{hl}} a_{hij} v_{x_i} v_{x_j}$$

just as we did in the case of an elliptic equation. Extending v and $\dot{u}_h$ by zero on the layer $E^n \times [0, T]$, we arrive at the inequality

$$\begin{aligned} &\int_{\Omega_t} |\tilde{v} - \tilde{u}_h|^2 dx + 2 \int_{Q_t} \tilde{a}_{hij} (\tilde{v}_{x_j} - \tilde{u}_{hx_j})(\tilde{v}_{x_i} - \tilde{u}_{hx_i})\, dxdt \\ &\quad \leqslant \Big| \int_{\Omega_0} |\tilde{u}_{h0}|^2 dx - 2 \int_{\Omega_t} \tilde{v}\tilde{u}_h dx + \int_{\Omega_t} |\tilde{u}_h|^2 dx - 4 \int_{Q_t} \tilde{a}_{hij} \tilde{v}_{x_i} \tilde{u}_{hx_j} dxdt \\ &\quad + 2 \int_{Q_t} \tilde{a}_{hij} \tilde{u}_{hx_i} \tilde{u}_{hx_j} dxdt - 2 \int_{Q_t} \tilde{a}_{hi} \tilde{v} \tilde{v}_{x_i} dxdt - 2 \int_{Q_t} \tilde{b}_{hi} \tilde{v}_{x_i} \tilde{v}\, dxdt \\ &\quad - 2 \int_{Q_t} \tilde{a}_h \tilde{v}^2 dxdt - 2 \int_{Q_t} \tilde{F}_{hi} \tilde{v}_{x_i} dxdt - 2 \int_{Q_t} \tilde{f}_h \tilde{v}\, dxdt \Big|. \end{aligned} \tag{3.22}$$

Here $t = m\tau$ and the term $\tau \int_{Q_t} |\tilde{v}_t|^2 dxdt$ is discarded, which only strengthens the inequality. Here, in order to avoid the appearance of integrals with limits which change with a change in τ, let us define a piecewise constant extension thus:

$$\tilde{\varphi}_h(x, t)\,|_{q^h(k, l-1)} = \varphi_h(k, l).$$

As is not difficult to see, the limit functions and nature of convergence are preserved. We shall further use the following: the weak convergence of

$\tilde{v}$ to u in $L_2(\Omega)$, $\tilde{v}_{x_i}$ to $\partial u/\partial x_i$ in $L_2(Q_T)$, the strong convergence of $\tilde{u}_h$ to $u(x,t)$ in $L_2(\Omega)$, $\tilde{u}_{hx_i}$ to $\partial u/\partial x_i$ in $L_2(Q_T)$, $\tilde{u}_h$ to u in $L_q(Q_T)$ with $(1+n/2)/q = n/4$, $\sum_{i=1}^n \tilde{a}_{hi}^2$, $\sum_{i=1}^n \tilde{b}_{hi}^2$, $\tilde{a}_h$ to $\sum_{i=1}^n a_i^2$, $\sum_{i=1}^n b_i^2$, a in $L_q(Q_T)$ with $(1+n/2)/q<1$, $\widetilde{F}_{hi}$ to F_i in $L_2(Q_T)$ and f_h to f in $L_{q_1}(Q_T)$ with $(1+n/2)/q_1 < 1+n/4$.

Here, under the limiting process in the terms

$$\int_{Q_T} (\tilde{a}_{hj}\tilde{v}\tilde{v}_{x_j} + \tilde{b}_{hi}\tilde{v}\tilde{v}_{x_i} + \tilde{a}_h\tilde{v}^2)\, dxdt$$

we shall use the strong convergence of $\tilde{v}$ to u in $L_q(Q_T)$ with $(1+n/4)/q > n/4$, by virtue of which one can show, for example, that $\tilde{a}_{hi}\tilde{v}$ converges strongly in $L_2(Q_T)$ to $a_i u$. The remaining two terms are considered in an analogous manner.

In the terms which are even linear in $\tilde{v}$, the possibility of the limiting process is obvious.

As a result of the limiting process in (3.22), we obtain that

$$\lim_{h,\tau\to 0} \int_{\Omega_t} |\tilde{v} - \tilde{u}_h|^2\, dx + 2\int_{Q_t} \tilde{a}_{hij}(\tilde{v}_{x_i} - \tilde{u}_{hx_i})(\tilde{v}_{x_j} - \tilde{u}_{hx_j})\, dxdt = 0, \tag{3.23}$$

since the expression on the right side has the limit

$$\Bigg| \int_{\Omega_0} u_0^2 dx - \int_{\Omega_t} u^2 dx - 2\int_{Q_t}\Big[a_{ij}\frac{\partial u}{\partial x_i}\frac{\partial u}{\partial x_j} + a_j u\frac{\partial u}{\partial x_j} + b_i\frac{\partial u}{\partial x_i}u + au^2 + F_i\frac{\partial u}{\partial x_i} + fu\Big] dxdt \Bigg| = 0.$$

For the reason for this relation cf. [1], Chapter III, §2.

From (3.23) we obtain

$$\int_{Q_T} |\tilde{v}_x - u_x|^2\, dxdt \xrightarrow[h,\tau\to 0]{} 0, \tag{3.24}$$

$$\int_{\Omega_t} |\tilde{v} - u|^2\, dx \xrightarrow[h,\tau\to 0]{} 0, \quad t = m\tau. \tag{3.25}$$

It is easy to conclude from (3.25) that

$$\int_{\Omega_t} |\tilde{v}(x,t) - u(x,t)|^2\, dx \xrightarrow[h,\tau\to 0]{} 0, \quad t\in[0,T]. \tag{3.26}$$

THEOREM 3.1. *The solutions of the difference scheme* (3.9) – (3.12) *converge to a solution of the problem* (3.1) – (3.3) *so that*

$$\int_{Q_T} |\tilde{v}_x - u_x|^2\, dxdt \xrightarrow[h,\tau\to 0]{} 0,$$

$$\int_{\Omega_t} |\tilde{v}(x,t) - u(x,t)|^2\, dx \xrightarrow[h,\tau\to 0]{} 0, \quad t\in[0,T],$$

if the coefficients and the free terms of the equation (3.1) *satisfy the conditions* (3.4) – (3.7).

§4. The case of a hyperbolic equation

We now pass to the study of the initial-boundary value problem for a hyperbolic equation and the difference scheme which corresponds to it. For simplicity we shall consider the case of the wave equation. The reasoning remains the same for a general hyperbolic equation.

We consider the problem

$$\frac{\partial^2 u}{\partial t^2} - \sum_{i=1}^{n} \frac{\partial^2 u}{\partial x^2} = f(x, t), \tag{4.1}$$

$$u|_{t=0} = \varphi(x), \quad u_t|_{t=0} = \psi(x), \tag{4.2}$$

$$u|_{\Gamma_{T_0}} = 0. \tag{4.3}$$

Let us assume that the free terms and the initial data satisfy the following conditions:

$$f(x, t) \in L_2(Q_{T_0}), \quad \varphi(x) \in \overset{\circ}{W}{}_2^1(\Omega), \quad \psi(x) \in L_2(\Omega). \tag{4.4}$$

In [3, 5] the existence and uniqueness of a generalized solution $u(x,t) \in \overset{\circ}{W}{}_2^1(Q_{T_0})$ is proved for this problem, and also for the case of a general hyperbolic equation. A stronger assertion is also proved there: under the satisfaction of (4.4) the derivatives $\partial u/\partial t$ and $\partial u/\partial x_i$ are elements of $L_2(\Omega)$ which continuously depend on t. However, this result is not used in the sequel.

A solution of (4.1) – (4.3) satisfies the identity

$$\int_{Q_{T_0}} \left[-\frac{\partial u}{\partial t}\frac{\partial \Phi}{\partial t} + \frac{\partial u}{\partial x_i}\frac{\partial \Phi}{\partial x_i} \right] dxdt - \int_{\Omega} \psi \Phi(x, 0)\, dx = \int_{Q_{T_0}} f \Phi dxdt \tag{4.5}$$

with any $\Phi(x,t) \in W_2^1(Q_{T_0})$ which is equal to zero on Γ_{T_0} and for $t = T_0$.

For a solution of (4.1) – (4.3) an "energy relation" is true, from which by integration with respect to t we obtain the equality

$$\int_{Q_t} \left[\left(\frac{\partial u}{\partial t}\right)^2 + \left(\frac{\partial u}{\partial x}\right)^2 \right] dxdt = t \int_{\Omega} \left[\left(\frac{\partial u}{\partial t}\right)^2 + \left(\frac{\partial u}{\partial x}\right)^2 \right] \Big|_{t=0} dx + 2 \int_0^t dt' \int_{Q_{t'}} f \frac{\partial u}{\partial t''}\, dxdt'', \tag{4.6}$$

which is valid for all $t \in [0, T_0]$.

In [3] various difference schemes corresponding to a mixed problem for hyperbolic equations are cited, and the weak convergence of the solutions of these schemes to the exact solution in $W_2^1(Q_{T_0})$ is proved.

Let us consider the following difference scheme:

$$v_{t\bar{t}} - v_{x_i \bar{x}_i} = f_h(k, l), \tag{4.7}$$

$$v|_{t=0} = \dot{\varphi}_h(k), \quad v_t|_{t=0} = \dot{\psi}_h(k), \tag{4.8}$$

$$v|_{S_T^h} = 0, \tag{4.9}$$

where $\dot{\varphi}_h(k)$ and $\dot{\psi}_h(k)$ are averages of the initial data $\varphi(x)$ and $\psi(k)$ over the variable x.

The equation (4.7) is assumed to be satisfied at the points of Q_T^h.

A solution v of the system (4.7) – (4.9) is uniquely determined. For it the equality

$$(4.10)\quad h^h \sum_{\Omega_{hl}} (v_t^2 + v_x^2 + \tau v_{xt} v_x)\,|_{l=0}^{l=m} = \tau \sum_{l=1}^{m} h^h \sum_{\Omega_{hl}} f_h [v_t + v_{\bar{t}}], \quad m = 1, 2, \ldots, N,$$

is true; it is obtained after multiplying (4.7) by $v_t + v_{\bar{t}}$, summing over the cylinder, and transforming the left-hand terms by means of "summation by parts".

With the aid of (4.10) one proves the possibility of taking the limit as $h, \tau \to 0$ if only the condition

$$(4.11)\qquad 2n\frac{\tau^2}{h^2} \leqslant 1 - \varepsilon, \quad \varepsilon > 0,$$

is satisfied.

With the satisfaction of (4.11) one obtains the following estimate for solutions of the difference schemes (4.7) – (4.9):

$$(4.12)\qquad \max_{1 \leqslant m \leqslant N} \Big\{ h^n \sum_{\Omega_{hm}} (v_t^2 + v_x^2) \Big\} \leqslant C \{ \| \varphi_x \|_{2,\Omega}^2 + \| \psi \|_{2,\Omega}^2 + \| f \|_{2,Q_{T_0}}^2 \}.$$

By means of (4.12) one proves that $\tilde{v}$, $\tilde{v}_x$ and $\tilde{v}_t$ converge weakly in $L_2(Q_{T_0})$ to some function u and its derivatives $\partial u/\partial x$ and $\partial u/\partial t$ respectively. One further proves that $u(x,t)$ is a generalized solution from $\overset{\circ}{W}{}_2^1(Q_{T_0})$ of the problem (4.1) – (4.3). Note that, by the uniqueness theorem, the entire sequence converges to $u(x,t)$.

Below we shall prove the strong convergence of $\tilde{v}$, $\tilde{v}_x$, $\tilde{v}_t$ to u, $\partial u/\partial x$, $\partial u/\partial t$ in $L_2(Q_T)$.

Let us consider the expression

$$(4.13)\quad \tau \sum_{m=1}^{N} h^n \sum_{\Omega_{hm}} [| v_t - \dot{u}_{ht} |^2 + | v_x - \dot{u}_{hx} |^2 + \tau (v_{xt} - \dot{u}_{hxt})(v_x - \dot{u}_{hx}).$$

In it, as we have repeatedly done above, after multiplying out parentheses, we substitute terms which are quadratic in v with the aid of equality (4.10), which has been summed over layers, i.e.

$$\tau \sum_{m=1}^{N} \Big[h^n \sum_{\Omega_{hm}} (v_t^2 + v_x^2 + \tau v_{xt} v_x) \Big] = (\tau N) \Big[h^n \sum_{\Omega_{h0}} (v_t^2 + v_x^2 + \tau v_{xt} v_x) + \tau \sum_{m=1}^{N} \tau \sum_{l=1}^{m} h^n \sum_{\Omega_{hl}} f_h (v_t + v_{\bar{t}}).$$

Taking also into account the fact that with the ratio of the steps h and τ satisfying (4.11), the expression (4.17) is equivalent to

$$\tau \sum_{m=1}^{N} h^n \sum_{\Omega_{hm}} (| v_t - \dot{u}_{ht} |^2 + | v_x - \dot{u}_{hx} |^2),$$

we arrive at the inequality

$$\tau \sum_{m=1}^{N} h^n \sum_{\Omega_{hm}} (| v_t - \dot{u}_{ht} |^2 + | v_x - \dot{u}_{hx} |^2) \leqq$$

$$
\begin{aligned}
&\leqslant C\Bigg\{\Bigg|(N\tau)h^n\sum_{\bar{\Omega}_{h0}}\tau v_{xt}v_x\Bigg|+\Bigg|\tau\sum_{m=1}^{N}h^n\sum_{\bar{\Omega}_{hm}}\tau\dot{u}_{hxt}\dot{u}_{hx}\Bigg|\\
&+\Bigg|\tau\sum_{m=1}^{N}h^n\sum_{\bar{\Omega}_{hm}}\tau v_{xt}\dot{u}_{hx}\Bigg|+\Bigg|\tau\sum_{m=1}^{N}h^n\sum_{\bar{\Omega}_{hm}}\tau\dot{u}_{hxt}\dot{u}_{hx}\Bigg|\\
&+\Bigg|(\tau N)\sum_{\bar{\Omega}_{h0}}(v_t^2+v_x^2)-2\tau\sum_{m=1}^{N}h^n\sum_{\bar{\Omega}_{hm}}v_t\dot{u}_{ht}+\tau\sum_{m=1}^{N}h^n\sum_{\bar{\Omega}_{hm}}\dot{u}_{ht}^2\\
&-2\tau\sum_{m=1}^{N}h^n\sum_{\bar{\Omega}_{hm}}v_x\dot{u}_{hx}+\tau\sum_{m=1}^{N}h^n\sum_{\bar{\Omega}_{hm}}\dot{u}_{hx}^2+\tau\sum_{m=1}^{N}\tau\sum_{l=1}^{m}h^n\sum_{\bar{\Omega}_{hl}}f_h(v_t+v_{\bar{t}})\Bigg|\Bigg\}.
\end{aligned}
\tag{4.14}
$$

Let us write (4.14) in integral form, for which we pass to piecewise continuous extensions with respect to x as described above, and with respect to t we construct an extension "downwards", i.e. we assume that

$$\tilde{v}(x,\ t)\big|_{q^h(k,\ l-1)}=v(k,\ l).$$

Note that the constant C in (4.14) is determined only by the quotient τ/h. Limiting processes will be implemented with a constant quotient τ/h. Taking into account the fact that $\tau N=T$, we obtain

$$
\begin{aligned}
&\int_{Q_T}(|\tilde{v}_t-\tilde{u}_{ht}|^2+|\tilde{v}_x-\tilde{u}_{hx}|^2)\,dxdt\leqslant CT\Bigg|\int_{\Omega_0}\tau\tilde{v}_{xt}\tilde{v}_x dx\Bigg|\\
&+C\Bigg|\int_{Q_T}\tau\tilde{u}_{hxt}\tilde{u}_{hx}dxdt\Bigg|+C\Bigg|\int_{Q_T}\tau\tilde{v}_{xt}\tilde{u}_{hx}dxdt\Bigg|+C\Bigg|\int_{Q_T}\tau\tilde{u}_{hxt}\tilde{u}_{hx}dxdt\Bigg|\\
&+C\Bigg|T\int_{\Omega_0}(\tilde{v}_t^2+\tilde{v}_x^2)\,dx+\int_{Q_T}[-2\tilde{v}_t\tilde{u}_{ht}+\tilde{u}_{ht}^2-2\tilde{v}_x\tilde{u}_{hx}+\tilde{u}_{hx}^2]\,dxdt\\
&\qquad+\tau\sum_{m=1}^{N}\int_{Q_{tm}}\tilde{f}_h(\tilde{v}_t+\tilde{v}_{\bar{t}})\,dxdt\Bigg|.
\end{aligned}
\tag{4.15}
$$

Let us pass to the limit in (4.15) as $h,\tau\to 0$, and let us preserve a constant quotient which satisfies (4.11).

We consider the first term on the right side of (4.15):

$$
\begin{aligned}
\tau\int_{\Omega_0}\tilde{v}_{xt}\tilde{v}_x dx&=\frac{\tau}{h}\sum_{i=1}^{n}\int_{\Omega_0}[\tilde{v}_t(k_i+1)-\tilde{v}_t(k_i)]\,\tilde{v}_{x_i}dx\\
&=\frac{\tau}{h}\sum_{i=1}^{n}\int_{\Omega_0}[\tilde{\psi}_h(k_i+1)-\tilde{\psi}_h(k_i)]\,\tilde{\varphi}_{hx_i}dx.
\end{aligned}
$$

Having written it in this manner, we obtain that its limit as $h,\tau\to 0$ is 0 by virtue of the fact that $\tilde{\psi}_h(k_i)$ and $\tilde{\psi}_h(k_i+1)$ have a limit $\psi(x)$, where they converge to $\psi(x)$ strongly in $L_2(\Omega)$, and $\tilde{\varphi}_{hx_i}$ converges strongly in $L_2(\Omega)$ to $\partial\varphi/\partial x_i$.

We consider yet another term of this type:

$$\int_{Q_T}\tau\tilde{v}_{xt}\tilde{u}_{hx}dxdt=\frac{\tau}{h}\sum_{i=1}^{n}\int_{Q_T}[\tilde{v}_t(k_i+1)-\tilde{v}_t(k_i)]\,\tilde{u}_{hx_i}dxdt=$$

$$=\frac{\tau}{h}\sum_{i=1}^{n}\left[\int\limits_{Q_T}\tilde{v}_t(k_i+1)\,\tilde{u}_{hx_i}dxdt-\int\limits_{Q_T}\tilde{v}_t(k_i)\,\tilde{u}_{hx_i}dxdt\right].$$

Using the fact that $\tilde{u}_{hx_i}$ converges strongly in $L_2(Q_T)$ to $\partial u/\partial x_i$, and both $\tilde{v}_t(k_i+1)$ and $\tilde{v}_t(k_i)$ converge weakly in $L_2(Q_T)$ to one and the same limit, which is equal to zero, we deduce that the limit of this term is 0.

The remaining two terms of the same type on the right side of (4.15) also have the limit 0.

In the last term it is not difficult to conclude that it has the relation (4.6) as a limit by using the weak convergence of $\tilde{v}_t$, $\tilde{v}_{\bar{t}}$, $\tilde{v}_x$ in $L_2(Q_T)$ to $\partial u/\partial t$, $\partial u/\partial x$ and the strong convergence of $\tilde{v}_t$ and $\tilde{v}_x$ with $t=0$ to $\psi(x)$ and $\tilde{\varphi}_x$ in $L_2(\Omega)$, and also the strong convergence in $L_2(Q_T)$ of $\tilde{u}_{ht}, \tilde{u}_{hx}$ and $\tilde{f}_h$ to $\partial u/\partial t$, u_x and f, respectively.

THEOREM 4.1. *The solutions of the difference schemes* (4.7) – (4.9) *converge to a solution of the problem* (4.1) – (4.3) *under the condition* (4.4) *so that*

$$\int\limits_{Q_T}\left(\left|\tilde{v}_t^2-\frac{\partial u}{\partial t}\right|^2+\left|\tilde{v}_x-\frac{\partial u}{\partial x}\right|^2\right)dxdt\underset{h,\tau\to 0}{\longrightarrow}0.$$

In conclusion the author is grateful to O. A. Ladyženskaja for formulating the problem and for assistance, and to V. Ja. Rivkind, who read the manuscript and gave much useful advice.

Bibliography

[1] O. A. Ladyženskaja, V. A. Solonnikov and N. N. Ural′ceva, *Linear and quasilinear equations of parabolic type*, "Nauka", Moscow, 1967; English transl., Transl. Math. Monographs, vol. 23, Amer. Math. Soc., Providence, R. I., 1968. MR **39** #3159a,b.

[2] O. A. Ladyženskaja and N. N. Ural′ceva, *Linear and quasilinear equations of elliptic type*, "Nauka", Moscow, 1964; English transl., Academic Press, New York, 1968. MR **35** #1955; **39** #5941.

[3] O. A. Ladyženskaja, *The mixed problem for a hyperbolic equation*, GITTL, Moscow, 1953. (Russian) MR **17**, 160.

[4] ———, *Mathematical problems in the dynamics of a viscous incompressible fluid*, 2nd rev. aug. ed., "Nauka", Moscow, 1970; English transl. of 1st ed., *The mathematical theory of viscous incompressible flow*, Gordon and Breach, New York, 1963; rev. 1969. MR **40** #7610; **27** #5034b; **42** #6442.

[5] ———, *On non-stationary operator equations and their applications to linear problems of mathematical physics*, Mat. Sb. **45 (87)** (1958), 123-158. (Russian) MR **22** #12290.

[6] L. A. Oganesjan, *Numerical computation of plates*, Solution of Engineering Problems on Electronic Computers, Central Bureau of Technical Information, Leningrad, 1963, pp. 84-97. (Russian) RŽMat. **1964** #3Б450.

[7] V. Ja. Rivkind, *Estimates of the rate of convergence of solutions of difference equations with discontinuous coefficients and a numerical method of solving the Dirichlet problem*, Dokl. Akad. Nauk SSSR **149** (1963), 1264-1267 = Soviet Math. Dokl. **4** (1963), 546-549. MR **26** #6561.

[8] Ju. K. Dem′janovič, *The net method for some problems in mathematical physics*, Dokl. Akad. Nauk SSSR **159** (1964), 250-253 = Soviet Math. Dokl. **5** (1964), 1452-1456. MR **32** #8517.

Translated by
J. D. FABREY

Trudy Mat. Inst. Steklov.
125 (1973)

Proc. Steklov Inst. Math.
125 (1973)

ON SOLUTIONS OF THE EQUATION $D_{x_i}(|Du|^{p-2}D_{x_i}u)=0$ WITH A SINGULARITY AT A BOUNDARY POINT

I. N. KROL′

Abstract. Special solutions of the equation $D_{x_i}(|Du|^{p-2}D_{x_i}u) = 0$ in a spherical cone of vertex half-angle l in R^n that are equal to zero on its boundary are constructed. These solutions have the form $u(x) = |x|^{\lambda}f_{\lambda}(x|x|^{-1})$, $\lambda < 0$, and have a singularity at the vertex of the cone. An asymptotic representation of $\lambda(l)$ for $l \to \pi$ is found.

Bibliography: 4 items.

In the papers [1, 2] of J. Serrin, which are devoted to a study of the behavior of the solutions of elliptic equations of the form

$$D_{x_i}[a_i(x,u,Du)] = a(x,u,Du)$$

in a neighborhood of an interior point of a domain, it is shown, in particular, that the relation

$$u(x) \sim |x|^{\frac{p-n}{p-1}},$$

where $p < n$, is valid for a solution having an isolated point singularity.

In this article we construct solutions of a Dirichlet problem in a cone for the special equation

$$D_{x_i}(|Du|^{p-2}D_{x_i}u) = 0 \tag{0.1}$$

that have the singularity $|x|^{(p-n)/(p-1)}$ at the vertex of the cone.

These solutions have the form

$$u(x) = |x|^{\lambda}f_{\lambda}(x|x|^{-1}), \quad \lambda < 0, \tag{0.2}$$

and satisfy equation (0.1) in a closed spherical cone $K(l)$ in R^n. They are equal to zero on its boundary and have a singularity at its vertex. We assume that $1 < p \leqq n-1$.

The solutions of equation (0.1) of form (0.2) were constructed in [3, 4] for $0 < \lambda \ll 1$ and $|\lambda| \gg 1$. With their help it was shown that a solution of a Dirichlet problem for equation (0.1) with continuous boundary conditions in a domain whose boundary contains a "null reentrant point" cannot belong to any Hölder class and can turn out to be discontinuous in a neighborhood of a boundary point.

In the author's note [4] the solutions of form (0.2) were considered for $|\lambda| \gg 1$. It was shown that a solution of a Dirichlet problem for equation (0.2) in a domain whose boundary contains a "null salient point" tends to its boundary value at the singular point in question at greater than power rate.

AMS (MOS) subject classifications (1970). Primary 35Q99; Secondary 34B10, 34B15, 34B30, 35J25.

Let $K(l)=\{x:0\leq\theta\leq l\}$, $\cos\theta=x_n|x|^{-1}$, and let l be a number sufficiently close to π. We write equation (0.1) in spherical coordinates:

$$\sin^{n-2}\theta\left\{\left[u_\rho^2+\frac{1}{\rho^2}u_\theta^2\right]^{\frac{p-2}{2}}\rho^{n-1}u_\rho\right\}'_\rho+\rho^{n-3}\left\{\left[u_\rho^2+\frac{1}{\rho^2}u_\theta^2\right]^{\frac{p-2}{2}}u_\theta\sin^{n-2}\theta\right\}'_\theta=0$$

and seek a solution of it in the form

$$u(\rho,\theta)=\rho^\lambda f_\lambda(\theta).$$

After a separation of variables, we find that the function $f_\lambda(\theta)$ must satisfy the ordinary differential equation

$$\begin{gathered}\left\{[\lambda^2f^2(\theta)+f'^2(\theta)]^{\frac{p-2}{2}}f'(\theta)\sin^{n-2}\theta\right\}'_\theta\\ +\lambda[\lambda(p-1)+n-p][\lambda^2f^2(\theta)+f'^2(\theta)]^{\frac{p-2}{2}}f(\theta)\sin^{n-2}\theta=0\end{gathered}\tag{0.3}$$

and the boundary conditions

$$f'(0)=0,\qquad f(l)=0.\tag{0.4}$$

We normalize $f(\theta)$ by requiring that $f(0)=1$.

It is obvious that when $\lambda\in[(p-n)/(p-1),0]$ the Cauchy problem for equation (0.3) with the initial data

$$f'(0)=0,\qquad f(0)=1\tag{0.5}$$

does not have solutions vanishing at some point of the interval $(0,\pi)$. Therefore the "exponent of growth" $\lambda=\lambda(l)$ of a solution of (0.1) must satisfy the inequality $\lambda(l)<(p-n)/(p-1)$.

In §1 of this article we give bilateral estimates of $|f'(\theta)|$ on the interval $(0,\pi)$ that will be used to prove the solvability of the boundary value problem (0.3), (0.4) for any l sufficiently close to π. In §2 we find an asymptotic representation of $\lambda(l)$ for $l\to\pi$.

§1. Construction of the "increasing" solution

In this section we construct and study a solution of equation (0.1) of form (0.2) in the cone $K(l)$ that is equal to zero on $\partial K(l)$ and is not bounded near the point 0.

LEMMA 1. *Suppose* $\mu=p-n-\lambda(p-1)\in(0,a(p-1))$, *where* a *is a sufficiently small positive number. Suppose, further,* $f(\theta)$ *is a solution of the Cauchy problem* (0.3), (0.5) *from the class* $C^2[0,\pi)$ *and* $[0,l]$ *is the interval of nonnegativity of* $f(\theta)$. *Then*

$$|f'(\theta)|\sim\begin{cases}\mu\theta(\pi-\theta)^{2-n}, & \text{if}\quad \theta\in\left[0,\ \pi-\mu^{\frac{1}{n-2}}\right],\\ [\mu(\pi-\theta)^{2-n}]^{\frac{1}{p-1}}, & \text{if}\quad \theta\in\left(\pi-\mu^{\frac{1}{n-2}},\ \pi\right),\end{cases}\tag{1.1}$$

$$(\pi-l)\sim\begin{cases}\mu^{\frac{1}{n-p-1}}, & \text{if}\quad p<n-1,\\ e^{-\mu^{\frac{1}{2-n}}}, & \text{if}\quad p=n-1.\end{cases}\tag{1.2}$$

Proof. We write equation (0.3) in the form

$$(1.3)\quad [\lambda^2 f^2(\theta)+f'^2(\theta)]^{\frac{p-2}{2}} f'(\theta)\sin^{n-2}\theta=\lambda\mu\int_0^\theta [\lambda^2 f^2(\tau)+f'^2(\tau)]^{\frac{p-2}{2}} f(\tau)\sin^{n-2}\tau d\tau$$

and first consider the case $\theta\in[0,\min\{(\pi-\alpha),l\}]$, where α is a sufficiently small positive number. From (1.3) it follows that

$$(1.4)\quad [\lambda^2 f^2(\theta)+f'^2(\theta)]^{\frac{p-2}{2}} |f'(\theta)|\leqslant c\mu\int_0^\theta [\lambda^2 f^2(\tau)+f'^2(\tau)]^{\frac{p-2}{2}} f(\tau)\,d\tau.$$

Let M denote the maximum of $|f'(\theta)|$ on $[0,\min\{(\pi-\alpha),l\}]$. Since

$$(1.5)\quad f'(\theta)\leqq 0 \quad\text{and}\quad f(\theta)\leqq 1,$$

when $p\geqq 2$ we obtain

$$M^{p-1}\leqq c\mu(\lambda^2+M^2)^{(p-2)/2},$$

i.e.

$$(1.6)\quad M\leqq c\mu^{1/(p-1)}.$$

Hence when $\theta\in[0,\min\{(\pi-\alpha),l\}]$ and $p\geqq 2$

$$(1.7)\quad f(\theta)\geqq 1-c\mu^{1/(n-2)}>\tfrac{1}{2}.$$

On the other hand, if $p<2$, we see from (1.4) that

$$(1.8)\quad |f'(\theta)|\leqslant c\mu[\lambda^2 f^2(\theta)+f'^2(\theta)]^{\frac{2-p}{2}}\int_0^\theta f^{p-1}(\tau)\,d\tau.$$

Consequently

$$M\leqq c\mu(\lambda^2+M^2)^{(2-p)/2},$$

from which we again obtain (1.6) and (1.7). Thus $l>\pi-\alpha$.

Suppose $p\geqq 2$. By virtue of (1.4)

$$|\lambda|^{p-2} f^{p-2}(\theta)|f'(\theta)|\leqslant c\mu\int_0^\theta [\lambda^2+f'^2(\tau)]^{\frac{p-2}{2}}\,d\tau,$$

and since (1.6) and (1.7) are satisfied,

$$(1.9)\quad |f'(\theta)|\leqq c\mu\theta.$$

When $p<2$ we again obtain (1.9) from (1.6) and (1.8).

We now estimate $|f'(\theta)|$ from below on $[0,\pi-\alpha]$. By virtue of (1.3)

$$(1.10)\quad [\lambda^2 f^2(\theta)+f'^2(\theta)]^{\frac{p-2}{2}} |f'(\theta)|\geqslant c\mu\int_{c\theta}^\theta f^{p-1}(\tau)\,d\tau\geqslant c\mu\theta.$$

If $p<2$, we see from (1.3) that

$$|f'(\theta)|\geqslant c\mu[\lambda^2 f^2(\theta)+f'^2(\theta)]^{\frac{2-p}{2}}\int_{c\theta}^\theta \frac{f(\tau)\,d\tau}{[\lambda^2 f^2(\tau)+f'^2(\tau)]^{\frac{2-p}{2}}}$$

Using (1.6), (1.7) and (1.9) to estimate the right side of the latter inequality, we arrive at (1.10).

Suppose $\theta \in (\pi - \alpha, \pi - \mu^{1/(n-2)}]$ and $p \geqq 2$. From (1.3) and (1.5) it follows that

$$|f'(\theta)|^{p-1} \sin^{n-2}\theta \leqslant c\mu \int_0^\theta [\lambda^2 + f'^2(\tau)]^{\frac{p-2}{2}} \sin^{n-2}\tau d\tau.$$

Since $\sin^{n-2}\theta \geqq c\mu$, the latter inequality can be rewritten in the form

$$|f'(\theta)|^{p-1} \leqslant c\left[1 + \int_0^\theta |f'(\tau)|^{p-2} d\tau\right]. \tag{1.11}$$

Integrating (1.11) from 0 to θ and applying Hölder's inequality, we get

$$\int_0^\theta |f'(\tau)|^{p-1} d\tau \leqslant c + \left[\int_0^\theta |f'(\tau)|^{p-1} d\tau\right]^{\frac{p-2}{p-1}},$$

so that

$$\int_0^\theta |f'(\tau)|^{p-1} d\tau \leqslant c,$$

which together with (1.11) implies

$$|f'(\theta)| \leqq c. \tag{1.12}$$

On the other hand, if $p < 2$, we see from (1.3) and (1.5) that

$$|f'(\theta)| \leqslant c[\lambda^2 f^2(\theta) + f'^2(\theta)]^{\frac{2-p}{2}} \int_0^\theta \frac{f(\tau)\, d\tau}{[\lambda^2 f^2(\tau) + f'^2(\tau)]^{\frac{2-p}{2}}} \leqslant c[\lambda^2 + f'^2(\theta)]^{\frac{2-p}{2}},$$

from which we again obtain (1.12).

By virtue of estimates (1.9) and (1.12) we get

$$1 - f(\theta) \leqslant \int_0^{\pi-\alpha} |f'(\tau)|\, d\tau + \int_{\pi-\alpha}^{\theta} |f'(\tau)|\, d\tau \leqslant c\mu + \alpha c,$$

so that for all $\theta \in [0, \pi - \mu^{1/(n-2)}]$

$$f(\theta) \geqq \text{const}, \quad l > \pi - \mu^{1/(n-2)}. \tag{1.13}$$

From (1.3) and (1.12) we obtain the inequality

$$f^{p-2}(\theta)\,|f'(\theta)|\sin^{n-2}\theta \leqq c\mu,$$

which together with (1.13) implies

$$|f'(\theta)| \leqq c\mu(\sin\theta)^{2-n}. \tag{1.14}$$

When $p < 2$ it follows from (1.3) and (1.12) that

$$|f'(\theta)|\sin^{n-2}\theta \leqslant c\mu[\lambda^2 f^2(\theta) + f'^2(\theta)]^{\frac{2-p}{2}} \int_0^\theta \frac{f(\tau)\sin^{n-2}\tau d\tau}{[\lambda^2 f^2(\tau) + f'^2(\tau)]^{\frac{2-p}{2}}}$$

$$\leqslant c\mu \int_0^\theta f^{p-1}(\tau) \sin^{n-2}\tau d\tau,$$

from which we again obtain (1.14).

We now estimate $|f'(\theta)|$ from below on $(\pi-\alpha, \pi-\mu^{1/(n-2)}]$. If $p \geqq 2$, we see from (1.3) that

$$[\lambda^2 f^2(\theta) + f'^2(\theta)]^{\frac{p-2}{2}} |f'(\theta)| \sin^{n-2}\theta \geqslant c\mu \int_0^\theta f^{p-1}(\tau) \sin^{n-2}\tau d\tau,$$

which by virtue of (1.12) and (1.13) implies

$$|f'(\theta)| \geqq c\mu(\sin\theta)^{2-n}. \tag{1.15}$$

Analogously, when $p < 2$ it follows from (1.3), (1.12) and (1.13) that

$$|f'(\theta)| \sin^{n-2}\theta \geqslant c\mu [\lambda^2 f^2(\theta) + f'^2(\theta)]^{\frac{p-2}{2}} \int_0^\theta \frac{f(\tau)\sin^{n-2}\tau d\tau}{[\lambda^2 f^2(\tau) + f'^2(\tau)]^{\frac{2-p}{2}}} \geqslant c\mu,$$

and (1.15) is again satisfied.

Suppose $\theta \in (\pi - \mu^{1/(n-2)}, l]$ and $p \geqq 2$. Inasmuch as $|f'(\theta)| \geqq$ const for $\theta > \pi - \mu^{1/(n-2)}$, we deduce from (1.3) that

$$\begin{aligned} |f'(\theta)|^{p-1} \sin^{n-2}\theta &\leqslant c\mu \int_0^\theta [\lambda^2 f^2(\tau) + f'^2(\tau)]^{\frac{p-2}{2}} f(\tau) \sin^{n-2}\tau d\tau \\ &\leqslant c\mu \left[\int_0^x |f'(\tau)|^{p-2} \sin^{n-2}\tau d\tau + 1\right]. \end{aligned} \tag{1.16}$$

Integrating this inequality from 0 to θ and applying Hölder's inequality, we get

$$\int_0^\theta |f'(\tau)|^{p-1} \sin^{n-2}\tau d\tau \leqslant c\mu.$$

From this result and (1.16) it follows that

$$|f'(\theta)| \leqq c[\mu(\sin\theta)^{2-n}]^{1/(p-1)}. \tag{1.17}$$

When $p < 2$ equality (1.3) implies the estimates

$$\begin{aligned} |f'(\theta)| \sin^{n-2}\theta &\leqslant c\mu [\lambda^2 f^2(\theta) + f'^2(\theta)]^{\frac{2-p}{2}} \int_0^\theta \frac{f(\tau)\sin^{n-2}\tau d\tau}{[\lambda^2 f^2(\tau) + f'^2(\tau)]^{\frac{2-p}{2}}} \\ &\leqslant c\mu |f'(\theta)|^{2-p} \int_0^\theta f^{p-1}(\tau) \sin^{n-2}\tau d\tau \leqslant c\mu |f'(\theta)|^{2-p}, \end{aligned}$$

from which we again obtain (1.17).

We now estimate $|f'(\theta)|$ from below on $(\pi - \mu^{1/(n-2)}, l]$. When $p \geqq 2$ equality (1.3) implies

$$\begin{aligned} [\lambda^2 f^2(\theta) + f'^2(\theta)]^{\frac{p-2}{2}} |f'(\theta)| \sin^{n-2}\theta &\geqslant c\mu \int_0^\theta [\lambda^2 f^2(\tau) + f'^2(\tau)]^{\frac{p-2}{2}} f(\tau) \sin^{n-2}\tau d\tau \\ &\geqslant c\mu \int_0^\theta f^{p-1}(\tau) \sin^{n-2}\tau d\tau \geqslant c\mu. \end{aligned} \tag{1.18}$$

Since the left side of the latter inequality does not exceed $c|f'(\theta)|^{p-1}\sin^{n-2}\theta$ for $\theta > \pi - \mu^{1/(n-2)}$, it follows from (1.18) that

$$|f'(\theta)| \geqq c[\mu(\sin\theta)^{2-n}]^{1/(p-1)}. \tag{1.19}$$

On the other hand, if $p < 2$,

$$|f'(\theta)|\sin^{n-2}\theta \geqslant c\mu[\lambda^2 f^2(\theta) + f'^2(\theta)]^{\frac{2-p}{2}} \int_0^\theta \frac{f(\tau)\sin^{n-2}\tau\,d\tau}{[\lambda^2 f^2(\tau) + f'^2(\tau)]^{\frac{2-p}{2}}}$$

$$\geqslant c\mu|f'(\theta)|^{2-p}\int_0^{\pi/2} f^{p-1}(\tau)\sin^{n-2}\tau\,d\tau \geqslant c\mu|f'(\theta)|^{2-p},$$

from which we again obtain (1.19).

From (1.17) and (1.19) we now deduce for $\theta \in (\pi - \mu^{1/(n-2)}, l)$ the relation

$$f\left(\pi - \mu^{\frac{1}{n-2}}\right) - f(\theta) \sim \begin{cases} [\mu(\pi - \theta)^{p-n+1}]^{\frac{1}{p-1}}, & \text{if} \quad p < n-1, \\ \mu^{\frac{1}{n-2}} \ln(\pi - \theta)^{-1}, & \text{if} \quad p = n-1, \end{cases} \tag{1.20}$$

which implies that $f(\theta)$ changes sign at some point of $(\pi - \mu^{1/(n-2)}, \pi)$. This point will be denoted below by l. Setting $\theta = l$ in (1.20) and using (1.13), we obtain relations (1.2).

To prove the validity of the second of relations (1.1) on the interval (l, π) we integrate equation (1.3) from l to θ, where $\theta < \pi$ if $f'(\theta) \neq 0$ on (l, π) or $\theta < \beta$ if $f'(\beta) = 0$, $\beta \in (l, \pi)$:

$$\begin{aligned} [\lambda^2 f^2(\theta) + f'^2(\theta)]^{\frac{p-2}{2}} |f'(\theta)|\sin^{n-2}\theta = |f'(l)|^{p-1}\sin^{n-2} l \\ - \lambda[\lambda(p-1) + n - p]\int_l^\theta [\lambda^2 f^2(\tau) + f'^2(\tau)]^{\frac{p-2}{2}} |f(\tau)|\sin^{n-2}\tau\,d\tau, \end{aligned} \tag{1.21}$$

whence by virtue of (1.17) we obtain

$$[\lambda^2 f^2(\theta) + f'^2(\theta)]^{\frac{p-2}{2}} |f'(\theta)|\sin^{n-2}\theta \leqslant |f'(l)|^{p-1}\sin^{n-2} l \leqslant c\mu. \tag{1.22}$$

When $p \geqq 2$ we obtain from the latter inequality

$$|f'(\theta)|^{p-1}\sin^{n-2}\theta \leqq c\mu,$$

which implies (1.17) for $\theta \in (l, \pi)$. When $p < 2$ inequality (1.22) can be rewritten in the form

$$|f'(\theta)|\sin^{n-2}\theta \leqslant c\mu[|f(\theta)|^{2-p} + |f'(\theta)|^{2-p}]. \tag{1.23}$$

Integrating the latter inequality from l to θ, where $\theta \leqq \min(\beta, \pi)$, and applying Hölder's inequality, we get

$$\sin^{n-2}\theta\int_l^\theta |f'(\tau)|\,d\tau \leqslant c\mu\left\{\int_l^\theta\left[\int_l^t |f'(\tau)|\,d\tau\right]^{2-p} dt\right.$$

$$\left. + \int_l^\theta |f'(\quad {}^{2-p}d\tau\right\} \leqslant c\mu(\pi - l)^{p-1}\left[\int_l^\theta |f'(\tau)|\,d\tau\right]^{2-p},$$

which implies

$$(1.24)\qquad |f(\theta)| = \int_l^\theta |f'(\tau)|\,d\tau \leqslant c(\pi - l)[\mu(\sin\theta)^{2-n}]^{\frac{1}{p-1}}.$$

From the latter estimate and (1.23) we again obtain (1.17) for $\theta \in (l, \pi)$.

We now estimate $|f'(\theta)|$ from below on $(l, \min\{\beta, \pi\})$. We begin with an estimation of the second summand in the right side of (1.21). Suppose $p \geqq 2$. By virtue of (1.17) and (1.24),

$$\lambda[\lambda(p-1)+n-p]\int_l^\theta [\lambda^2 f^2(\tau)+f'^2(\tau)]^{\frac{p-2}{2}}|f(\tau)|\sin^{n-2}\tau\,d\tau$$

$$\leqslant c\mu\int_l^\theta |f(\tau)|^{p-1}\sin^{n-2}\tau\,d\tau + c\mu\int_0^\theta |f'(\tau)|^{p-2}|f(\tau)|\sin^{n-2}\tau\,d\tau \leqslant c\mu^2.$$

When $p < 2$ it follows from (1.17) that

$$\lambda[\lambda(p-1)+n-p]\int_l^\theta \frac{f(\tau)\sin^{n-2}\tau\,d\tau}{[\lambda^2 f^2(\tau)+f'^2(\tau)]^{\frac{2-p}{2}}} \leqslant c\mu\int_l^\theta |f(\tau)|^{p-1}\sin^{n-2}\tau\,d\tau \leqslant c\mu^2.$$

Since

$$|f'(l)|^{p-1}\sin^{n-2} l \geqq c\mu,$$

we see from (1.21) that

$$(1.25)\qquad [\lambda^2 f^2(\theta)+f'^2(\theta)]^{\frac{p-2}{2}}|f'(\theta)|\sin^{n-2}\theta \geqslant c\mu.$$

Suppose $p \geqq 2$. By virtue of (1.17) and (1.24) the left side of the latter inequality does not exceed

$$c\mu^{\frac{p-2}{p-1}}(\sin\theta)^{\frac{n-2}{p-1}}|f'(\theta)|,$$

which together with (1.25) implies (1.19). When $p < 2$ it follows from (1.25) that

$$c\mu \leqslant \frac{|f'(\theta)|\sin^{n-2}\theta}{[\lambda^2 f^2(\theta)+f'^2(\theta)]^{\frac{2-p}{2}}} \leqslant |f'(\theta)|^{p-1}\sin^{n-2}\theta,$$

from which we again obtain (1.19).

Estimate (1.19) guarantees that $f'(\theta)$ does not vanish for all $\theta \in (l, \pi)$, and hence the estimates obtained by us hold for all $\theta < \pi$. The lemma is proved.

Lemma 2. *For any* $\lambda \in ((p-n)/(p-1) - a, (p-n)/(p-1))$ *(a is the constant in Lemma 1) there exists in a neighborhood of the point $\theta = 0$ one and only one solution $f(\theta) \in C^2[0, \pi)$ of the Cauchy problem for equation (0.3) with the initial data (0.5). This solution can be uniquely extended onto the interval $[0, \pi)$.*

Proof. It suffices to rewrite (0.3) in the equivalent form

$$f(\theta) = 1 - \lambda[\lambda(p-1)+n-p]\int_0^\theta \frac{d\tau}{[\lambda^2 f^2(\tau)+f'^2(\tau)]^{\frac{p-2}{2}}\sin^{n-2}\tau} \tag{1.26}$$
$$\times \int_0^\tau [\lambda^2 f^2(t)+f'^2(t)]^{\frac{p-2}{2}} f(t)\sin^{n-2}t\,dt,$$

where $\theta \in [0,\epsilon]$, and note that the right side of this equality is a contraction from the set

$$M = \left\{f : f \in C^2[0,\varepsilon],\ f(0)=1,\ f'(0)=0,\ -\frac{1}{2}\leqslant f'(\theta)\leqslant 0\right\}$$

into itself for small $\epsilon > 0$.

The possibility of extending $f(\theta)$ onto any interval $[0,\alpha]$, $\alpha < \pi$, follows from relations (1.1) of Lemma 1.

LEMMA 3. *The first zero* $l(\lambda)$ *of the solution of the Cauchy problem* (0.3), (0.5) *is a continuous function on* $((p-n)/(p-1)-a,(p-n)/(p-1))$, *where* a *is the constant in Lemma* 1.

PROOF. Let λ be an arbitrary number in

$$((p-n)/(p-1)-a,(p-n)/(p-1))$$

and let $f_\lambda(\theta)$ be the solution of the Cauchy problem (0.3), (0.5) corresponding to λ. Further, let $\{\lambda_n\}_1^\infty$ be an arbitrary sequence of numbers in this same interval, that converges to λ, and let $f_n(\theta)$, $n=1,2,\cdots$, be the solution of the Cauchy problem (0.3), (0.5) corresponding to λ_n.

By virtue of (1.1) the sequence $\{f_n(\theta)\}_1^\infty$ is compact in $C^1[0,\pi-\epsilon]$, where ϵ is any sufficiently small positive number. We distinguish a subsequence $\{f_{n_k}\}$ that uniformly converges to a function $f(\theta)$ on $[0,\pi-\epsilon]$. Passing to the limit in the integral identity (0.3) and using relation (1.1), we get that $f(\theta)$ is the solution of the Cauchy problem (0.3), (0.5) corresponding to the parameter λ. Since this solution is unique, the whole sequence $\{f_n(\theta)\}_1^\infty$ converges to $f_\lambda(t)$ in $C^1[0,\pi-\epsilon]$.

We now note that

$$|l-l_n| = |f_n(l)|\,|f_n'(\tau_n)|^{-1},$$

where τ_n is a point between l and l_n. Since $|f_n'(\tau_n)| \geqq \text{const}$ by virtue of (1.1), we have

$$|l-l_n| \leqslant c\,|f_n(l)| \xrightarrow[n\to\infty]{} 0.$$

By the eigenvalues of problem (0.3), (0.4) we will mean those values of the parameter λ for which this problem has a nontrivial solution.

From Lemmas 2 and 3 we immediately obtain the following assertion.

THEOREM 1. *There exists an eigenvalue* $\lambda(l) < (p-n)/(p-1)$ *of problem* (0.3), (0.4) *to which there corresponds a unique eigenfunction that is positive on* $[0,l)$.

§2. Asymptotic behavior for $l \to \pi$ of the singularity of the "growing" solution

THEOREM 2. *Suppose* $\lambda(l) < (p-n)/(p-1)$ *and* $l \to \pi$. *Then*

$$\lambda(l) = \frac{p-n}{p-1} - \frac{\Gamma\left(\frac{n}{2}\right)}{\sqrt{\pi}\,(p-1)\,\Gamma\left(n-1\right)}\left(\frac{n-p-1}{n-p}\right)^{p-1}(\pi - l)^{n-p-1}$$

$$\times\left\{1 + O\left[(\pi - l)^{\frac{n-p-1}{n-2}}\right]\right\}$$

for $p < n-1$ *and*

$$\lambda(l) = \frac{1}{2-n} - \frac{\Gamma\left(\frac{n}{2}\right)}{\sqrt{\pi}\,(n-2)^{3-n}\,\Gamma\left(\frac{n-1}{2}\right)}[\ln(\pi - l)^{-1}]^{2-n}\left\{1 + O\left[\frac{1}{\ln(\pi - l)^{-1}}\right]\right\}$$

for $p = n-1$.

PROOF. We set $f'(\theta) = \lambda g(\theta)$ and rewrite equation (0.3) in the form

$$\sin^{n-2}\theta\,[f^2(\theta) + g^2(\theta)]^{\frac{p-2}{2}}\,g(\theta) = \mu\int_0^\theta [f^2(\tau) + g^2(\tau)]^{\frac{p-2}{2}} f(\tau)\sin^{n-2}\tau\,d\tau. \tag{2.1}$$

Let

$$z(\theta) = \int_0^\theta [f^2(\tau) + g^2(\tau)]^{\frac{p-2}{2}}\sin^{n-2}\tau\,d\tau,$$

$$\xi(\theta) = \int_0^\theta [f^2(\tau) + g^2(\tau)]^{\frac{p-2}{2}}[1 - f(\tau)]\sin^{n-2}\tau\,d\tau.$$

By virtue of (2.1) the function $z(\theta)$ satisfies the equation

$$[z'^2(\theta)]^{p-1} = \sin^{2(n-2)}\theta\,\{\mu^2[z(\theta) - \xi(\theta)]^2 + f^2(\theta)\,z'^2(\theta)\}^{p-2}. \tag{2.2}$$

Consider the function

$$[0, \infty) \ni t \to \psi(t, \alpha, \beta) = t^{2(p-1)}(\alpha^2 + \beta^2 t^2)^{2-p} - 1, \tag{2.3}$$

where $\alpha, \beta \geqq 0$, $\alpha + \beta > 0$. Since $\psi(0, \alpha, \beta) = -1$, $\psi(\infty, \alpha, \beta) = \infty$ and $\psi_t'(t, \alpha, \beta) > 0$ for $t \in (0, \infty)$, the equation $\psi(t, \alpha, \beta) = 0$ has a unique positive solution $t = \Phi(\alpha, \beta)$.

We also note that when $t = \Phi(\alpha, \beta)$

$$\frac{\partial\Phi}{\partial\alpha} = \frac{(p-2)\,\alpha t}{(p-1)\,\alpha^2 + \beta^2 t^2}, \quad \frac{\partial\Phi}{\partial\beta} = \frac{(p-2)\,\beta t^3}{(p-1)\,\alpha^2 + \beta^2 t^2}. \tag{2.4}$$

We introduce the notation

$$\left.\begin{aligned} t_z &= z'(\theta)\,(\theta^{p-2}\sin\theta)^{\frac{2-n}{p-1}}, \\ \alpha_z &= \mu\,[z(\theta) - \xi(\theta)]\,\theta^{2-n}, \quad \beta_z = f(\theta)\,(\theta^{-1}\sin\theta)^{\frac{n-2}{p-1}} \end{aligned}\right\} \tag{2.5}$$

and write equation (2.2) in the form $\psi(t_z, \alpha_z, \beta_z) = 0$, where ψ is the function (2.3). The latter equation can be solved for t_z, whence

$$z'(\theta) = (\theta^{p-2}\sin\theta)^{\frac{n-2}{p-1}}\Phi(\alpha_z, \beta_z). \tag{2.6}$$

Further, let

$$y(\theta)=\int_0^\theta \sin^{n-2}\tau\,d\tau. \tag{2.7}$$

We introduce the notation

$$t_y=y'(\theta)[\theta^{p-2}\sin\theta]^{\frac{2-n}{p-1}},\quad \alpha_y=0,\quad \beta_y=[(\theta^{-1}\sin\theta]^{\frac{n-2}{p-1}} \tag{2.8}$$

and write (2.7) in the form $\psi(t_y,\alpha_y,\beta_y)=0$, where ψ is the function (2.3). Solving the latter equation for t_y, we get

$$y'(\theta)=(\theta^{p-2}\sin\theta)^{\frac{n-2}{p-1}}\Phi(\alpha_y,\beta_y). \tag{2.9}$$

Using (1.1) and (2.4), we establish that

$$\left|\frac{\partial\Phi}{\partial\alpha}\right|,\quad \left|\frac{\partial\Phi}{\partial\beta}\right|\leqslant \text{const}$$

on the line segment joining the points $[\alpha_y(\theta),\beta_y(\theta)]$ and $[\alpha_z(\theta),\beta_z(\theta)]$. From (2.6) and (2.9) we now get

$$|y(\pi)-z(\pi)|\leqslant c\int_0^\pi(\theta^{p-2}\sin\theta)^{\frac{n-2}{p-1}}[|\alpha_z(\theta)-\alpha_y(\theta)|+|\beta_z(\theta)-\beta_y(\theta)|]\,d\theta. \tag{2.10}$$

From (2.5) and (2.8) we see that the differences in the integrand here have the form

$$\alpha_z(\theta)-\alpha_y(\theta)=\mu\theta^{2-n}[z(\theta)-\xi(\theta)],$$

$$\beta_z(\theta)-\beta_y(\theta)=-(\theta^{-1}\sin\theta)^{\frac{n-2}{p-1}}[1-f(\theta)].$$

When $\theta\in[0,\pi)$ we have by virtue of (1.1) the inequalities

$$\begin{gathered}|1-f(\theta)|\leqslant c\mu^{\frac{1}{n-2}}\left[1+(\pi-\theta)^{\frac{p-n+1}{p-1}}\right],\\ z(\theta)\leqslant c\theta^{n-1},\quad |\xi(\theta)|\leqslant c\mu^{\frac{1}{n-2}}\theta^{n-1},\end{gathered} \tag{2.11}$$

and hence the inequalities

$$|\alpha_y(\theta)-\alpha_z(\theta)|\leqslant c\mu^{\frac{1}{n-2}},\quad |\beta_y(\theta)-\beta_z(\theta)|\leqslant c\mu^{\frac{1}{n-2}},$$

which together with (2.10) imply the inequality

$$|y(\pi)-z(\pi)|\leqslant c\mu^{\frac{1}{n-2}}. \tag{2.12}$$

We note that by virtue of the second of relations (1.1)

$$\begin{aligned}z(\pi)-z(\theta)&=\int_\theta^\pi[f^2(\tau)+g^2(\tau)]^{\frac{p-2}{2}}\sin^{n-2}\tau\,d\tau\\ &\leqslant c\mu^{\frac{p-2}{p-1}}\int_\theta^\pi(\pi-\tau)^{\frac{n-2}{p-1}}d\tau\leqslant c[\mu^{p-2}(\pi-\theta)^{n+p-3}]^{\frac{1}{p-1}}\end{aligned} \tag{2.13}$$

for all $\theta\in(\pi-\mu^{1/(n-2)},\pi)$.

It will be convenient in the sequel to write (2.1) in the form

$$(2.14)\quad [f^2(\theta)+g^2(\theta)]^{\frac{p-2}{2}} g(\theta)\sin^{n-2}\theta = \mu\{y(\pi)+[z(\pi)-y(\pi)] - [z(\pi)-z(\theta)]-\xi(\theta)\} = \mu[y(\pi)+\gamma(\theta)],$$

where

$$\gamma(\theta)=[z(\pi)-y(\pi)]-[z(\pi)-z(\theta)]-\xi(\theta).$$

As follows from (2.11) – (2.13), the function $\gamma(\theta)$ satisfies the estimate

$$(2.15)\quad |\gamma(\theta)|\leqslant c\mu^{\frac{1}{n-2}}+c\mu^{\frac{p-2}{p-1}}(\pi-\theta)^{\frac{n+p-3}{p-1}}.$$

We write (2.14) in the form

$$g(\theta)=\{\mu[y(\pi)+\gamma(\theta)]\}^{\frac{1}{p-1}}\sin^{\frac{2-n}{p-1}}\theta\left[1+\frac{f^2(\theta)}{g^2(\theta)}\right]^{\frac{2-p}{2(p-1)}}$$

Since $|\lambda|\int_0^l g(\theta)d\theta=\int_0^l|f'(\theta)|d\theta=1$ and by virtue of (1.1)

$$\int_0^{\pi-\mu^{\frac{1}{n-2}}}|f'(\theta)|\,d\theta=O\left(\mu^{\frac{1}{n-2}}\right),$$

we get

$$\frac{1}{|\lambda|}=\mu^{\frac{1}{p-1}}\int_{\pi-\mu^{\frac{1}{n-2}}}^{\pi}[y(\pi)+\gamma(\theta)]^{\frac{1}{p-1}}\left[1+\frac{f^2(\theta)}{y^2(\theta)}\right]^{\frac{2-p}{2(p-1)}}\sin^{\frac{2-n}{p-1}}\theta d\theta+O\left(\mu^{\frac{1}{n-2}}\right).$$

For all $\theta\in(\pi-\mu^{1/(n-2)},\pi)$ we have

$$\frac{|f(\theta)|}{g(\theta)}\sim\left[\frac{\sin^{n-2}\theta}{\mu}\right]^{\frac{1}{p-1}},$$

which implies

$$\frac{1}{|\lambda|\mu^{\frac{1}{p-1}}}=\int_{\pi-\mu^{\frac{1}{n-2}}}^{\pi}[y(\pi)+\gamma(\theta)]^{\frac{1}{p-1}}\sin^{\frac{2-n}{p-1}}\theta d\theta+O\left[\mu^{\frac{p-n+1}{(n-2)(p-1)}}\right].$$

By virtue of (2.15)

$$\left|[y(\pi)+\gamma(\theta)]^{\frac{1}{p-1}}-[y(\pi)]^{\frac{1}{p-1}}\right|\leqslant c\gamma(\theta)\leqslant c\mu^{\frac{1}{n-2}}+c[\mu^{p-2}(\pi-\theta)^{n+p-3}]^{\frac{1}{p-1}}.$$

From the latter inequality and (1.2) we see that

$$\left|\left(|\lambda|\mu^{\frac{1}{p-1}}\right)^{-1}-y^{\frac{1}{p-1}}(\pi)\int_{\pi-\mu^{\frac{1}{n-2}}}^{l}\sin^{\frac{2-n}{p-1}}\theta d\theta\right|\leqslant c\mu^{\frac{p-n+1}{(n-2)(p-1)}}$$
$$+c\mu^{\frac{p-2}{p-1}}\int_{\pi-\mu^{\frac{1}{n-2}}}^{l}(\pi-\theta)^{\frac{n+p-3}{p-1}-\frac{n-2}{p-1}}d\theta\leqslant c\mu^{\frac{p-n+1}{(n-2)(p-1)}}.$$

Hence, if $p < n-1$,

$$\mu^{-\frac{1}{p-1}} = |\lambda|[y(\pi)]^{\frac{1}{p-1}} \int_{\pi-\mu^{\frac{1}{n-2}}}^{l} \sin^{\frac{2-n}{p-1}}\theta d\theta + O\left[\mu^{\frac{p-n+1}{(n-2)(p-1)}}\right]$$

$$= \frac{n-p}{n-p-1}[y(\pi)]^{\frac{1}{p-1}}(\pi-l)^{\frac{p-n+1}{p-1}} + O\left[\mu^{\frac{p-n+1}{(n-2)(p-1)}}\right].$$

If $p = n-1$,

$$\mu^{-\frac{1}{n-2}} = \frac{1}{n-2}[y(\pi)]^{\frac{1}{n-2}} \ln(\pi-l)^{-1} + O(1).$$

It remains to note that

$$y(\pi) = \int_0^\pi \sin^{n-2}\theta d\theta = \sqrt{\pi}\,\frac{\Gamma\left(\frac{n-1}{2}\right)}{\Gamma\left(\frac{n}{2}\right)}.$$

The theorem is proved.

Remark. The solutions of form (0.1) in the cone $K(l)$ that are equal to zero on $\partial K(l)$ of the quasilinear elliptic equation

$$D_{x_i}(|x|^\beta |Du|^{p-2} D_{x_i}u) = 0$$

where $p \leq \min(n-1, n+\beta)$, are studied in exactly the same way.

We note that when $\beta \neq 0$ the asymptotic representations for $l \to \pi$ of $\lambda(l) < (p-n-\beta)/(p-1)$ analogous to those obtained in Theorem 2 are

$$\lambda(l) = \frac{p-n-\beta}{p-1} - \frac{\Gamma\left(\frac{n}{2}\right)}{\sqrt{\pi}(p-1)\Gamma\left(\frac{n-1}{2}\right)}\left(\frac{n-p-1}{n-p+\beta}\right)^{p-1}(\pi-l)^{n-p-1} \times \left\{1 + O\left[(\pi-l)^{\frac{n-p-1}{n-2}}\right]\right\},$$

for $p < n-1$ and

$$\lambda(l) = \frac{1+\beta}{2-n} - \frac{\Gamma\left(\frac{n}{2}\right)}{\sqrt{\pi}(n-2)^{3-n}\Gamma\left(\frac{n-1}{2}\right)}[\ln(\pi-l)^{-1}]^{2-n}\left\{1 + O\left[\frac{1}{\ln(\pi-l)^{-1}}\right]\right\},$$

for $p = n-1$.

Bibliography

[1] J. Serrin, *Local behavior of solutions of quasi-linear equations,* Acta Math. **111** (1964), 247-302. MR **30** # 337.

[2] ———, *Isolated singularities of solutions of quasi-linear equations,* Acta Math. **113** (1965), 219-240. MR **31** # 494.

[3] I. N. Krol′ and V. G. Maz′ja, *The lack of continuity and Hölder continuity of the solutions of a certain quasilinear equation,* Zap. Naučn. Sem. Leningrad. Otdel. Mat. Inst. Steklov. (LOMI) **14** (1969), 89-91 = Sem. Math. V. A. Steklov Math. Inst. Leningrad **14** (1969), 44-45. MR **43** # 3659.

[4] I. N. Krol′, *On barriers for a quasilinear equation,* Abstracts of Reports, Jubilee Scientific-Technological Conf., Leningrad Shipbuilding Inst., Leningrad, 1969, pp. 88-89. (Russian)

Translated by
S. SMITH

Trudy Mat. Inst. Steklov.
125 (1973)

Proc. Steklov Inst. Math.
125 (1973)

ON THE BEHAVIOR OF THE SOLUTIONS OF A QUASILINEAR EQUATION NEAR NULL SALIENT POINTS OF THE BOUNDARY

I. N. KROL′

Abstract. A solution $u(x)$ of the equation $D_{x_i}(|Du|^{p-2}D_{x_i}u) = 0$ (1) in a spherical cone $K(l)$ of vertex half-angle l of the form $u(x) = |x|^\lambda f_\lambda(x_n|x|^{-1})$ (2) that is equal to zero on $\partial K(l)$ is constructed. It is shown that the exponent $\lambda = \lambda(l)$ satisfies the asymptotic formula $\lambda(l) = \pm Ll^{-1} + O(1)$ for $l \to 0$, where L is the first zero of the solution of a Cauchy problem for a certain ordinary differential equation. By using the solution (2) as a barrier it is proved that the solution of the Dirichlet problem for equation (1) in a domain Ω tends to its boundary value at a "null salient point" of $\partial\Omega$ at a greater than power rate.

In this note we consider the Dirichlet problem for the equation

$$D_{x_i}(|Du|^{p-2}D_{x_i}u) = 0, \quad p > 1, \tag{1}$$

in a convex cone $K(l) = \{x : 0 \leq \theta \leq l\}$, $\cos\theta = x_n|x|^{-1}$, of vertex half-angle l for $l \to 0$.[1)] We will construct a solution of equation (1) of the form

$$u(x) = |x|^\lambda f_\lambda(\theta)$$

that is equal to zero on $\partial K(l)$, and show that for $l \to 0$

$$\lambda(l) = \pm Ll^{-1} + O(1).$$

By using this solution as a barrier, we prove that the solution of a Dirichlet problem for equation (1) in a domain Ω with a boundary function h that vanishes on $\partial\Omega$ near a "null salient point" of $\partial\Omega$ tends to its boundary value at this point at greater than power rate.

§1. Construction of a special solution

We will seek a nonnegative solution of equation (1) of the form

$$u(\rho,\theta) = \rho^\lambda f_\lambda(\theta)$$

that is equal to zero on $\partial K(l)$, where $|\lambda| \gg 1$ and $f_\lambda(\theta) \in C^2[0,\pi)$. Equation (1) for the function $u(\rho,\theta)$ has the form

$$\sin^{n-2}\theta\left\{\left[u_\rho^2 + \frac{1}{\rho^2}u_\theta^2\right]^{\frac{p-2}{2}}\rho^{n-1}u_\rho\right\}_\rho' + \rho^{n-3}\left\{\left[u_\rho^2 + \frac{1}{\rho^2}u_\theta^2\right]^{\frac{p-2}{2}}u_\theta\sin^{n-2}\theta\right\}_\theta' = 0.$$

Separating the variables ρ and θ, we find that the function $f_\lambda(\theta)$ must satisfy the ordinary differential equation

$$\begin{aligned}&\left\{[\lambda^2 f^2(\theta) + f'^2(\theta)]^{\frac{p-2}{2}} f'(\theta)\sin^{n-2}\theta\right\}_\theta \\ &\qquad + \lambda[\lambda(p-1) + n - p][\lambda^2 f^2(\theta) + f'^2(\theta)]^{\frac{p-2}{2}} f(\theta)\sin^{n-2}\theta = 0\end{aligned} \tag{2}$$

AMS (MOS) subject classifications (1970). Primary 35Q99; Secondary 34B10, 34B15, 34B30, 35J25.

[1)] The results contained in this note were announced under the same title in the abstracts of the Jubilee Scientific and Technological Conference of the Leningrad Institute of Naval Architecture, 1969.

and the boundary conditions

$$f'(0) = 0, \qquad f(l) = 0. \tag{3}$$

We normalize $f(\theta)$ by requiring that $f(0) = 1$.

Lemma 1. *Suppose $|\lambda| \in (a, \infty)$, where a is a sufficiently large positive number, suppose $f(\theta) \in C^2[0,\pi)$ is a solution of the Cauchy problem for equation (2) with the initial data*

$$f(0) = 1, \qquad f'(0) = 0 \tag{4}$$

and suppose $[0, l]$ is the interval of nonnegativity of $f(\theta)$. Then[2]

$$|f'(\theta)| \sim \lambda^2\theta \quad \text{and} \quad l \sim |\lambda|^{-1}. \tag{5}$$

Proof. We make the change of variable $\tau = |\lambda|\theta$ in equation (2) and consider the Cauchy problem with initial data (4) for the resultant equation

$$\begin{aligned} &\left\{[\varphi^2(\tau) + \varphi'^2(\tau)]^{\frac{p-2}{2}} \varphi'(\tau) \sin^{n-2}(\tau|\lambda|^{-1})\right\}' \\ &\quad + \left[p - 1 + \frac{n-p}{\lambda}\right][\varphi^2(\tau) + \varphi'^2(\tau)]^{\frac{p-2}{2}} \varphi(\tau) \sin^{n-2}(\tau|\lambda|^{-1}) = 0. \end{aligned} \tag{6}$$

We integrate equation (6) from 0 to τ:

$$\begin{aligned} &-[\varphi^2(\tau) + \varphi'^2(\tau)]^{\frac{p-2}{2}} \varphi'(\tau) \sin^{n-2}(\tau|\lambda|^{-1}) \\ &\quad = [p - 1 + (n-p)\lambda^{-1}] \int_0^\tau [\varphi^2(t) + \varphi'^2(t)]^{\frac{p-2}{2}} \varphi(t) \sin^{n-2}(t|\lambda|^{-1})\, dt, \end{aligned} \tag{7}$$

and estimate the modulus of the derivative of the solution on the interval $[0, |\lambda| l]$ where $\varphi(\tau) \geqq 0$. We note that for all $\tau \in [0, |\lambda| l]$

$$\varphi'(\tau) \leqq 0 \quad \text{and} \quad \varphi(\tau) \leqq 1. \tag{8}$$

Let M denote the maximum of $|\varphi'(\tau)|$ on $[0, \min(|\lambda|, |\lambda| l)]$. It follows from (7) for $p \geqq 2$ that

$$M^{p-1} \leqq c(1 + M^2)^{(p-2)/2},$$

from which we get $M \leqq c$. On the other hand, it follows from (7) for $p < 2$ that

$$M \leqslant c(1 + M^2)^{\frac{2-p}{2}} \int_0^\tau \frac{\varphi(t)\, dt}{[\varphi^2(t) + \varphi'^2(t)]^{\frac{2-p}{2}}} \leqslant c(1 + M^2)^{\frac{2-p}{2}},$$

from which we again get $M \leqq c$.

Since $M \leqq c$, we have

$$\varphi(\tau) \geqq 1 - c\alpha\tau > \tfrac{1}{2} \tag{9}$$

---

[2] Here and below $a \sim b$ means that $c_1 \leqq ab^{-1} \leqq c_2$, where c_1 and c_2 are positive constants depending only on n and p.

for all $\tau \in [0, \alpha]$, where α is a sufficiently small positive number depending only on n and p.

It now follows from (7) for $p \geqq 2$ that

$$[\varphi(\tau)]^{p-2} |\varphi'(\tau)| \leqslant \int_0^\tau (1 + M^2)^{\frac{p-2}{2}} d\tau \leqslant c\tau,$$

which together with (9) implies the estimate

$$|\varphi'(\tau)| \leqq c\tau, \quad \tau \in [0, \alpha].$$

And if $p < 2$, it follows at once that

$$|\varphi'(\tau)| \leqslant c(1 + M^2)^{\frac{2-p}{2}} \int_0^\tau \varphi^{p-1}(\tau)\, d\tau \leqslant c\tau$$

for all $\tau \in [0, \min(|\lambda|, |\lambda| l)]$.

When $p \geqq 2$ we get from (7) that

$$(1 + M^2)^{\frac{p-2}{2}} |\varphi'(\tau)| \geqslant c \int_{c\tau}^{\tau} \varphi^{p-1}(t)\, dt \geqslant c\tau,$$

for $\tau \in [0, \alpha]$, which implies $|\varphi'(\tau)| \geqq c\tau$.

Analogously, when $p < 2$

$$|\varphi'(\tau)| \geqslant c[\varphi^2(\tau) + \varphi'^2(\tau)]^{\frac{2-p}{2}} \int_{c\tau}^{\tau} \frac{\varphi(t)\, dt}{(1 + M^2)^{\frac{2-p}{2}}} \geqslant c\tau.$$

Thus the relation

$$|\varphi'(\tau)| \sim \tau \tag{10}$$

is valid for all $\tau \in [0, \alpha]$.

Suppose $\tau \in [\alpha, \min(|\lambda|, |\lambda| l)]$. Since $M \leqq c$, the estimate $|\varphi'(\tau)| \leqq c\tau$ holds for all $\tau \in [0, \min(|\lambda|, |\lambda| l)]$. The reverse inequality obviously follows from (7) and (9).

We note that $l < 1$ for sufficiently large $|\lambda|$. For otherwise we would have

$$1 = \int_0^{|\lambda| l} |\varphi'(\tau)|\, d\tau > \int_0^{|\lambda|} |\varphi'(\tau)|\, d\tau \geqslant c|\lambda|^2.$$

Therefore (10) is proved on the whole interval of nonnegativity of $\varphi(\tau)$. Integrating $|\varphi'(\tau)|$ from 0 to τ with the use of (10), we get

$$1 - \varphi(\tau) = \int_0^\tau |\varphi'(\tau)|\, d\tau \geqslant c\tau^2.$$

Since this inequality is not satisfied for $\tau = |\lambda|$, the function $\varphi(\tau)$ vanishes at the point $|\lambda| l$ of the interval $[0, |\lambda|]$. Finally, from (10) we get

$$1 = \int_0^{|\lambda| l} |\varphi'(\tau)|\, d\tau \sim (|\lambda| l)^2,$$

which implies $l \sim |\lambda|^{-1}$.

LEMMA 2. *For any* λ, $|\lambda| \in (a, \infty)$, *where* a *is the constant in Lemma* 1, *there exists in a neighborhood of the point* $\theta = 0$ *one and only one solution* $f(\theta) \in C^2[0, \pi)$ *of the Cauchy problem for equation* (2) *with initial data* (4). *This solution can be uniquely extended onto the interval of nonnegativity of* $f(\theta)$.

PROOF. It suffices to rewrite (2) in the equivalent form

$$f(\theta) = 1 - \lambda[\lambda(p-1) + n - p] \int_0^\theta \frac{d\tau}{[\lambda^2 f^2(\tau) + f'^2(\tau)]^{\frac{p-2}{2}} \sin^{n-2}\tau} \times \int_0^\tau [\lambda^2 f^2(t) + f'^2(t)]^{\frac{p-2}{2}} f(t) \sin^{n-2} t\, dt, \tag{11}$$

where $\theta \in [0, \epsilon]$, and note that the right side of this equality is a contraction from the set

$$M = \left\{ f : f \in C^2[0, \varepsilon],\ f(0) = 1,\ f'(0) = 0,\ -\frac{1}{2} \leqslant f'(\theta) \leqslant 0 \right\}$$

into itself for small $\epsilon > 0$.

The possibility of extending $f(\theta)$ onto the interval of nonnegativity follows from (5).

LEMMA 3. *The first zero* $l(\lambda)$ *of the solution* $f(\theta)$ *of the Cauchy problem* (2), (4) *is a continuous function of* λ *on* $(-\infty, -a)$ *and* (a, ∞), *where* a *is the constant in Lemma* 1.

PROOF. We will prove the continuity of $l(\lambda)$ on (a, ∞). Let λ be an arbitrary number in (a, ∞) and let $f_\lambda(\theta)$ be the solution of the Cauchy problem (2), (4) corresponding to λ.

Further, let $\{\lambda_n\}_1^\infty$ be an arbitrary sequence of numbers in (a, ∞) that converges to λ and let $f_n(\theta)$, $n = 1, 2, \cdots$, be the solution of the Cauchy problem (2), (4) corresponding to λ_n. By virtue of (5) the sequence $\{f_n(\theta)\}_1^\infty$ is compact in $C^1[0, \epsilon]$, where ϵ is any sufficiently small positive number. We distinguish a subsequence $\{f_{n_k}\}$ that uniformly converges to some function $f(\theta)$ on $[0, \epsilon]$. Passing to the limit in the integral identity (2) and using (5), we get that $f(\theta)$ is the solution of the Cauchy problem (2), (4) corresponding to the parameter λ. Since this solution is unique, the whole sequence converges to $f_\lambda(\theta)$ in $C^1[0, \epsilon]$.

We now note that

$$|l - l_n| = |f_n(l)|\, |f_n'(\tau_n)|^{-1},$$

where τ_n is a point between l and l_n. Since $|f_n'(\tau_n)| \geqq \text{const}$ by virtue of (5), we have

$$|l - l_n| \leqslant c\, |f_n(l)| \xrightarrow[n \to \infty]{} 0.$$

The continuity of $l(\lambda)$ on $(-\infty, -a)$ is proved in the same way.

By the eigenvalues of problem (2), (3) we will mean those values of the parameter λ for which this problem has a nontrivial solution.

THEOREM 1. *There exists an eigenvalue* $\lambda(l)$, $|\lambda(l)| > a$, *where* a *is the constant in Lemma* 1, *of problem* (2), (3) *to which there corresponds a unique eigenfunction that is positive on* $[0, l)$.

§2. Asymptotic formula for $\lambda(l)$ for $l \to 0$

THEOREM 2. *The following equality holds for* $l \to 0$:

$$\lambda(l) = \pm L l^{-1} + O(1), \tag{12}$$

where L is the first zero of the solution of the Cauchy problem for the equation

$$\left\{[g^2(\tau) + g'^2(\tau)]^{\frac{p-2}{2}} g'(\tau)\,\tau^{n-2}\right\}'_\tau + (p-1)[g^2(\tau) + g'^2(\tau)]^{\frac{p-2}{2}} g(\tau)\,\tau^{n-2} = 0 \tag{13}$$

with the initial data

$$g'(0) = 0, \qquad g(0) = 1. \tag{14}$$

PROOF. We first note that the relations

$$|g(\tau)| \sim \rho, \quad L \sim 1 \tag{15}$$

are established by a verbatim repetition of the reasoning in the proof of Lemma 1.

Next, let $A_f(\tau) = [f^2(\tau) + f'^2(\tau)]^{(p-2)/2}$.

Multiplying equation (13) by $|\lambda|^{2-n}$, integrating (2) and (13) from 0 to τ and subtracting one from the other, we get

$$\begin{aligned} |f'(\tau) - g'(\tau)| \leqslant (p-1)\Bigg| &\frac{\int_0^\tau A_f(t) f(t) \sin^{n-2}(t|\lambda|^{-1})\,dt}{A_f(\tau)\sin^{n-2}(\tau|\lambda|^{-1})} \\ &- \frac{\int_0^\tau A_g(t) g(t)(t|\lambda|^{-1})^{n-2}\,dt}{A_g(\tau)(\tau|\lambda|^{-1})^{n-2}}\Bigg| + \left|\frac{n-p}{\lambda}\right|\left|\frac{\int_0^\tau A_f(t) f(t)\sin^{n-2}(t|\lambda|^{-1})\,dt}{A_f(\tau)\sin^{n-2}(\tau|\lambda|^{-1})}\right|. \end{aligned} \tag{16}$$

We note that by virtue of (5) and (15)

$$A_f(\tau) \sim 1, \qquad A_g(\tau) \sim 1. \tag{17}$$

Using (17) to estimate the residual terms in (16), we get

$$\begin{aligned} \max_{\tau\in[0,\alpha]} |f'(\tau) - g'(\tau)| &\leqslant c\tau \max_{\tau\in[0,\alpha]} \{|A_f(\tau) - A_g(\tau)| + |f(\tau) - g(\tau)|\} \\ &+ O(|\lambda|^{-1}) \leqslant c\tau \max_{\tau\in[0,\alpha]} |f'(\tau) - g'(\tau)| + O(|\lambda|^{-1}), \end{aligned}$$

where α is a sufficiently small positive number.

From the latter inequality it follows that for all $\tau \in [0,\alpha]$

$$|f'(\tau) - g'(\tau)| \leqq O(|\lambda|^{-1}). \tag{18}$$

Let $B_f(\tau) = [f^2(\tau) + (p-1)f'^2(\tau)]^{(p-2)/2}$.

Carrying out the differentiation in equations (2) and (13) and subtracting one from the other, we get

$$\begin{aligned} f''(\tau) - g''(\tau) + (n-2)&\left\{|\lambda|^{-1}\cot(\tau|\lambda|^{-1}) f'(\tau)\left[\frac{A_f(\tau)}{B_f(\tau)}\right]^{\frac{2}{p-2}} - \tau^{-1} g'(\tau)\left[\frac{A_g(\tau)}{B_g(\tau)}\right]^{\frac{2}{p-2}}\right\} \\ &+ [p-1+(n-p)|\lambda|^{-1}]\left[\frac{A_f(\tau)}{B_f(\tau)}\right]^{\frac{2}{p-2}} f(\tau) - (p-1)\left[\frac{A_g(\tau)}{B_g(\tau)}\right]^{\frac{2}{p-2}} g(\tau) \\ &+ (p-2)\left\{\left[\frac{f'^2(\tau)}{B_f^{\frac{2}{p-2}}(\tau)}\right] f(\tau) - \left[\frac{g'^2(\tau)}{B_g^{\frac{2}{p-2}}(\tau)}\right] g(\tau)\right\} = 0. \end{aligned}$$

Integrating the resultant equality from α to τ and applying (5) and (15) to estimate the residual terms, we get

$$|f'(\tau)-g'(\tau)|\leqslant c\int_0^{\tau}|f'(t)-g'(t)|\,dt+O(|\lambda|^{-1}),$$

from which we again obtain (18). Integrating (18) from 0 to $|\lambda|l$, we get

$$|f(|\lambda|l)-g(|\lambda|l)|\leqq O(|\lambda|^{-1}).$$

It remains to estimate the difference of the zeros of the solutions of the Cauchy problems (2), (4) and (13), (14), which by virtue of the latter inequality and relation (15) does not exceed

$$|L-|\lambda|l|\leqq c|f(|\lambda|l)-g(|\lambda|l)|\leqq O(|\lambda|^{-1}).$$

Q.E.D.

REMARK. In the case $n=2$ the eigenvalues of the boundary value problem (2), (3) can be written out in explicit form. In fact, in this case equation (2) can be written in the form

$$\left\{[\lambda^2f^2(\theta)+f'^2(\theta)]^{\frac{p-2}{2}}f'(\theta)\right\}'+\lambda[\lambda(p-1)+2-p][\lambda^2f^2(\theta)+f'^2(\theta)]^{\frac{p-2}{2}}f(\theta)=0$$

or, using different notation,

$$z'(\theta)[\lambda^2+(p-1)z^2(\theta)]+[\lambda^2+z^2(\theta)]\{\lambda[\lambda(p-1)+2-p]+(p-1)z^2(\theta)\}=0,$$

where $z(\theta)=f'(\theta)/f(\theta)$.

We rewrite the latter equation in the form

$$\frac{\lambda dz}{\lambda^2+z^2(\theta)}-\frac{(\lambda-1)\,dz}{\lambda^2+z^2(\theta)+\dfrac{\lambda(2-p)}{p-1}}+d\theta=0,$$

which is convenient for integration, and integrate from 0 to l. We get

$$1-\frac{\lambda-1}{\sqrt{\lambda^2-\lambda\dfrac{p-2}{p-1}}}=\frac{2l}{\pi},$$

from which we obtain the quadratic equation

$$\lambda^2\left[1-\left(1-\frac{2l}{\pi}\right)^2\right]-\left[2-\frac{p-2}{p-1}\left(1-\frac{2l}{\pi}\right)^2\right]\lambda+1=0$$

for the determination of λ.

The latter equation readily implies the asymptotic formula

$$\lambda=\pm\frac{\pi p}{4l(p-1)}+O(1)$$

for $l\to 0$.

Let Ω denote a bounded open subset of R^n with boundary $\partial\Omega$. We place the origin 0 on $\partial\Omega$. Let Π_ρ denote the projection of $D_\rho\cap\Omega$, where D_ρ is the ball of radius ρ centered at the origin, from the point 0 onto ∂D_1, and let $\epsilon(\rho)$ denote the greatest distance calculated on ∂D_1 between the points of Π_ρ.

We will say that the boundary $\partial\Omega$ of Ω has a "null salient point" at 0 if $\epsilon(\rho)\to 0$ as $\rho\to 0$.

We direct the x_n axis toward the interior of Ω. Let φ denote the distance from a point $x \in R^n$ to the origin 0 and let θ denote the angle between the radius vector of a point x and the x_n axis. Let K_ρ denote the following closed spherical cone in R^n with vertex at 0: $K_\rho = \{x : 0 \leq \theta \leq \epsilon(\rho)/2\}$, and let ∂K_ρ denote its boundary.

A function $u \in L_p^1(\Omega)$ is said to be a solution of the Dirichlet problem for equation (1) if for all $\varphi \in L_p^1(\Omega)$

$$\int_\Omega |Du|^{p-2} Du D\varphi dx = 0, \quad (u - h) \in \mathring{L}_p^1(\Omega), \tag{19}$$

where $h \in L_p^1(\Omega)$.

Theorem 3. *Suppose $\partial\Omega$ has a null salient point at* 0, $h(x) = 0$ *on $\partial\Omega$ in a neighborhood of* 0 *and $u(x)$ is a solution of the Dirichlet problem* (19). *Then for any $N > 0$*

$$\lim_{x \to 0} |x|^{-N} u(x) = 0.$$

Proof. Let

$$w(x) = \frac{\max\limits_{x \in \Omega \cap \partial D_1} |u(x)|}{\min\limits_{x \in \Omega \cap \partial D_1} v(x)} v(x),$$

where $v(x)$ is a barrier for equation (1) in the cone K_ρ. We compare the functions $u(x)$ and $w(x)$ in the domain $\Omega \cap D_\rho$ with boundary

$$\partial(\Omega \cap D_\rho) = (\partial\Omega \cap D_\rho) \cup (\Omega \cap \partial D_\rho).$$

Clearly

$$0 = u(x) \leq w(x), \quad x \in \partial\Omega \cap D_\rho.$$

On the other hand, by virtue of the choice of $w(x)$

$$|u(x)| \leq w(x), \quad x \in \Omega \cap \partial D_\rho. \tag{20}$$

By virtue of the maximum principle inequality (20) holds for all $x \in \Omega \cap D_\rho$, and hence

$$|u(x)| \leqslant c|x|^{\lambda(\rho)} \max_{x \in \Omega \cap \partial D_1} |u(x)|.$$

The theorem follows from the latter inequality and (12).

Translated by
S. SMITH

Trudy Mat. Inst. Steklov.
125 (1973)

Proc. Steklov Inst. Math.
125 (1973)

ON THE ASYMPTOTIC BEHAVIOR OF SOLUTIONS OF CERTAIN SYSTEMS WITH A SMALL PARAMETER APPROXIMATING THE SYSTEM OF NAVIER-STOKES EQUATIONS

A. P. OSKOLKOV

Abstract. Elliptic and parabolic systems with a small parameter ϵ in front of the highest order derivatives in one of the equations are considered which degenerate as $\epsilon \to 0$ into stationary and nonstationary systems of Navier-Stokes equations. The asymptotic behavior of the solutions of these systems (mainly of the regularized presure) for $\epsilon \to 0$ is studied.

Bibliography: 5 items.

In the present paper quasilinear systems with a small parameter in front of the highest order derivatives in one of the equations are considered which for $\epsilon \to 0$ degenerate into a stationary or nonstationary system of Navier-Stokes equations, and the asymptotic behavior of the solutions of these systems, mainly of the "pressure" p^ϵ, is studied as $\epsilon \to 0$. The nonstationary systems that we study were introduced in [1] as examples of systems which regularize the nonstationary system of Navier-Stokes equations in the sense that they make it possible to construct for it a scheme of variable directions, concrete from the point of view of convergence and stability, which are most economical from the viewpoint of minimizing the volume of computations. The results we obtain are also of purely theoretical interest, since at the present time questions of the degeneration of quasilinear elliptic and parabolic systems have not been studied in sufficient detail.

The asymptotic behavior of the solutions of the systems we consider coincides with the standard asymptotics for solutions of a single elliptic or parabolic equation with a small parameter in front of the highest order derivatives.

The paper consists of three sections. In the first we study the asymptotic behavior of solutions of a nonlinear parabolic system which degenerates into the nonstationary, nonlinear system of Navier-Stokes equations. In the second we study the asymptotic behavior of solutions of a linear parabolic system which degenerates into the Stokes linearization of the nonstationary system of Navier-Stokes equations. In the last section a nonlinear elliptic system is considered which degenerates into the stationary nonlinear system of Navier-Stokes equations.

We remark that in all three of the parabolic or elliptic systems with a small parameter ϵ that we consider it is possible to eliminate the "pressure" p^ϵ and reduce matters to the investigation of a system of integro-differential equations with a large parameter $1/\epsilon$ which contain only the "velocity" $\mathbf{v}^\epsilon$; this reduction is achieved by means of the Green's function of the first or second initial-boundary-value problem for the heat equation in the stationary

---

AMS (MOS) subject classifications (1970). Primary 35B40, 35K55, 35K10, 35J70, 35Q10; Secondary 76D05.

case and by means of the Green's function of the Dirichlet problem or the Neumann function of the Neumann problem for the Laplace equation in the stationary case (this will be discussed in more detail below); the integral operators which hereby arise afford additional information on the asymptotic behavior of the "pressure" p^ϵ and $\operatorname{div} \mathbf{v}^\epsilon$. This state of affairs may be useful in the practical solution of problems on the motion of a viscous fluid especially if from the very formulation of the problem it is necessary to find only the velocity of the flow $\mathbf{v}$.

In presenting the results of this work we shall make essential use of our paper [1], in which theorems on the existence of generalized solutions of the problems with a small parameter which we consider are proved and preliminary information on the asymptotic behavior of solutions of these systems for $\epsilon \to 0$ is obtained.

Throughout the following Ω is a bounded domain in two-dimensional or three-dimensional Euclidean space, $\partial\Omega$ is its boundary, $Q_T = \Omega \times [0, T]$, $0 < T < \infty$, $\partial Q_T = \partial\Omega \times [0, T]$ is the lateral surface of Q_T, $\Omega_t \equiv \Omega$ is the cross-section of Q_T cut by the plane $t = t$, $0 \leq t \leq T$, and $\omega(\epsilon)$, $\epsilon \geqq 0$, is an arbitrary positive function which decreases monotonically to zero as $\epsilon \to 0$.

The results of the present work were reported in June, 1971 at the Jubilee Inter-Institutional Scientific Conference at the Kabarda-Balkar State University (in the city of Nalchik) in honor of the fiftieth anniversary of the formation of the Kabarda-Balkar ASSR.

§1. In the cylinder Q_T we consider the following quasilinear system with a small parameter $\epsilon > 0$:

$$(1) \qquad \begin{gathered} \frac{\partial \mathbf{v}^\varepsilon}{\partial t} - \nu\Delta \mathbf{v}^\varepsilon + v_k^\varepsilon \frac{\partial \mathbf{v}^\varepsilon}{\partial x_k} + \frac{1}{2} \mathbf{v}^\varepsilon \operatorname{div} \mathbf{v}^\varepsilon + \operatorname{grad} p^\varepsilon = \mathbf{f}(x, t), \\ \varepsilon\left(\frac{\partial p^\varepsilon}{\partial t} - \Delta p^\varepsilon\right) + \operatorname{div} \mathbf{v}^\varepsilon = 0, \quad \varepsilon > 0. \end{gathered}$$

For each $\epsilon > 0$ this system is obviously parabolic in the sense of Petrovskiĭ, and for $\epsilon = 0$ it degenerates into the nonstationary nonlinear system of Navier-Stokes equations

$$(2) \qquad \frac{\partial \mathbf{v}}{\partial t} - \nu\Delta \mathbf{v} + v_k \frac{\partial \mathbf{v}}{\partial x_k} + \operatorname{grad} p = \mathbf{f}(x, t), \ \operatorname{div} \mathbf{v} = 0.$$

We shall consider two initial-boundary-value problems for the system (1); namely we shall seek a solution $\mathbf{v}^\epsilon(x,t)$, $p^\epsilon(x,t)$ in the cylinder Q_T which satisfies the initial conditions

$$(3) \qquad \mathbf{v}^\varepsilon |_{t=0} = \mathbf{a}(x), \ \operatorname{div} \mathbf{a}(x) = 0, \ p^\varepsilon |_{t=0} = P(x), \ x \in \Omega,$$

and one of the following boundary conditions:

$$(4) \qquad \mathbf{v}^\varepsilon |_{\partial Q_T} = 0, \quad p^\varepsilon |_{\partial Q_T} = 0,$$

$$(5) \qquad \mathbf{v}^\varepsilon |_{\partial Q_T} = 0, \quad \left.\frac{\partial p^\varepsilon}{\partial n}\right|_{\partial Q_T} = 0.$$

We shall say that the functions $\mathbf{u}(x,t)$ and $q(x,t)$ belong to the space $V_2^\epsilon(Q_T)$, $\epsilon \geqq 0$, if for them the following norm is defined and finite:

(6)
$$|\mathbf{u}, q|_{V_2^\varepsilon(Q_T)} \equiv \sup_{0\leqslant t\leqslant T} \|\mathbf{u}(x, t)\|_{L_2(\Omega)} + \|\mathbf{u}_x\|_{L_2(Q_T)} + \varepsilon^{1/2}\left\{\sup_{0\leqslant t\leqslant T}\|q(x, t)\|_{L_2(\Omega)} + \|q_x\|_{L_2(Q_T)}\right\} < \infty.$$

We further let $\overset{\circ}{V}{}_2^\epsilon(Q_T)$ denote the set of functions $\mathbf{u}$, $q \in V_2^\epsilon(Q_T)$ satisfying the boundary conditions (4), and $\underset{\circ}{V}{}_2^\epsilon(Q_T)$ the set of functions $\mathbf{u}$, $q \in V_2^\epsilon(Q_T)$ satisfying the boundary condition $\mathbf{u}|_{\partial Q_T} = 0$.

Following [1], we shall call a *strong generalized solution* (a Ladyženskaja solution) of problem (1), (3), (4) or problem (1), (3), (5) a solution $\mathbf{v}^\epsilon(x,t)$, $p^\epsilon(x,t)$ of these problems which belongs to the space $\overset{\circ}{V}{}_2^\epsilon(Q_T) \cap L_{q,r}(Q_T)$ or respectively the space $\underset{\circ}{V}{}_2^\epsilon(Q_T) \cap L_{q,r}(Q_T)$ with indices q and r satisfying the conditions

(7)
$$1/r + n/2q = \tfrac{1}{2},$$

where $r \in [2, \infty)$, $q \in (n, \infty]$ or $r = \infty$, $q > n$, $n = 2, 3$, and which satisfies the integral identities

(8)
$$\iint_{Q_T}\left[-\mathbf{v}^\varepsilon\Phi_t + \nu\mathbf{v}^\varepsilon_x\Phi_x + \left(v^\varepsilon_k\frac{\partial \mathbf{v}^\varepsilon}{\partial x_k} + \frac{1}{2}\mathbf{v}^\varepsilon \operatorname{div}\mathbf{v}^\varepsilon + \operatorname{grad} p^\varepsilon\right)\Phi\right]dQ + \int_\Omega \mathbf{v}^\varepsilon\Phi|_{t=T}\,dx - \int_\Omega \mathbf{a}(x)\Phi|_{t=0}\,dx = \iint_{Q_T}\mathbf{f}\Phi\,dQ,$$

(9)
$$\iint_{Q_T}(-\varepsilon p^\varepsilon\psi_t + \varepsilon p^\varepsilon_x\psi_x + \psi\operatorname{div}\mathbf{v}^\varepsilon)\,dQ + \varepsilon\left[\int_\Omega p^\varepsilon\psi|_{t=T}\,dx - \int_\Omega P(x)\psi|_{t=0}\,dx\right] = 0$$

for any $\Phi(x,t) \in \overset{\circ}{W}{}_2^{1,1}(Q_T)$ and $\psi(x,t) \in \overset{\circ}{W}{}_2^{1,1}(Q_T)$ or respectively $\psi(x,t) \in W_2^{1,1}(Q_T)$ and for which, in addition, $(\mathbf{v}^\epsilon)^2$, $(p^\epsilon)^2$, $\mathbf{v}^\epsilon_x$, p^ϵ_x, $\mathbf{v}^\epsilon_t$, p^ϵ_t, $\mathbf{v}^\epsilon_{xt}$, $p^\epsilon_{xt} \in L_2(Q_T)$.

It is easily seen that because of the assumptions $\mathbf{v}^\epsilon_t$, $p^\epsilon_t \in L_2(Q_T)$ the integral identities (8) and (9) may be transformed to the form

(8′)
$$\iint_{Q_T}\left[\mathbf{v}^\varepsilon_t\Phi + \nu\mathbf{v}^\varepsilon_x\Phi_x + \left(v^\varepsilon_k\frac{\partial\mathbf{v}^\varepsilon}{\partial x_k} + \frac{1}{2}\mathbf{v}^\varepsilon\operatorname{div}\mathbf{v}^\varepsilon + \operatorname{grad}p^\varepsilon\right)\Phi\right]dQ = \iint_{Q_T}\mathbf{f}\Phi\,dQ,$$

(9′)
$$\varepsilon\iint_{Q_T}(p^\varepsilon_t\psi + p^\varepsilon_x\psi_x)\,dQ + \iint_{Q_T}\psi\operatorname{div}\mathbf{v}^\varepsilon\,dQ = 0.$$

It is shown in [1] (cf. Theorems 6–8) that under specific conditions on the data of problems (1), (3), (4) and (1), (3), (5) (that is, on the initial data $\mathbf{a}(x)$ and $P(x)$, the free term $\mathbf{f}(x,t)$, the coefficient of viscosity ν, and the dimensions of the cylinder Q_T) problems (1), (3), (4) and (1), (3), (5) have a unique strong generalized solution $\mathbf{v}^\epsilon(x,t)$, $p^\epsilon(x,t)$, and for this solution there is the following a priori estimate:

(10)
$$\sup_{0\leqslant t\leqslant T}\left(\|\mathbf{v}^\varepsilon\|^2_{L_2(\Omega)} + \|\mathbf{v}^\varepsilon_x\|^2_{L_2(\Omega)} + \|\mathbf{v}^\varepsilon_t\|^2_{L_2(\Omega)}\right) + \varepsilon\sup_{0\leqslant t\leqslant T}\left(\|p^\varepsilon\|^2_{L_2(\Omega)} + \|p^\varepsilon_x\|^2_{L_2(\Omega)} + \|p^\varepsilon_t\|^2_{L_2(\Omega)}\right) + \|\mathbf{v}^\varepsilon_{xt}\|^2_{L_2(Q_T)} + \varepsilon\|p^\varepsilon_{xt}\|^2_{L_2(Q_T)} \leqslant C_1,$$

where the constant C_1 is determined only by the aforementioned data of the problems in question. It is also shown there (Theorem 9) that for $\epsilon \to 0$ the strong generalized solution $\mathbf{v}^\epsilon$, p^ϵ of problem (1), (3), (4) and of problem (1), (3), (5) goes over into a strong generalized solution $\mathbf{v}$, p (the Ladyženskaja solution; see [**2**], Chapter 4) of the first initial-boundary-value problem for the system of Navier-Stokes equations (2):

$$\frac{\partial \mathbf{v}}{\partial t} - \nu\Delta\mathbf{v} + v_k\frac{\partial \mathbf{v}}{\partial x_k} + \operatorname{grad} p = \mathbf{f}(x, t), \ \operatorname{div}\mathbf{v} = 0, \tag{11}$$
$$\mathbf{v}\,|_{t=0} = \mathbf{a}(x), \quad \mathbf{v}\,|_{\partial Q_T} = 0,$$

where the solution $\mathbf{v}^\epsilon$, p^ϵ goes over to the solution $\mathbf{v}$, p of problem (11) in the sense that the integral identity (9) is canceled and the integral identity (8) on solenoidal $\Phi(x,t) \in \overset{\circ}{W}{}_2^{1,1}(Q_T)$ goes over into the integral identity

$$\iint_{Q_T}\left[-\mathbf{v}\Phi_t + \nu\mathbf{v}_x\Phi_x + v_k\frac{\partial \mathbf{v}}{\partial x_k}\Phi\right]dQ + \int_\Omega \mathbf{v}\Phi\,|_{t=T}\,dx \tag{12}$$
$$- \int_\Omega \mathbf{a}(x)\Phi\,|_{t=0}\,dx = \iint_{Q_T}\mathbf{f}\Phi\,dQ,$$

defining a strong solution of problem (11). The rate of convergence of the solutions $\mathbf{v}^\epsilon$, p^ϵ to the solution $\mathbf{v}$, p in the energy metric (6) is also studied (cf. Theorems 11, 12). In the present section, on the basis of these results (especially inequality (10)), we continue the investigation of the asymptotic behavior as $\epsilon \to 0$ of the regularized pressure $p^\epsilon(x,t)$ and $\operatorname{div}\mathbf{v}^\epsilon(x,t)$.

We assume that the boundary $\partial\Omega$ belongs to the class C^2, and we suppose that $P(x) \in W_{2,0}^2(\Omega)$ or respectively $P(x) \in W_2^2(\Omega)$. In the integral identity (9) we consider $\operatorname{div}\mathbf{v}^\epsilon$ as the free term which for each $t \in [0, T]$ belongs to $L_2(\Omega)$ by inequality (10). From the theory of generalized solutions of parabolic equations (cf. [**3**], Chapter III) we then find easily that $p^\epsilon(x,t) \in W_2^{2,1}(Q_T)$ almost everywhere in Q_T satisfies the equation

$$\varepsilon\left(\frac{\partial p^\varepsilon}{\partial t} - \Delta p^\varepsilon\right) + \operatorname{div}\mathbf{v}^\varepsilon = 0 \tag{13}$$

and as an element of $W_2^{2,1}(Q_T)$ it satisfies the initial condition (3) and the boundary condition (4) or (5) respectively.

Next, in equation (13) we consider $\epsilon\,\partial p^\epsilon/\partial t + \operatorname{div}\mathbf{v}^\epsilon$ as the free term. By (10) this free term for each $t \in [0, T]$ belongs to $L_2(\Omega)$, and then $p^\epsilon(x,t)$ for each $t \in [0, T]$ almost everywhere in the domain Ω satisfies the Poisson equation

$$\Delta p^\varepsilon = \frac{\partial p^\varepsilon}{\partial t} + \frac{1}{\varepsilon}\operatorname{div}\mathbf{v}^\varepsilon \tag{14}$$

with a free term in $L_2(\Omega)$ and one of the boundary conditions (4), (5). From this it follows immediately that for each $t \in [0, T]$ we have $p^\epsilon(x,t) \in W_{2,0}^2(\Omega)$ or respectively $p^\epsilon(x,t) \in W_2^2(\Omega)$, and it satisfies one of the following two inequalities (the second basic inequality for the Dirichlet problem or the Neumann problem; cf. [**4**], Chapter III):

$$\|p^\varepsilon\|_{W_2^2(\Omega)}^2 \leqslant C_2\left(\left\|\frac{\partial p^\varepsilon}{\partial t}\right\|_{L_2(\Omega)}^2 + \frac{1}{\varepsilon^2}\|\operatorname{div}\mathbf{v}^\varepsilon\|_{L_2(\Omega)}^2\right), \tag{15}$$

if $p^\epsilon(x,t)$ satisfies the Dirichlet boundary condition (4) and

(16) $$\|p^\varepsilon\|^2_{W_2^2(\Omega)} \leqslant C_3\left(\left\|\frac{\partial p^\varepsilon}{\partial t}\right\|^2_{L_2(\Omega)} + \frac{1}{\varepsilon^2}\|\operatorname{div}\mathbf{v}^\varepsilon\|^2_{L_2(\Omega)} + \|p^\varepsilon\|^2_{W_2^1(\Omega)}\right),$$

if $p^\epsilon(x,t)$ satisfies the Neumann boundary condition (5), where the constants C_2 and C_3 depend only on the norm of the boundary $\partial\Omega$ in C^2.

By means of (10) we obtain from (15) and (16) the following estimate for the "pressure" $p^\epsilon(x,t)$:

(17) $$\sup_{0\leqslant t\leqslant T}\int_\Omega \|p^\varepsilon_{xx}\|^2\,dx \leqslant \frac{C_4(C_1,\,\partial\Omega)}{\varepsilon^2},$$

and from this inequality and (10), using Sobolev's imbedding theorem, we deduce the following estimates for the Hölder norm of $p^\epsilon(x,t)$: in the two-dimensional case for any $\beta\in(0,1)$

(18) $$\sup_{0\leqslant t\leqslant T}|p^\varepsilon|_\Omega^{(\beta)} \equiv \sup_{0\leqslant t\leqslant T}\left\{\sup_\Omega|p^\varepsilon(x,\,t)| + \sup_{x,\,x'\in\Omega}\frac{|p^\varepsilon(x,\,t)-p^\varepsilon(x',\,t)|}{|x-x'|^\beta}\right\} \leqslant \frac{C_5(C_4,\,\beta)}{\varepsilon},$$

and in the three-dimensional case

(19) $$\sup_{0\leqslant t\leqslant T}|p^\varepsilon|_\Omega^{(1/2)} \leqslant \frac{C_6(C_4)}{\varepsilon}.$$

It is obvious that, for each $t\in[0,T]$, $p^\epsilon(x,t)$ satisfies the integral identity

(20) $$\varepsilon\int_\Omega p^\varepsilon_t\psi\,dx + \varepsilon\int_\Omega p^\varepsilon_x\psi_x\,dx + \int_\Omega\psi\operatorname{div}\mathbf{v}^\varepsilon dx = 0$$

for any $\psi(x)\in W_2^1(\Omega)$. As above, let $\omega(\epsilon)$, $\epsilon\geqq 0$, be an arbitrary positive function which tends monotonically to zero as $\epsilon\to 0$. From (10) it follows that for $\epsilon\to 0$

(21) $$\omega(\varepsilon)\,\varepsilon^{1/2}\left[\int_\Omega p^\varepsilon_t\psi\,dx + \int_\Omega p^\varepsilon_x\psi_x\,dx\right]\to 0$$

for any $\psi(x)\in W_2^1(\Omega)$, and hence from the integral identity (20) it follows that for any $\psi(x)\in W_2^1(\Omega)$

(22) $$\frac{\omega(\varepsilon)}{\varepsilon^{1/2}}\int_\Omega\psi(x)\operatorname{div}\mathbf{v}^\varepsilon(x,\,t)\,dx\to 0,\quad \varepsilon\to 0,\quad 0\leqslant t\leqslant T.$$

It follows further from (10) that for $\epsilon\to 0$

(23) $$\omega(\varepsilon)\,\varepsilon^{1/2}\int_\Omega p^\varepsilon_t\psi(x)\,dx\to 0,\quad 0\leqslant t\leqslant T,$$

for any $\psi(x)\in L_2(\Omega)$, and then from the integral identity (20), on integrating by parts in the middle term and using (22) and (23), we obtain as $\epsilon\to 0$ for any $\psi(x)\in \overset{\circ}{W}{}_2^1(\Omega)$ the limit relation

(24) $$\omega(\varepsilon)\,\varepsilon^{1/2}\int_\Omega\Delta p^\varepsilon(x,\,t)\,\psi(x)\,dx\to 0,\quad 0\leqslant t\leqslant T.$$

Suppose, as previously, that the boundary $\partial\Omega$ belongs to the class C^2, and let $G(x,y;t,\tau)$ be the Green's function of the first or second initial-boundary-value problem for the heat equation (cf., for example, [4], Chapter IV). Then, considering $-\epsilon^{-1}\operatorname{div}\mathbf{v}^\epsilon$ as the free term in (13), we can write the solution $p^\epsilon(x,t)$ of (13) which satisfies the initial condition (3) and the boundary condition (4) or respectively (5) in the following form:

$$p^\varepsilon(x,\ t) = -\frac{1}{\varepsilon}\iint_{Q_t} G(x,\ y;\ t,\ \tau)\operatorname{div}\mathbf{v}^\varepsilon dy d\tau + \int_\Omega G(x,\ y;\ t,\ 0)\,P(y)\,dy. \tag{25}$$

Making use of the representation (25), we can eliminate the "pressure" $p^\epsilon(x,t)$ from the first (vector) equation of the system (1) and obtain for the regularized velocity $\mathbf{v}^\epsilon(x,t)$ the first initial-boundary-value problem for the following system of integro-differential equations with the large parameter $1/\epsilon$:

$$\frac{\partial \mathbf{v}^\varepsilon}{\partial t} - \nu\Delta\mathbf{v}^\varepsilon + v_k^\varepsilon\frac{\partial \mathbf{v}^\varepsilon}{\partial x_k} + \frac{1}{2}\mathbf{v}^\varepsilon\operatorname{div}\mathbf{v}^\varepsilon - \frac{1}{\varepsilon}\iint_{Q_t}\operatorname{grad}_x G(x,\ y;\ t,\ \tau)\operatorname{div}\mathbf{v}^\varepsilon dy d\tau = \mathbf{f}(x,\ t) - \int_\Omega \operatorname{grad}_x G(x,\ y;\ t,\ 0)\,P(y)\,dy, \tag{26}$$

$$\mathbf{v}^\varepsilon|_{t=0} = \mathbf{a}(x),\quad \mathbf{v}^\varepsilon|_{\partial Q_T} = 0. \tag{27}$$

We set

$$G_1(\operatorname{div}\mathbf{v}^\varepsilon) \equiv \iint_{Q_t}\operatorname{grad}_x G(x,\ y;\ t,\ \tau)\operatorname{div}\mathbf{v}^\varepsilon dy d\tau, \tag{28}$$

$$G_2(P) \equiv \int_\Omega \operatorname{grad}_x G(x,\ y;\ t,\ 0)\,P(y)\,dy. \tag{29}$$

It is obvious that $\mathbf{v}^\epsilon(x,t)$ for any $t\in[0,T]$ satisfies the integral identity

$$\int_\Omega \mathbf{v}_t^\varepsilon\Phi dx + \nu\int_\Omega \mathbf{v}_x^\varepsilon\Phi_x dx + \int_\Omega\left(v_k^\varepsilon\mathbf{v}_{x_k}^\varepsilon + \frac{1}{2}\mathbf{v}^\varepsilon\operatorname{div}\mathbf{v}^\varepsilon\right)\Phi dx = \frac{1}{\varepsilon}\int_\Omega \Phi G_1(\operatorname{div}\mathbf{v}^\varepsilon)\,dx + \int_\Omega(\mathbf{f} - G_2(P))\,\Phi dx \tag{30}$$

for any $\Phi(x)\in W_2^1(\Omega)$. In (30) all the integrals, except for $\epsilon^{-1}\int_\Omega\Phi(x)\,G_1(\operatorname{div}\mathbf{v}^\epsilon)\,dx$, by virtue of (10) and Sobolev's imbedding theorems are bounded uniformly with respect to $\epsilon\geqq 0$ for any $\Phi(x)\in W_2^1(\Omega)$, and hence from (30) it follows that as $\epsilon\to 0$

$$\frac{\omega(\varepsilon)}{\varepsilon}\int_\Omega G_1(\operatorname{div}\mathbf{v}^\varepsilon)\,\Phi(x)\,dx \to 0,\quad \omega(\varepsilon)\downarrow 0,\quad \varepsilon\to 0, \tag{31}$$

for any $\Phi(x)\in W_2^1(\Omega)$ and for any $t\in[0,T]$.

Finally, since

$$\frac{1}{\varepsilon}G_1(\operatorname{div}\mathbf{v}^\varepsilon) = -\operatorname{grad} p^\varepsilon - G_2(P),$$

where $G_2(P)$ does not depend on ϵ and therefore $\omega(\epsilon)G_2(P)$ converges strongly to zero in $L_2(\Omega)$ as $\epsilon\to 0$, we also obtain the following result: for any $t\in[0,T]$

$$\omega(\varepsilon)\int_\Omega \Phi(x)\operatorname{grad} p^\varepsilon(x,\ t)\,dx \to 0,\quad \varepsilon\to 0 \tag{32}$$

for any $\Phi(x) \in W_2^1(\Omega)$, where $\omega(\epsilon)$ is an arbitrary positive function which tends monotonically to zero as $\epsilon \to 0$.

We summarize the results obtained on the asymptotic behavior for $\epsilon \to 0$ of the "pressure" $p^\epsilon(x,t)$ and $\operatorname{div} \mathbf{v}^\epsilon(x,t)$ in the form of the following theorem.

THEOREM 1. *Let $\mathbf{v}^\epsilon$, p^ϵ be the strong generalized solution of problem* (1), (3), (4) *or of problem* (1), (3), (5), *the initial data of which are such that inequality* (10) *holds for $\mathbf{v}^\epsilon$ and p^ϵ. Suppose that the boundary $\partial\Omega \in C^2$, and let $\omega(\epsilon)$, $\epsilon \geq 0$, be an arbitrary positive function which tends monotonically to zero as $\epsilon \to 0$. Then as $\epsilon \to 0$ the following asymptotic estimates hold:*

(33) $$\sup_{0\leqslant t\leqslant T} (\| p^\varepsilon \|^2_{W_2^1(\Omega)} + \| p_t^\varepsilon \|^2_{L_2(\Omega)}) \leqslant \frac{C_1}{\varepsilon},$$

(34) $$\sup_{0\leqslant t\leqslant T} [\| p^\varepsilon_{xx} \|^2_{L_2(\Omega)} + (| p^\varepsilon |_\Omega^{(1/2)})^2] \leqslant \frac{C_9}{\varepsilon^2},$$

(35) $$\omega(\varepsilon)\, \varepsilon^{1/2} \int_\Omega \Delta p^\varepsilon (x, t)\, \psi(x)\, dx \to 0 \quad \forall \psi(x) \in \overset{\circ}{W}{}_2^1(\Omega), \ 0 \leqslant t \leqslant T,$$

(36) $$\omega(\varepsilon) \int_\Omega \operatorname{grad} p^\varepsilon (x, t)\, \Phi(x)\, dx \to 0 \quad \forall \Phi(x) \in W_2^1(\Omega), \ 0 \leqslant t \leqslant T,$$

(37) $$\frac{\omega(\varepsilon)}{\varepsilon^{1/2}} \int_\Omega \operatorname{div} \mathbf{v}^\varepsilon (x, t)\, \psi(x)\, dx \to 0 \quad \forall \psi(x) \in W_2^1(\Omega), \ 0 \leqslant t \leqslant T,$$

(38) $$\frac{\omega(\varepsilon)}{\varepsilon} \int_\Omega G_1 (\operatorname{div} \mathbf{v}^\varepsilon)\, \Phi(x)\, dx \to 0, \ 0 \leqslant t \leqslant T \quad \forall \Phi(x) \in W_2^1(\Omega),$$

where the operator G_1 is defined by (28) *and $G(x,y;t,\tau)$ is the Green's function of the first or respectively the second initial-boundary-value problem for the heat equation.*

REMARK. The limit relation (35) – (38) can be given the following equivalent form. Let A_0 and A be the Riesz operators taking the space $L_2(\Omega)$ into the spaces $\overset{\circ}{W}{}_2^1(\Omega)$ and $W_2^1(\Omega)$ respectively and acting by the Riesz theorem on the general form of a linear functional in a Hilbert space on any element $\varphi(x) \in L_2(\Omega)$ according to the following formulas:

$$\int_\Omega \varphi(x)\, \psi(x)\, dx = (A_0(\varphi), \psi)_{\overset{\circ}{W}{}_2^1(\Omega)}, \quad \forall \psi(x) \in \overset{\circ}{W}{}_2^1(\Omega),$$

$$\int_\Omega \varphi(x)\, \psi(x)\, dx = (A(\varphi), \psi)_{W_2^1(\Omega)}, \quad \forall \psi(x) \in W_2^1(\Omega).$$

Then the limit relations (35) – (38) can be represented as follows: for $\epsilon \to 0$ and for any $t \in [0, T]$

(35′) $$\omega(\varepsilon)\, \varepsilon^{1/2} A_0 (\Delta p^\varepsilon) \rightharpoonup 0 \ \text{ in } \overset{\circ}{W}{}_2^1(\Omega),$$

(36′) $$\omega(\varepsilon)\, A(\operatorname{grad} p^\varepsilon) \rightharpoonup 0 \ \text{ in } W_2^1(\Omega),$$

(37′) $$\frac{\omega(\varepsilon)}{\varepsilon^{1/2}} A(\operatorname{div} \mathbf{v}^\varepsilon) \rightharpoonup 0 \ \text{ in } W_2^1(\Omega),$$

(38′) $$\frac{\omega(\varepsilon)}{\varepsilon} A(G_1(\operatorname{div} \mathbf{v}^\varepsilon)) \rightharpoonup 0 \ \text{ in } W_2^1(\Omega),$$

where the symbol $\rightharpoonup$, as usual, denotes weak convergence.

The asymptotic relations (33) – (34) which we have obtained for the regularized pressure $p^\epsilon(x,t)$ coincide with the standard asymptotic expressions for solutions of a single elliptic or parabolic equation with a small parameter in front of the highest order derivatives (cf., for example, [5]).

§2. In the cylinder Q_T we consider the following linear, nonstationary system with a small parameter $\epsilon > 0$:

$$\frac{\partial \mathbf{v}^\varepsilon}{\partial t} - \nu\Delta \mathbf{v}^\varepsilon + \operatorname{grad} p^\varepsilon = \mathbf{f}(x, t), \quad \varepsilon\left(\frac{\partial p^\varepsilon}{\partial t} - \Delta p^\varepsilon\right) + \operatorname{div} \mathbf{v}^\varepsilon = 0, \quad \varepsilon > 0. \tag{39}$$

For each $\epsilon > 0$ this system is parabolic in the sense of Petrovskiĭ, and for $\epsilon = 0$ it degenerates into the linearized nonstationary system of Navier-Stokes equations

$$\frac{\partial \mathbf{v}}{\partial t} - \nu\Delta \mathbf{v} + \operatorname{grad} p = \mathbf{f}(x, t), \quad \operatorname{div} \mathbf{v} = 0. \tag{40}$$

For the system (39) we shall consider the same two initial-boundary-value problems as for the system (1), i.e. in the cylinder Q_T we seek a solution $\mathbf{v}^\epsilon, p^\epsilon$ which satisfies the initial conditions (3) and one of the boundary conditions (4) or (5).

We shall call a *strong generalized solution* of problem (39), (3), (4) or respectively of problem (39), (3), (5) a solution $\mathbf{v}^\epsilon(x,t)$, $p^\epsilon(x,t)$ of these problems such that $\mathbf{v}^\epsilon$, $\mathbf{v}^\epsilon_x$, $\mathbf{v}^\epsilon_t$, $\mathbf{v}^\epsilon_{xt}$, p^ϵ, p^ϵ_x, p^ϵ_t, $p^\epsilon_{xt} \in L_2(Q_T)$ and which satisfies the integral identities

$$\iint_{Q_T} (\mathbf{v}^\varepsilon_t \Phi + \nu \mathbf{v}^\varepsilon_x \Phi_x + \operatorname{grad} p^\varepsilon \Phi)\, dQ = \iint_{Q_T} \mathbf{f}\Phi dQ, \tag{41}$$

$$\varepsilon \iint_{Q_T} (p^\varepsilon_t \psi + p^\varepsilon_x \psi_x)\, dQ + \iint_{Q_T} \psi \operatorname{div} \mathbf{v}^\varepsilon dQ = 0 \tag{42}$$

for any $\Phi(x,t) \in \overset{\circ}{W}{}_2^{1,0}(Q_T)$ and $\psi(x,t) \in \overset{\circ}{W}{}_2^{1,0}(Q_T)$ or respectively $\psi(x,t) \in \overset{\circ}{W}{}_2^{1,0}(Q_T)$.

The linearized system (39) represents a particular case of the nonlinear system (1), and therefore for the problems (39), (3), (4) and (39), (3), (5) the same existence theorems for a strong solution $\mathbf{v}^\epsilon$, p^ϵ and the consequent asymptotic estimates clearly hold as for problems (1), (3), (4), and (1), (3), (5). Moreover, just as for the linearized system of Navier-Stokes equations (40) (see in this connection [2], Chapters 4 and 6), these existence theorems and the estimates (33) – (37) which they imply hold in both the two-dimensional and three-dimensional cases without any smallness restrictions on the initial data of the problems in question. In other words, the following theorem is valid.

THEOREM 2. *Let* $\mathbf{v}^\epsilon|_{t=0} = \mathbf{a}(x) \in W^2_{2,0}(\Omega)$, $p^\epsilon|_{t=0} = P(x) \in W^2_{2,0}(\Omega)$ *or respectively* $P(x) \in W^2_2(\Omega)$ *and* $\partial P/\partial n|_{\partial\Omega} = 0$, $\mathbf{f}(x,t) \in L_2(Q_T)$ *and* $\mathbf{f}_t \in L_{2,1}(Q_T)$. *Then for any* $\epsilon > 0$ *problems* (39), (3), (4) *and* (39), (3), (5) *have a unique strong generalized solution* $\mathbf{v}^\epsilon$, p^ϵ, *and for this solution inequality* (10) *and the asymptotic estimates* (33) – (38) *hold; as* $\epsilon \to 0$ *this solution goes over into a strong generalized solution* $\mathbf{v}$, p *of the first initial-boundary-value problem for the linearized system of Navier-Stokes equations* (40).

In addition, for solutions of the linearized problems (39), (3), (4) and (39), (3), (5) it is possible to obtain information on the asymptotic behavior for $\epsilon \to 0$ of the derivatives $\partial^{k+1}p^\epsilon/\partial t^{k+1}$, $\partial^{k+1}p^\epsilon/\partial x\partial t^k$, $\partial^{k+2}p^\epsilon/\partial x^2\partial t^k$ and also of $\operatorname{div}(\partial^k \mathbf{v}^\epsilon/\partial t^k)$, $k = 1, 2, \cdots$, by imposing additional smoothness conditions on the initial data $\mathbf{a}(x)$, $P(x)$ and the free term $\mathbf{f}(x,t)$. Indeed, we shall prove the following theorem.

THEOREM 3. *Suppose that for $k = 1, 2, \cdots$ we have $\mathbf{v}^\epsilon|_{t=0} = \mathbf{a}(x) \in \overset{\circ}{W}{}^{2+2k}_{2,0}(\Omega)$, $p^\epsilon|_{t=0} = P(x) \in \overset{\circ}{W}{}^{2+2k}_{2,0}(\Omega)$ or respectively $P(x) \in W_2^{2+2k}(\Omega)$, and $\partial P/\partial n|_{\partial\Omega} = 0$, $\partial^l f/\partial t^l \in L_2(Q_T)$, $l = 0, 1, \cdots, k+1$, and that the boundary $\partial\Omega$ belongs to the class C^2; suppose moreover that the necessary compatibility conditions for the initial conditions $\mathbf{a}(x)$ and $P(x)$, the boundary conditions, and the free term $\mathbf{f}(x,t)$ are satisfied. Then for the strong generalized solution $\mathbf{v}^\epsilon$, p^ϵ of problems (39), (3), (4) and (39), (3), (5) the following inequalities are valid:*

$$\sum_{l=0}^{k+1} \sup_{0\leq t\leq T} \left\| \frac{\partial^l \mathbf{v}^\varepsilon}{\partial t^l} \right\|^2_{L_2(\Omega)} + \sum_{l=0}^{k} \sup_{0\leq t\leq T} \left\| \frac{\partial^{l+1} \mathbf{v}^\varepsilon}{\partial x \partial t^l} \right\|^2_{L_2(\Omega)} + \left\| \frac{\partial^{k+2} \mathbf{v}^\varepsilon}{\partial x \partial t^{k+1}} \right\|^2_{L_2(Q_T)}$$

$$+\varepsilon \left[\sum_{l=0}^{k+1} \sup_{0\leq t\leq T} \left\| \frac{\partial^l p^\varepsilon}{\partial t^l} \right\|^2_{L_2(\Omega)} + \sum_{l=0}^{k} \sup_{0\leq t\leq T} \left\| \frac{\partial^{l+1} p^\varepsilon}{\partial x \partial t^l} \right\|^2_{L_2(\Omega)} + \left\| \frac{\partial^{k+2} p^\varepsilon}{\partial x \partial t^{k+1}} \right\|^2_{L_2(Q_T)} \right] \leq C_{10}, \tag{43}$$

$$\sum_{l=0}^{k} \sup_{0\leq t\leq T} \left\| \frac{\partial^{l+2} p^\varepsilon}{\partial x^2 \partial t^l} \right\|^2_{L_2(\Omega)} + \sum_{l=0}^{k-1} \sup_{0\leq t\leq T} \left(\left| \frac{\partial^l p^\varepsilon}{\partial t^l} \right|^{(1/2)}_{\Omega} \right)^2 \leq \frac{C_{11}}{\varepsilon^2}, \tag{44}$$

and as $\epsilon \to 0$ for $l = 0, 1, \cdots, k$ and $0 \leq t \leq T$ the following limit relations hold:

$$\omega(\varepsilon)\, \varepsilon^{1/2} \int_\Omega \frac{\partial^l}{\partial t^l} \Delta p^\varepsilon(x,\ t)\, \psi(x)\, dx \to 0 \qquad \textit{for any} \quad \psi(x) \in \overset{\circ}{W}{}^1_2(\Omega),$$

$$\omega(\varepsilon) \int_\Omega \operatorname{grad} \frac{\partial^l p^\varepsilon}{\partial t^l}\, \mathbf{\Phi}(x)\, dx \to 0 \qquad \textit{for any} \quad \mathbf{\Phi}(x) \in W_2^1(\Omega), \tag{45}$$

$$\frac{\omega(\varepsilon)}{\varepsilon^{1/2}} \int_\Omega \operatorname{div} \frac{\partial^l \mathbf{v}^\varepsilon}{\partial t^l}\, \psi(x)\, dx \to 0 \qquad \textit{for any} \quad \psi(x) \in W_2^1(\Omega), \tag{46}$$

$$\frac{\omega(\varepsilon)}{\varepsilon} \int_\Omega G_1\left(\operatorname{div} \frac{\partial^l \mathbf{v}^\varepsilon}{\partial t^l}\right) \mathbf{\Phi}(x)\, dx \to 0 \qquad \textit{for any} \quad \mathbf{\Phi}(x) \in W_2^1(\Omega), \tag{47}$$

where the constants C_{10} and C_{11} depend only on the initial data of the problem in question and on the boundary $\partial\Omega$; $\omega(\epsilon)$, $\epsilon \geq 0$, is an arbitrary positive function which tends monotonically to zero as $\epsilon \to 0$, and $G(x,y;t,\tau)$ is the Green's function of the first or respectively the second initial-boundary-value problem for the heat equation.

Inequalities (43), (44) and the asymptotic relations (45)–(47) show that with smooth initial data of order $k = 1, 2, \cdots$ the asymptotic estimates (10), (33)–(37) guaranteed by Theorem 2 for p^ϵ and $\operatorname{div} \mathbf{v}^\epsilon$ also hold for their derivatives with respect to t: $\partial^l p^\epsilon/\partial t^l$ and $\partial^l \operatorname{div} \mathbf{v}^\epsilon/\partial t^l$, $l = 1, \cdots, k$.

In proceeding to the proof of Theorem 3 we remark first of all that as in the proof of Theorem 1 it is possible to go over from the integral identity (42) to the equation

$$\frac{\partial p^\varepsilon}{\partial t} - \Delta p^\varepsilon = -\frac{1}{\varepsilon}\operatorname{div}\mathbf{v}^\varepsilon \tag{48}$$

with initial condition (3) and one of the boundary conditions (4), (5), where (48) is satisfied for each $t \in [0, T]$ almost everywhere in Ω and for each $t \in [0, T]$ we have p_t^ϵ, $p_{xx}^\epsilon \in L_2(\Omega)$. After this, considering $\operatorname{grad} p^\epsilon$ as the free term in (41), we may consider $\mathbf{v}^\epsilon(x, t)$ to be a generalized solution in $\overset{\circ}{W}{}_2^{1,1}(Q_T)$ of the linear parabolic system

$$\mathbf{v}_t^\varepsilon - \nu\Delta\mathbf{v}^\varepsilon = \mathbf{f}(x, t) - \operatorname{grad} p^\varepsilon \tag{49}$$

with initial and boundary conditions

$$\mathbf{v}^\varepsilon|_{t=0} = \mathbf{a}(x), \quad \mathbf{v}^\varepsilon|_{\partial Q_T} = 0. \tag{50}$$

Since it is known that $(\mathbf{f} - \operatorname{grad} p^\epsilon) \in L_2(Q_T)$, it follows from the theory of generalized solutions of linear parabolic systems (cf., for example, [**3**], Chapter VII) that $\mathbf{v}^\epsilon(x, t) \in \overset{\circ}{W}{}_2^{2,1}(Q_T)$ and satisfies the system (49) almost everywhere in Q_T. The strong generalized solution $\mathbf{v}^\epsilon$, p^ϵ of problems (39), (3), (4) and (39), (3), (5) therefore satisfies (39) almost everywhere in Q_T and also the initial conditions (3) and one of the boundary conditions (4), (5).

Further, inequality (10) holds for the strong generalized solution $\mathbf{v}^\epsilon$, p^ϵ of problems (39), (3), (4) and (39), (3), (5), and this means that the "free terms" $\epsilon^{-1}\operatorname{div}\mathbf{v}^\epsilon$ and $\mathbf{f} - \operatorname{grad} p^\epsilon$ have generalized derivatives with respect to t in $L_2(Q_T)$; using the smoothness of the initial data and of the boundary $\partial\Omega$, we now find from standard theorems of the theory of linear Petrovskiĭ-parabolic systems ([**3**], Chapter VIII) that the solution $\mathbf{v}^\epsilon$, $p^\epsilon \in W_2^{2,2}(Q_T)$. Applying this operation of successively increasing the smoothness of the solution $\mathbf{v}^\epsilon$, p^ϵ in the variable t k times and so exhausting the smoothness of the initial conditions $\mathbf{a}(x)$ and $P(x)$, the free term $\mathbf{f}(x, t)$, and the boundary $\partial\Omega$ stipulated in Theorem 3, we find that the solution $\mathbf{v}^\epsilon$, $p^\epsilon \in W_2^{2,k+1}(Q_T)$. After this, to establish the asymptotic estimates (43)–(47) for each $l = 1, \cdots, k$, it suffices to differentiate the system (39) l times in the variable t and apply (10) and the asymptotic estimates (33)–(37), using the fact that the derivatives $\partial^l\mathbf{v}^\epsilon/\partial t^l$ and $\partial^l p^\epsilon/\partial t^l$ constitute a solution of a problem of the type (39), (3), (4).

For solutions of the linearized problems (39), (3), (4) and (39), (3), (5) it is possible to obtain still another asymptotic estimate which, in general, does not hold for solutions of the nonlinear problems considered in the preceding section. Namely, let $G_\nu(x, y; t, \tau)$ be the Green's function of the first initial-boundary-value problem for the equation $\partial v/\partial t - \nu\Delta\mathbf{v} = \mathbf{f}(x, t)$. Then, considering $\mathbf{f} - \operatorname{grad} p^\epsilon$ as the free term in the system (49), we can write the solution of the system (49) satisfying the initial and boundary conditions (50) in the following form:

$$\begin{aligned}\mathbf{v}^\varepsilon(x, t) = &-\iint_{Q_t} G_\nu(x, y; t, \tau)\operatorname{grad} p^\varepsilon dy d\tau \\ &+ \iint_{Q_t} G_\nu \mathbf{f} dy d\tau + \int_\Omega G_\nu(x, y; t, 0)\mathbf{a}(y)\, dy.\end{aligned} \tag{51}$$

From the representation (51) for solutions of problems (39), (3), (4) and (39), (3), (5) and the asymptotic estimate (46) it follows that the following estimate holds as $\epsilon \to 0$ for the "pressure" $p^\epsilon(x,t)$ for each $t \in [0,T]$:

$$\int_\Omega \frac{\omega(\varepsilon)}{\varepsilon^{1/2}} \left[\iint_{Q_t} \frac{\partial G_\nu}{\partial x_i}(x,\ y;\ t,\ \tau) \frac{\partial p^\varepsilon}{\partial y_i}\, dy d\tau \right] \psi(x)\, dx \to 0, \quad \forall \psi(x) \in W_2^1(\Omega). \tag{52}$$

It is easily seen that if the initial data has smoothness of order $k = 1, 2, \cdots$, the asymptotic estimate (52) continues to hold also for the derivatives $\partial^l p^\epsilon / \partial t^l$, $l = 1, \cdots, k$.

We remark finally that by using the representation (25) for the "pressure" $p^\epsilon(x,t)$ we are able, in the general nonlinear case, to eliminate $p^\epsilon(x,t)$ from the system (39) and obtain for $\mathbf{v}^\epsilon(x,t)$ the first initial-boundary-value problem for the following nonlinear, nonstationary system of integro-differential equations with a large parameter $1/\epsilon$:

$$\begin{aligned}\frac{\partial \mathbf{v}^\varepsilon}{\partial t} - \nu\Delta \mathbf{v}^\varepsilon - \frac{1}{\varepsilon} \iint_{Q_t} \operatorname{grad}_x G(x,\ y;\ t,\ \tau) \operatorname{div} \mathbf{v}^\varepsilon dy d\tau \\ = \mathbf{f}(x,\ t) - \int_\Omega \operatorname{grad}_x G(x,\ y;\ t,\ 0) P(y)\, dy,\end{aligned} \tag{53}$$

$$\mathbf{v}^\varepsilon|_{t=0} = \mathbf{a}(x), \quad \mathbf{v}^\varepsilon|_{\partial Q_T} = 0. \tag{54}$$

§3. In the domain Ω we consider the following stationary nonlinear system with the small parameter $\epsilon > 0$:

$$\begin{aligned}&\nu\Delta \mathbf{v}^\varepsilon - v_k^\varepsilon \frac{\partial \mathbf{v}^\varepsilon}{\partial x_k} - \frac{1}{2} \mathbf{v}^\varepsilon \operatorname{div} \mathbf{v}^\varepsilon - \operatorname{grad} p^\varepsilon = \mathbf{f}(x), \\ &\varepsilon\Delta p^\varepsilon - \operatorname{div} \mathbf{v}^\varepsilon = 0, \quad \varepsilon > 0.\end{aligned} \tag{55}$$

For each $\epsilon > 0$ this system is elliptic in the sense of Petrovskiĭ, and for $\epsilon = 0$ it degenerates into the stationary nonlinear system of Navier-Stokes equations:

$$\nu\Delta \mathbf{v} - v_k \frac{\partial \mathbf{v}}{\partial x_k} - \operatorname{grad} p = \mathbf{f}(x), \quad \operatorname{div} \mathbf{v} = 0. \tag{56}$$

For the elliptic system (55) we consider two boundary value problems; namely, we seek a solution $\mathbf{v}^\epsilon(x)$, $p^\epsilon(x)$ satisfying one of the two boundary conditions

$$\mathbf{v}^\varepsilon|_{\partial\Omega} = 0, \quad p^\varepsilon|_{\partial\Omega} = 0, \tag{57}$$

$$\mathbf{v}^\varepsilon|_{\partial\Omega} = 0, \quad \left.\frac{\partial p^\varepsilon}{\partial n}\right|_{\partial\Omega} = 0. \tag{58}$$

We shall say that the functions $\mathbf{u}(x)$, $q(x)$ belong to the space $H^\epsilon(\Omega)$, $\epsilon \geq 0$, if for them the following norm is defined and finite:

$$|\mathbf{u},\ q|^2_{H^\varepsilon(\Omega)} \equiv \int_\Omega |\mathbf{u}_x|^2\, dx + \frac{\varepsilon}{\nu} \int_\Omega |q_x|^2\, dx < \infty. \tag{59}$$

The norm (59) is obviously generated by the scalar product

$$[\mathbf{u},\ q;\ \mathbf{v},\ p]_{H^{\varepsilon}(\Omega)} = \int_{\Omega} \mathbf{u}_x \mathbf{v}_x dx + \frac{\varepsilon}{\nu} \int_{\Omega} q_x p_x dx. \tag{60}$$

We further denote by $\overset{\circ}{H}{}^{\epsilon}(\Omega)$ the set of functions $\mathbf{u}$, q of $H^{\epsilon}(\Omega)$ satisfying the boundary conditions (57), and by $\underset{\circ}{H}{}^{\epsilon}(\Omega)$ the set of functions $\mathbf{u}(x)$, $q(x) \in \overset{\circ}{H}{}^{\epsilon}(\Omega)$ satisfying the boundary conditions (58).

As in [1], we call a *generalized solution* of problem (55), (57) or of problem (55), (58), respectively, in the space $H^{\epsilon}(\Omega)$ a solution $\mathbf{v}^{\epsilon}(x)$, $p^{\epsilon}(x)$ of these problems which belongs to the space $\overset{\circ}{H}{}^{\epsilon}(\Omega)$ or $\underset{\circ}{H}{}^{\epsilon}(\Omega)$ and satisfies the integral identities

$$\nu \int_{\Omega} \mathbf{v}^{\varepsilon}_x \Phi_x dx + \int_{\Omega} \left(v^{\varepsilon}_k \frac{\partial \mathbf{v}^{\varepsilon}}{\partial x_k} + \frac{1}{2} \mathbf{v}^{\varepsilon} \operatorname{div} \mathbf{v}^{\varepsilon} + \operatorname{grad} p^{\varepsilon} \right) \Phi dx = \int_{\Omega} \mathbf{f}(x) \Phi(x)\, dx, \tag{61}$$

$$\varepsilon \int_{\Omega} p^{\varepsilon}_x \psi_x dx + \int_{\Omega} \psi \operatorname{div} \mathbf{v}^{\varepsilon} dx = 0 \tag{62}$$

for any $\Phi(x) \in \overset{\circ}{W}{}_2^1(\Omega)$ and $\psi(x) \in \overset{\circ}{W}{}_2^1(\Omega)$ or $\psi(x) \in W_2^1(\Omega)$ respectively.

It is shown in [1] (cf. Theorem 13) that if the free term $\mathbf{f}(x)$ of the system (55) is such that the integral $\int_{\Omega} \mathbf{f}(x)\Phi(x)\,dx$ represents a bounded linear functional of $\Phi(x) \in \overset{\circ}{W}{}_2^1(\Omega)$, then problems (55), (57) and (55), (58) have at least one generalized solution $\mathbf{v}^{\epsilon}(x)$, $p^{\epsilon}(x)$ of the class $H^{\epsilon}(\Omega)$, and for this solution the following a priori estimate holds:

$$\| \mathbf{v}^{\varepsilon},\ p^{\varepsilon} \|_{H^{\varepsilon}(\Omega)} \leqslant \frac{\| \mathbf{f} \|}{\nu}. \tag{63}$$

It is also shown there that for $\epsilon \to 0$ the generalized solutions $\mathbf{v}^{\epsilon}$, p^{ϵ} of the space $H^{\epsilon}(\Omega)$ of problems (55), (57) and (55), (58) go over into generalized solutions $\mathbf{v}$, p of the class $H(\Omega)$ (cf. [2], Chapters 1, 3 and 5) of the first boundary value problem for the system of Navier-Stokes equations (56):

$$\nu \Delta \mathbf{v} - v_k \frac{\partial \mathbf{v}}{\partial x_k} - \operatorname{grad} p = \mathbf{f}(x), \quad \operatorname{div} \mathbf{v} = 0, \quad \mathbf{v}|_{\partial\Omega} = 0, \tag{64}$$

where $(\mathbf{v}^{\epsilon}, p^{\epsilon}) \to (\mathbf{v}, p)$ in the sense that in the limit the integral identity (62) is canceled, while the integral identity (61) on solenoidal $\Phi(x) \in \overset{\circ}{W}{}_2^1(\Omega)$, i.e. on $\Phi(x) \in H(\Omega)$, goes over into the integral identity

$$\nu \int_{\Omega} \mathbf{v}_x \Phi_x dx + \int_{\Omega} v_k \frac{\partial \mathbf{v}}{\partial x_k} \Phi dx = \int_{\Omega} \mathbf{f}(x) \Phi(x)\, dx, \tag{65}$$

defining a generalized solution of problem (64) in the space $H(\Omega)$ (cf. [2], Chapter 4). In the present section we continue the study of the asymptotic properties of the "pressure" $p^{\epsilon}(x)$ and $\operatorname{div} \mathbf{v}^{\epsilon}(x)$ on the basis of the estimate (63).

We shall assume that the boundary $\partial\Omega$ belongs to the class C^2. In the integral identity (62) we consider $\epsilon^{-1} \operatorname{div} \mathbf{v}^{\epsilon}$ as the free term which for each $\epsilon > 0$ belongs to $L_2(\Omega)$ by (63). Then from the theory of generalized solutions of linear elliptic equations (cf. [4], Chapter III) it follows that $p^{\epsilon}(x) \in \overset{\circ}{W}{}_2^2(\Omega)$ almost everywhere in Ω satisfies the equation

$$\Delta p^{\varepsilon} = \frac{1}{\varepsilon} \operatorname{div} \mathbf{v}^{\varepsilon}, \tag{66}$$

and also one of the boundary conditions (57) or (58). From this it follows immediately that for $p^\epsilon(x)$ one of the following two inequalities holds (the second basic inequality for the Laplace operator: cf. [2], Chapter 3, §§1 and 4):

$$\|p^\varepsilon\|^2_{W_2^2(\Omega)} \leqslant \frac{C_{12}}{\varepsilon^2}\|\operatorname{div} \mathbf{v}^\varepsilon\|^2_{L_2(\Omega)}, \tag{67}$$

if $p^\epsilon(x)$ satisfies the Dirichlet boundary condition (57) and

$$\|p^\varepsilon\|^2_{W_2^2(\Omega)} \leqslant C_{13}\left(\frac{1}{\varepsilon^2}\|\operatorname{div} \mathbf{v}^\varepsilon\|^2_{L_2(\Omega)} + \|p^\varepsilon\|^2_{W_2^1(\Omega)}\right), \tag{68}$$

if $p^\epsilon(x)$ satisfies the Neumann boundary condition (58), where the constants C_{12} and C_{13} depend only on the norm of the boundary $\partial\Omega$ in C^2.

Further, from the integral identity (62), which holds for any $\psi(x) \in W_2^1(\Omega)$, and (63) it follows easily that if $\omega(\epsilon)$, $\epsilon \geq 0$, is any positive function tending monotonically to zero for $\epsilon \to 0$, then as $\epsilon \to 0$

$$\frac{\omega(\varepsilon)}{\varepsilon^{1/2}}\int_\Omega \psi(x)\operatorname{div} \mathbf{v}^\varepsilon(x)\,dx \to 0 \tag{69}$$

for any $\psi(x) \in W_2^1(\Omega)$; after this, from the integral identity (62), on integrating by parts in the first term, we find that for $\epsilon \to 0$

$$\omega(\varepsilon)\,\varepsilon^{1/2}\int_\Omega \Delta p^\varepsilon(x)\,\psi(x)\,dx \to 0 \tag{70}$$

for all $\psi(x) \in \mathring{W}_2^1(\Omega)$.

From (63) and (67) we obtain the following estimate for the "pressure" $p^\epsilon(x)$ satisfying the Dirichlet boundary condition (57):

$$\|p^\varepsilon\|^2_{W_2^2(\Omega)} \leqslant \frac{C_{14}(C_{12}, \nu, \|\mathbf{f}\|)}{\varepsilon^2}, \tag{71}$$

and from this inequality we deduce via Sobolev's imbedding theorems an estimate for the Hölder norm of $p^\epsilon(x)$ with exponent 1/2 (we recall that $n = 2$ or 3):

$$|p^\varepsilon|_\Omega^{(1/2)} \leqslant \frac{C_{15}(C_{14})}{\varepsilon}. \tag{72}$$

Further, from (63) and (68) we obtain for the "pressure" $p^\epsilon(x)$ satisfying the Neumann boundary condition (58) the following estimate:

$$\|p^\varepsilon\|^2_{W_2^2(\Omega)} \leqslant \frac{C_{16}(C_{13}, \nu, \|\mathbf{f}\|)}{\varepsilon^2} + C_{17}(C_{13})\|p^\varepsilon\|^2_{L_2(\Omega)}, \tag{73}$$

and from this we obtain the following estimate for the Hölder norm of $p^\epsilon(x)$ with exponent 1/2:

$$|p^\varepsilon|_\Omega^{(1/2)} \leqslant C_{18}(C_{16}, C_{17})\left(\|p^\varepsilon\|^2_{L_2(\Omega)} + \frac{1}{\varepsilon^2}\right). \tag{74}$$

Let $N(x,y)$ be the Neumann function of the Neumann problem for the Laplace equation. Since for the Neumann problem

$$\Delta p^\varepsilon = \frac{1}{\varepsilon}\operatorname{div} \mathbf{v}^\varepsilon, \quad \left.\frac{\partial p^\varepsilon}{\partial n}\right|_{\partial\Omega} = 0, \tag{75}$$

in which $\epsilon^{-1}\operatorname{div} \mathbf{v}^\epsilon$ is considered as the free term, the necessary condition for solvability

$$\int_{\partial\Omega} \frac{\partial p^\varepsilon}{\partial n} dS = \frac{1}{\varepsilon} \int_\Omega \operatorname{div} \mathbf{v}^\varepsilon dx = \frac{1}{\varepsilon} \int_{\partial\Omega} v_n^\varepsilon dS = 0 \tag{76}$$

is satisfied, the pressure $p^\epsilon(x)$ of problem (75) can be represented in the following form:

$$p^\varepsilon(x) = \frac{1}{\varepsilon} \int_\Omega N(x, y) \operatorname{div} \mathbf{v}^\varepsilon(y)\, dy + C_{19}, \tag{77}$$

where the arbitrary constant C_{19}—a general solution of the homogeneous problem (75)—does not depend on ϵ. Since the function $N(x,y)$ is square summable over the domain Ω, the representation (77) implies the following inequality:

$$\|p^\varepsilon\|^2_{L_2(\Omega)} \leqslant \frac{2}{\varepsilon^2} \int_\Omega \int_\Omega |N(x, y)|^2 dx dy \|\operatorname{div} \mathbf{v}^\varepsilon\|^2_{L_2(\Omega)} + 2C^2_{19} \operatorname{meas}^2\Omega, \tag{78}$$

and it then follows from (70), (73), (74) and (78) that asymptotic estimates of the type of (71) and (72) also hold for the "pressure" $p^\epsilon(x)$ satisfying the Neumann boundary condition (58).

For brevity we further denote by $G(x,y)$ the Green's function of the Dirichlet problem or the Neumann function of the Neumann problem for the Laplace equation. Then the solution of equation (66), in which $\epsilon^{-1}\operatorname{div} \mathbf{v}^\epsilon(x)$ is considered the free term, which satisfies one of the boundary conditions (57) or (58) can be represented in the form

$$p^\varepsilon(x) = \frac{1}{\varepsilon} \int_\Omega G(x, y) \operatorname{div} \mathbf{v}^\varepsilon(y)\, dy + C_{19}, \tag{79}$$

where if $p^\epsilon(x)$ satisfies the Dirichlet boundary condition (57) then $C_{19} = 0$. Using the representation (79), we can eliminate the "pressure" $p^\epsilon(x)$ from the system (55) and obtain for the regularized velocity $\mathbf{v}^\epsilon(x)$ the first boundary value problem for the stationary, nonlinear system of integro-differential equations with a large parameter $1/\epsilon$

$$\nu\Delta\mathbf{v}^\varepsilon - v_k^\varepsilon \frac{\partial \mathbf{v}^\varepsilon}{\partial x_k} - \frac{1}{2} \mathbf{v}^\varepsilon \operatorname{div} \mathbf{v}^\varepsilon - \frac{1}{\varepsilon} \int_\Omega \operatorname{grad}_x G(x, y) \operatorname{div} \mathbf{v}^\varepsilon dy = \mathbf{f}(x), \quad \mathbf{v}^\varepsilon|_{\partial\Omega} = 0. \tag{80}$$

We set

$$G(\operatorname{div} \mathbf{v}^\varepsilon) \equiv \int_\Omega \operatorname{grad}_x G(x, y) \operatorname{div} \mathbf{v}^\varepsilon(y)\, dy. \tag{81}$$

It is obvious that the solution $\mathbf{v}^\epsilon(x)$ of problem (80) satisfies for any $\mathbf{\Phi}(x) \in \overset{\circ}{W}{}_2^1(\Omega)$ the integral identity

$$\nu \int_\Omega \mathbf{v}^\varepsilon_x \mathbf{\Phi}_x dx + \int_\Omega \left(v_k^\varepsilon \frac{\partial \mathbf{v}^\varepsilon}{\partial x_k} + \frac{1}{2} \mathbf{v}^\varepsilon \operatorname{div} \mathbf{v}^\varepsilon\right) \mathbf{\Phi} dx + \frac{1}{\varepsilon} \int_\Omega G(\operatorname{div} \mathbf{v}^\varepsilon) \mathbf{\Phi} dx = \int_\Omega \mathbf{f}(x) \mathbf{\Phi}(x)\, dx. \tag{82}$$

In this identity the integrals, except for the third integral on the left, by virtue of (63) and Sobolev's imbedding theorems are bounded uniformly for $\epsilon \geqq 0$ for any $\Phi(x) \in W_2^1(\Omega)$, and it then follows from (82) that for $\epsilon \to 0$

$$\frac{\omega(\varepsilon)}{\varepsilon} \int_\Omega \Phi(x)\, G(\operatorname{div} \mathbf{v}^\varepsilon)\, dx \to 0 \tag{83}$$

for any $\Phi(x) \in W_2^1(\Omega)$; since by (79)

$$\operatorname{grad} p^\varepsilon = \frac{1}{\varepsilon} G(\operatorname{div} \mathbf{v}^\varepsilon), \tag{84}$$

it also follows that

$$\omega(\varepsilon) \int_\Omega \operatorname{grad} p^\varepsilon \Phi(x)\, dx \to 0 \qquad \forall \Phi(x) \in W_2^1(\Omega), \tag{85}$$

where, as above, $\omega(\epsilon)$ is an arbitrary positive function tending monotonically to zero as $\epsilon \to 0$.

We summarize the results obtained on the asymptotic behavior as $\epsilon \to 0$ of $p^\epsilon(x)$ and $\operatorname{div} \mathbf{v}^\epsilon(x)$ in the following theorem.

Theorem 4. *Let $\mathbf{v}^\epsilon(x)$, $p^\epsilon(x)$ be a generalized solution of problem* (55), (57) *or of problem* (55), (58) *of the class $H^\epsilon(\Omega)$, and let $\mathbf{f}(x)$ be such that $\int_\Omega \mathbf{f}(x)\Phi(x)\,dx$ is a bounded linear functional on $\Phi(x) \in \overset{\circ}{W}{}_2^1(\Omega)$; suppose that the boundary $\partial\Omega$ belongs to the class C^2, and let $\omega(\epsilon)$, $\epsilon \geqq 0$, be an arbitrary positive function which tends monotonically to zero as $\epsilon \to 0$. Then as $\epsilon \to 0$ the following asymptotic estimates hold*:

$$\| p^\varepsilon \|_{W_2^1(\Omega)} \leqslant \frac{C_{20}}{\varepsilon} \tag{86}$$

for the problem (55), (57),

$$\omega(\varepsilon) \| p^\varepsilon \|^2_{L_2(\Omega)} + \| p^\varepsilon_x \|^2_{L_2(\Omega)} \leqslant \frac{C_{21}}{\varepsilon} \tag{87}$$

for the problem (55), (58), *and for both problems*

$$\| p^\varepsilon_{xx} \|^2_{L_2(\Omega)} + (| p^\varepsilon |_\Omega^{1/2})^2 \leqslant \frac{C_{22}}{\varepsilon^2}, \tag{88}$$

$$\omega(\varepsilon)\, \varepsilon^{1/2} \int_\Omega \Delta p^\varepsilon(x)\, \psi(x)\, dx \to 0 \qquad \forall \psi(x) \in \overset{\circ}{W}{}_2^1(\Omega), \tag{89}$$

$$\omega(\varepsilon) \int_\Omega \operatorname{grad} p^\varepsilon(x)\, \Phi(x)\, dx \to 0 \qquad \forall \Phi(x) \in W_2^1(\Omega), \tag{90}$$

$$\frac{\omega(\varepsilon)}{\varepsilon^{1/2}} \int_\Omega \operatorname{div} \mathbf{v}^\varepsilon(x)\, \psi(x)\, dx \to 0 \qquad \forall \psi(x) \in W_2^1(\Omega), \tag{91}$$

$$\frac{\omega(\varepsilon)}{\varepsilon} \int_\Omega G(\operatorname{div} \mathbf{v}^\varepsilon)\, \Phi(x)\, dx \to 0 \qquad \forall \Phi(x) \in W_2^1(\Omega), \tag{92}$$

where the operator G is defined by (81), *$G(x,y)$ is the Green's function of the Dirichlet problem or respectively the Neumann function of the Neumann problem for the Laplace equation, and the constants $C_{20}-C_{22}$ depend only on $\|\mathbf{f}\|$, ν, the dimensions of Ω, and the norm of the boundary $\partial\Omega$ in C^2.*

Theorem 4 shows that as in the nonstationary case the asymptotic expression for the "pressure" $p^\epsilon(x)$ that we have found coincides with the standard asymptotic behavior for a single elliptic equation (cf. [5]).

In the domain Ω we now consider the following stationary linear system with a small parameter ϵ:

(93) $$\nu\Delta\mathbf{v}^{\varepsilon}-\operatorname{grad}p^{\varepsilon}=\mathbf{f}(x),\quad \varepsilon\Delta p^{\varepsilon}-\operatorname{div}\mathbf{v}^{\varepsilon}=0,\quad \varepsilon>0.$$

This system, just as the system (55), is for each $\epsilon>0$ an elliptic system in the sense of Petrovskiĭ, and for $\epsilon=0$ it degenerates into the stationary Stokes linearization of the system of Navier-Stokes equations:

(94) $$\nu\Delta\mathbf{v}-\operatorname{grad}p=\mathbf{f}(x),\quad \operatorname{div}\mathbf{v}=0.$$

For the system (93) it is possible to consider the same two boundary value problems as for the general nonlinear system (55), and for the solutions $\mathbf{v}^{\epsilon}(x)$, $p^{\epsilon}(x)$ of these problems—problems (93), (57) and (93), (58)—all the assertions of Theorem 4 hold. Moreover, for the solutions of the linearized problems (93), (57) and (93), (58) it is possible to obtain two more asymptotic estimates which in general do not hold for solutions of the nonlinear problems (55), (57) and (55), (58). Namely, let $G(x,y)$ be the Green's function of the Dirichlet problem for the Laplace equation. Then, considering the right side to be the free term in the system $\nu\Delta\mathbf{v}^{\epsilon}=\mathbf{f}+\operatorname{grad}p$, we can write the solution of this system satisfying the boundary condition $\mathbf{v}^{\epsilon}|_{\partial\Omega}=0$ in the following form:

(95) $$\mathbf{v}^{\varepsilon}(x)=\frac{1}{\nu}\int_{\Omega}G(x,y)\operatorname{grad}p^{\varepsilon}dy+\frac{1}{\nu}\int_{\Omega}G(x,y)\mathbf{f}(y)\,dy.$$

From the representation (95) for solutions of the problems (93), (57) and (93), (58) and the asymptotic estimates (91) and (92) it then follows that for the "pressure" $p^{\epsilon}(x)$ the following estimates are valid as $\epsilon\to0$:

(96) $$\frac{\omega(\varepsilon)}{\varepsilon^{1/2}}\int_{\Omega}\operatorname{div}\int_{\Omega}G(x,y)\operatorname{grad}p^{\varepsilon}dy\,\psi(x)\,dx\to0,\qquad \forall\psi(x)\in W_2^1(\Omega),$$

(97) $$\frac{\omega(\varepsilon)}{\varepsilon}\int_{\Omega}\left[\operatorname{grad}_x\int_{\Omega}G(x,y)\operatorname{div}_y\int_{\Omega}G(y,z)\operatorname{grad}p^{\varepsilon}(z)\,dz\,dy\right]\Phi(x)\,dx\to0,$$
$$\forall\Phi(x)\in W_2^1(\Omega).$$

We remark finally that using the representation (79) for the regularized pressure $p^{\epsilon}(x)$ we can eliminate it from the system (93) and obtain for the regularized velocity $\mathbf{v}^{\epsilon}(x)$ the first boundary value problem for the stationary linear system of integro-differential equations with a large parameter $1/\epsilon$:

(98) $$\nu\Delta\mathbf{v}^{\varepsilon}=\frac{1}{\varepsilon}\int_{\Omega}\operatorname{grad}_x G(x,y)\operatorname{div}\mathbf{v}^{\varepsilon}_{\cdot}(y)\,dy=\mathbf{f}(x),$$
$$\mathbf{v}^{\varepsilon}|_{\partial\Omega}=0.$$

Bibliography

[1] A. P. Oskolkov, *A certain quasilinear parabolic system with small parameter that approximates a system of Navier-Stokes equations*, Zap. Naučn. Sem. Leningrad. Otdel. Mat. Inst. Steklov. (LOMI) **21** (1971), 79-103 = J. Soviet Math. **1** (1973), 452-470. MR **45** # 7314.

[2] O. A. Ladyženskaja, *Mathematical problems in the dynamics of a viscous incompressible fluid,* 2nd rev. aug. ed., "Nauka", Moscow, 1970; English transl. of 1st ed., *The mathematical theory of viscous incompressible flow,* Gordon and Breach, New York, 1963; rev. 1969. MR **40** # 7610; **27** # 5034b; **42** # 6442.

[3] O. A. Ladyženskaja, V. A. Solonnikov and N. N. Ural′ceva, *Linear and quasilinear equations of parabolic type,* "Nauka", Moscow, 1967; English transl., Transl. Math. Monographs, vol. 23, Amer. Math. Soc., Providence, R. I., 1968. MR **39** # 3159a,b.

[4] O. A. Ladyženskaja and N. N. Ural′ceva, *Linear and quasilinear equations of elliptic type,* "Nauka", Moscow, 1964; English transl., Academic Press, New York, 1968. MR **35** # 1955; **39** # 5941.

[5] A. L. Treskunov, *Exact estimates in L_p for a certain class of second order degenerate equations of elliptic type,* Zap. Naučn. Sem. Leningrad. Otdel. Mat. Inst. Steklov. (LOMI) **5** (1967), 232-249 = Sem. Math. V. A. Steklov Math. Inst. Leningrad **5** (1967), 88-94. MR **37** # 1757.

Translated by
J. SCHULENBERGER

Trudy Mat. Inst. Steklov.
125 (1973)

Proc. Steklov Inst. Math.
125 (1973)

ON SOME CONVERGENT DIFFERENCE SCHEMES FOR THE NAVIER-STOKES EQUATIONS

A. P. OSKOLKOV

Abstract. We set forth criteria of convergence and stability of explicit finite-difference schemes approximating a quasilinear parabolic system with a small parameter degenerating in the limit to a nonstationary system of Navier-Stokes equations, as proposed in our earlier work.
Bibliography: 5 items.

In a note of the author [1] (see also [2]) two explicit finite-difference schemes were constructed, approximating a proposed quasilinear parabolic system with small parameter ϵ, which for $\epsilon=0$ degenerates into a nonstationary nonlinear system of Navier-Stokes equations; conditions were established under which solutions of these finite-difference problems converge to the exact (sufficiently smooth) solution of the Navier-Stokes equations. The conditions we obtained up to order ϵ are better than those under which O. A. Ladyženskaja [3] and R. Temam [4,5] proved convergence of solutions of finite-difference schemes approximating the systems with small parameter ϵ suggested by Temam [4,5], which for $\epsilon=0$ degenerate into a system of Navier-Stokes equations.

Let Ω be a bounded domain in two- or three-dimensional Euclidean space with boundary $\partial\Omega$ of class $C^{(2)}$, $Q_T=\Omega\times[0,T]$, and $\partial\Omega_T=\partial\Omega\times[0,T]$ the lateral surface of the cylinder Q_T. Consider the following quasilinear system in Q_T with small parameter $\epsilon>0$:

$$\begin{gathered}\frac{\partial \mathbf{v}^\varepsilon}{\partial t}-\nu\Delta\mathbf{v}^\varepsilon+v_k^\varepsilon\frac{\partial \mathbf{v}^\varepsilon}{\partial x_k}+\frac{1}{2}\mathbf{v}^\varepsilon\operatorname{div}\mathbf{v}^\varepsilon+\operatorname{grad}p^\varepsilon=\mathbf{f}(x,t),\\ \varepsilon\frac{\partial p^\varepsilon}{\partial t}-\varepsilon^\alpha\Delta p^\varepsilon+\operatorname{div}\mathbf{v}^\varepsilon=0,\quad \varepsilon>0,\end{gathered}\tag{1}$$

in which $\alpha\geqq 1$. For every $\epsilon>0$ this system is parabolic in the sense of Petrovskiĭ, and for $\epsilon=0$ it degenerates into the nonstationary nonlinear system of Navier-Stokes equations

$$\frac{\partial \mathbf{v}}{\partial t}-\nu\Delta\mathbf{v}+v_k\frac{\partial \mathbf{v}}{\partial x_k}+\operatorname{grad}p=\mathbf{f}(x,t),\quad \operatorname{div}\mathbf{v}=0.\tag{2}$$

For the system (1) let us consider the first initial-boundary problem, namely, to seek a solution $\mathbf{v}^\epsilon$, p^ϵ of (1) in the cylinder Q_T that satisfies the following initial and boundary conditions:

$$\mathbf{v}^\varepsilon|_{t=0}=\mathbf{a}(x),\ \operatorname{div}\mathbf{a}(x)=0,\ p^\varepsilon|_{t=0}=P(x),\ x\in\Omega,\tag{3}$$

$$\mathbf{v}^\varepsilon|_{\partial Q_T}=0,\ p^\varepsilon|_{\partial Q_T}=0.\tag{4}$$

The first boundary-value problem (1), (3), (4) for the system and the asymptotic behavior of its solution $\mathbf{v}^\epsilon$, p^ϵ when $\alpha=1$ have been studied in detail in the author's articles [1,2]. For arbitrary $\alpha\geqq 1$ theorems on the solvability of the problem (1), (3), (4) and on the relation of its solution

AMS (MOS) subject classifications (1970). Primary 65N10, 35Q10.

$\mathbf{v}^\epsilon$, p^ϵ as $\epsilon \to 0$ to that of the first initial-boundary problem for the system of Navier-Stokes equations

$$\frac{\partial \mathbf{v}}{\partial t} - \nu \Delta \mathbf{v} + v_k \frac{\partial \mathbf{v}}{\partial x_k} + \operatorname{grad} p = \mathbf{f}(x, t), \quad \operatorname{div} \mathbf{v} = 0, \tag{5}$$
$$\mathbf{v}\big|_{t=0} = \mathbf{a}(x), \quad \mathbf{v}\big|_{\partial Q_T} = 0$$

are completely analogous to the theorems in [1, 2]. For example, in the two-dimensional case, when a strong generalized solution (see [1, 2]) of problem (1), (3), (4) exists without any limitations of smallness on the data of the problem the following theorem is valid:

THEOREM 1. *Let Ω be an arbitrary plane region with boundary $\partial\Omega$ of class $C^{(2)}$, $\mathbf{a}(x) \in W^2_{2,0}(\Omega)$, $P(x) \in W^2_{2,0}(\Omega)$, $\mathbf{f}(x,t) \in L_2(Q_T)$ and $\mathbf{f}_t \in L_{2,1}(Q_T)$. Then problem* (1), (3), (4) *has a strong generalized solution $\mathbf{v}^\epsilon$, p^ϵ for every $\epsilon > 0$, that is moreover unique, satisfying the inequality*

$$\sup_{0\leq t\leq T} [\|\mathbf{v}^\varepsilon\|^2_{L_2(\Omega)} + \|\mathbf{v}^\varepsilon_x\|^2_{L_2(\Omega)} + \|\mathbf{v}^\varepsilon_t\|^2_{L_2(\Omega)}] + \varepsilon \sup_{0\leq t\leq T} [\|p^\varepsilon\|^2_{L_2(\Omega)} + \|p^\varepsilon_t\|^2_{L_2(\Omega)}] \tag{6}$$
$$+ \|\mathbf{v}^\varepsilon_{xt}\|^2_{L_2(Q_T)} + \varepsilon^\alpha \Big[\sup_{0\leq t\leq T} \|p^\varepsilon_x\|^2_{L_2(\Omega)} + \|p^\varepsilon_{xt}\|^2_{L_2(Q_T)}\Big] \leq C_1,$$

and as $\epsilon \to 0$ the following asymptotic estimates hold:

$$\varepsilon \sup_{0\leq t\leq T} \{\|p^\varepsilon_{xx}\|^2_{L_2(\Omega)} + (|p^\varepsilon|^{(1/2)}_\Omega)^2\} \leq \frac{C_2}{\varepsilon^{2\alpha}}, \tag{7}$$

$$\frac{\omega(\varepsilon)}{\varepsilon^{1/2}} \int_\Omega \operatorname{div} \mathbf{v}^\varepsilon(x, t)\, \psi(x)\, dx \to 0, \quad 0 \leq t \leq T, \quad \forall \psi(x) \in W^1_2(\Omega), \tag{8}$$

in which the constants C_1 and C_2 depend only on the original data of the problem, and $\omega(\epsilon)$, $\epsilon \geq 0$, is an arbitrary positive monotonically increasing function as $\epsilon \to 0$.

We shall partition the space of x and t into elementary cells by planes $x_i = k_i h$, $h > 0$, $k_i = 0, \pm 1, \pm 2, \cdots$, and $t_l = l\Delta t$, $\Delta t = T/N$, $l = 0, 1, \cdots, N$. Denote by $\Omega^l \equiv \Omega$ the section of the cylinder Q_T by the plane $t_l = l\Delta t$, and by $\partial\Omega^l \equiv \partial\Omega$ the boundary of Ω^l. Moreover, let Ω^l_h be the set of all lattice points in Ω^l, $\partial\Omega^l_h$ the boundary of Ω^l_h, $\overline{\Omega}^l_h = \Omega^l_h \cup \partial\Omega^l_h$, and $\partial\Omega^l_{ih}$ $(i = 1, 2, 3)$ the set of all points of $\partial\Omega^l_h$ which go into some point Ω^l_h under a translation by h in the direction of the x_i-axis. Let us put

$$\overset{\pm i}{u}(x, t) = u(x \pm h\mathbf{e}^i, t), \quad u_{x_i}(x, t) = \frac{1}{h}[\overset{+i}{u}(x, t) - u(x, t)],$$
$$u_{\bar{x}_i} = \frac{1}{h}[u(x, t) - \overset{-i}{u}(x, t)],$$

where $\mathbf{e}^i$ $(i = 1, 2, 3)$ is the unit vector along the x_i-axis. In the same way we define displacements $\overset{\pm 0}{u}(x, t)$ and the difference quotients $u_t(x, t)$ and $u_{\bar{t}}(x, t)$ with respect to the variable t.

Furthermore, let u^l_h be a function defined only on the lattice points in the layer t_l, $l = 0, 1, \cdots, N$; for $\mathbf{u}^l_h = (u^l_{ih})$ we introduce the following expression:

$$(\mathbf{u}^l_h, \mathbf{v}^l_h) = h^3 \sum_{\Omega^l_h} u^l_{ih} v^l_{ih}, \quad \|\mathbf{u}^l_h\|^2 = (\mathbf{u}^l_h, \mathbf{u}^l_h), \quad \|\mathbf{u}^l_{hx}\|^2 = h^3 \sum_{i=1}^3 \sum_{\Omega^l_h \cup \partial\Omega^l_{ih}} (u^l_{hx_i})^2. \tag{9}$$

Following Ladyženskaja ([3], Chapter VI), we shall write down two explicit finite-difference schemes approximating problem (1), (3), (4) — one asymmetric and one symmetric. The asymmetric scheme has the following form (for brevity we drop the index h on the functions $v_{ih}^{\epsilon,l}$ and $p_{ih}^{\epsilon,l}$):

$$(10)\qquad \begin{gathered} v_{it}^{\varepsilon,l} - \nu v_{ix_k\bar{x}_k}^{\varepsilon,l} + \frac{1}{2}\overset{+k}{v_k^{\varepsilon,l}} v_{ix_k}^{\varepsilon,l} + \frac{1}{2} v_k^{\varepsilon,l} v_{i\bar{x}_k}^{\varepsilon,l} + \frac{1}{2} v_i^{\varepsilon,l} v_{kx_k}^{\varepsilon,l} + p_{x_i}^{\varepsilon,l} = f_i^l, \\ i=1,\ 2,\ 3, \\ \varepsilon p_t^{\varepsilon,l} - \varepsilon^\alpha p_{x_k\bar{x}_k}^{\varepsilon,l} + v_{k\bar{x}_k}^{\varepsilon,l} = 0,\ l=1,\ 2,\ \ldots,\ N, \end{gathered}$$

and the symmetric scheme has the form

$$(11)\qquad \begin{gathered} v_{it}^{\varepsilon,l} - \nu v_{ix_k\bar{x}_k}^{\varepsilon,lk} + \frac{1}{4}(\overset{-k}{v_k^{\varepsilon,l}} + v_k^{\varepsilon,l})\, v_{i\bar{x}_k}^{\varepsilon,l} + \frac{1}{4}(v_k^{\varepsilon,l} + \overset{+k}{v_k^{\varepsilon,l}})\, v_{ix_k}^{\varepsilon,l} \\ + \frac{1}{4}(v_{k\bar{x}_k}^{\varepsilon,l} + v_{kx_k}^{\varepsilon,l})\, v_i^{\varepsilon,l} + \frac{1}{2}(p_{x_i}^{\varepsilon,l} + p_{\bar{x}_i}^{\varepsilon,l}) = f_i^l, \quad i=1,\ 2,\ 3, \end{gathered}$$

$$(12)\qquad \varepsilon p_t^{\varepsilon,l} - \varepsilon^\alpha p_{x_k\bar{x}_k}^{\varepsilon,l} + \frac{1}{2}(v_{kx_k}^{\varepsilon,l} + v_{k\bar{x}_k}^{\varepsilon,l}) = 0, \quad l=1,\ 2,\ldots,\ N.$$

To both difference schemes we annex the same initial and boundary conditions, which approximate the original conditions (3), (4):

$$(13)\qquad \mathbf{v}^{\varepsilon,0} = \mathbf{a}_h,\quad p^{\varepsilon,0} = P_h,$$

$$(14)\qquad \mathbf{v}^{\varepsilon,l}\big|_{\partial\Omega_h^l} = 0,\quad p^{\varepsilon,l}\big|_{\partial\Omega_h^l} = 0,\quad l=1,\ 2,\ \ldots,\ N.$$

It is easy to see that equations (10), (13), (14) and (11) — (14) uniquely determine $\mathbf{v}_h^{\epsilon,l}$ and $p_h^{\epsilon,l}$ for all $l = 0, 1, \cdots, N$. We shall show that the following theorem obtains:

Theorem 2. *Let Ω be a three-dimensional region with boundary $\partial\Omega$ of class $C^{(2)}$, and suppose that the original data of the initial-boundary problem* (5) *for the system of Navier-Stokes equations are such that problem* (5) *has a classical solution $\mathbf{v}(x,t)$, $p(x,t)$.*[1] *Moreover, let $\mathbf{v}_h^\epsilon$, p_h^ϵ be a solution of the difference problem* (10), (13), (14) *or problem* (11) — (14) *with $\alpha = 3/2$, and when ϵ, Δt, $h \to 0$, suppose the following conditions are satisfied*:

$$(15)\qquad \frac{h^2}{\varepsilon} \to 0,$$

$$(16)\qquad \frac{1}{2} - 12\nu\frac{\Delta t}{h^2} - \frac{2\Delta t}{\varepsilon^{3/2}} \equiv \delta_1 > 0,$$

$$(17)\qquad \nu - \frac{2\Delta t}{\varepsilon} - 2\sqrt{18}\left(\frac{4}{3}\right)^{3/4}\left[\|\mathbf{a}_h\| + 2\Delta t\sum_{k=0}^{N-1}\|\mathbf{f}_h^k\|\right]^2\frac{\Delta t}{h^3} \equiv \delta_2 > 0.$$

Then, as ϵ, Δt, $h \to 0$, the solutions $\mathbf{v}_h^\epsilon$, p_h^ϵ of the finite-difference problems (10), (13), (14) *and* (11) — (14) *with $\alpha = 3/2$ converge to the exact solution $\mathbf{v}$, p of problem* (5) *in the energy metric associated with system* (1).

Remark. For the assertions of Theorem 2 to be valid in the two-dimensional case, we must take $\alpha = 1$ in the systems (10) and (11), (12); replace $\epsilon^{3/2}$ by

[1] In this connection see [3], Chapter VI, §5.

ϵ in (16); replace h^3 by h^2 in (17); and replace the constant $2\sqrt{18}(4/3)^{3/4}$ by $4\sqrt{2}$ in (17).

The proof of Theorem 2 is carried out according to methods developed by Ladyženskaja ([**3**], Chapter VI) for analyzing the convergence of implicit and explicit finite-difference schemes, as well as variable-direction schemes for Navier-Stokes systems with small parameter and their approximations advanced by R. Temam [**4, 5**]. In particular, Ladyženskaja studied convergence of explicit difference schemes for the first of the two systems Temam proposed for approximating the Navier-Stokes system

$$\text{(18)}\qquad \begin{gathered} \frac{\partial \mathbf{v}^\varepsilon}{\partial t} - \nu\Delta\mathbf{v}^\varepsilon + v_k^\varepsilon \frac{\partial \mathbf{v}^\varepsilon}{\partial x_k} + \frac{1}{2}\mathbf{v}^\varepsilon \operatorname{div}\mathbf{v}^\varepsilon + \operatorname{grad} p^\varepsilon = \mathbf{f}(x, t), \\ \varepsilon p^\varepsilon + \operatorname{div}\mathbf{v}^\varepsilon = 0, \quad \varepsilon > 0. \end{gathered}$$

Theorem 2 shows that solutions of the difference problems (10), (13), (14) and (11) – (14) constructed according to (1), owing to the presence of the term $\epsilon^\alpha p^\epsilon_{x_i\bar{x}_i}$, converge to the exact (classical) solution of problem (5) as ϵ, Δt, $h \to 0$ for a better ratio between the parameter ϵ and the increments h and Δt of the independent variables than the ones occurring in the difference problems constructed according to (18) (compare condition (96) of Chapter VI in [**3**] and our condition (16)). We emphasize once more that not only the power of ϵ in (16) but also the exponent α in (10) – (12) must be assigned differently in the three- and two-dimensional cases; namely, $\alpha = 3/2$ for $n = 3$ and $\alpha = 1$ for $n = 2$.

The proof of Theorem 2 comprises two stages: first we obtain the basic a priori estimate (33) for the solutions of the difference problems (10), (13), (14) and (11) – (14), which represents the difference analogue of the energy inequality for problem (1), (3), (4) (see inequalities (11) and (12) of [**1**]), and then we study convergence of $\mathbf{v}_h^\epsilon$, p_h^ϵ to the classical solution of problem (5) in the energy metric (33). Since the two results for problems (10), (13), (14) and (11) – (14) are both proved in the same way, we will deal with problem (10), (13), (14) for definiteness.

That being the case, we multiply the ith equation of system (10) by $2\Delta t h^3 v_i^{\epsilon, l+1}$ and the last equation by $2\Delta t h^3 p^{\epsilon, l+1}$, afterwards summing over all points of the grid Ω_h^l. Then we get these equations:

$$\text{(19)}\qquad \begin{aligned} \|\mathbf{v}^{\varepsilon, l+1}\|^2 - \|\mathbf{v}^{\varepsilon, l}\|^2 + (\Delta t)^2\|\mathbf{v}_t^{\varepsilon, l}\|^2 + 2\nu\Delta t\|\mathbf{v}_x^{\varepsilon, l}\|^2 = 2\Delta t\,(\mathbf{f}^l, \mathbf{v}^{\varepsilon, l+1}) \\ - 2\nu(\Delta t)^2(\mathbf{v}_x^{\varepsilon, l}, \mathbf{v}_{xt}^{\varepsilon, l}) - 2\Delta t h^3 \sum_{\Omega_h^l} p_{x_i}^{\varepsilon, l} v_i^{\varepsilon, l+1} + j^l, \end{aligned}$$

$$\text{(20)}\qquad \begin{aligned} \varepsilon\|p^{\varepsilon, l+1}\|^2 - \varepsilon\|p^{\varepsilon, l}\|^2 + \varepsilon(\Delta t)^2\|p_t^{\varepsilon, l}\|^2 + 2\varepsilon^{3/2}\Delta t\|p_x^{\varepsilon, l}\|^2 \\ + 2\varepsilon^{3/2}(\Delta t)^2(p_x^{\varepsilon, l}, p_{xt}^{\varepsilon, l}) + 2\Delta t h^3 \sum_{\Omega_h^l} p^{\varepsilon, l} v_{k\bar{x}_k}^{\varepsilon, l} = 0, \end{aligned}$$

where

$$j^l = -\Delta t\left(\overset{+k}{v_k^{\varepsilon, l}} v_{ix_k}^{\varepsilon, l} + v_k^{\varepsilon, l} v_{i\bar{x}_k}^{\varepsilon, l} + v_{k\bar{x}_k}^{\varepsilon, l} v_i^{\varepsilon, l},\ v_i^{\varepsilon, l+1}\right).$$

Using the formula for summation by parts and applying simple transformations, we obtain

$$-2\Delta t h^3 \sum_{\Omega_h^l} p^{\varepsilon,l}_{x_i} v^{\varepsilon,l+1}_i = 2\Delta t h^3 \sum_{\Omega_h^l} p^{\varepsilon,l} v^{\varepsilon,l+1}_{i\bar{x}_i} = 2\Delta t\,(p^{\varepsilon,l}, v^{\varepsilon,l+1}_{i\bar{x}_i}),$$

$$\begin{aligned}(21)\qquad & 2\Delta t h^3 \sum_{\Omega_h^l} p^{\varepsilon,l+1} v^{\varepsilon,l}_{k\bar{x}_k} = 2\Delta t\left(p^{\varepsilon,l+1}, v^{\varepsilon,l}_{k\bar{x}_k}\right) = 2\Delta t\left(p^{\varepsilon,l+1} - p^{\varepsilon,l}, v^{\varepsilon,l}_{k\bar{x}_k}\right)\\ &+ 2\Delta t\left(p^{\varepsilon,l}, v^{\varepsilon,l}_{k\bar{x}_k}\right) = 2(\Delta t)^2\left(p^{\varepsilon,l}_t, v^{\varepsilon,l}_{k\bar{x}_k}\right) + 2\Delta t\left(p^{\varepsilon,l}, v^{\varepsilon,l}_{k\bar{x}_k} - v^{\varepsilon,l+1}_{k\bar{x}_k}\right)\\ &+ 2\Delta t\left(p^{\varepsilon,l}, v^{\varepsilon,l+1}_{k\bar{x}_k}\right) = 2(\Delta t)^2\left(p^{\varepsilon,l}_t, v^{\varepsilon,l}_{k\bar{x}_k}\right) - 2(\Delta t)^2\left(p^{\varepsilon,l}, v^{\varepsilon,l}_{k\bar{x}_k t}\right)\\ &\qquad + 2\Delta t\left(p^{\varepsilon,l}, v^{\varepsilon,l+1}_{k x_k}\right),\end{aligned}$$

$$(21')\qquad \Delta t\left(\overset{+k}{v^{\varepsilon,l}_k} v^{\varepsilon,l}_{ix_k} + v^{\varepsilon,l}_k v^{\varepsilon,l}_{i\bar{x}_k} + v^{\varepsilon,l}_{kx_k} v^{\varepsilon,l}_i, v^{\varepsilon,l}_i\right) = 0$$

and therefore

$$(22)\qquad j^l = (\Delta t)^2\left(\overset{+k}{v^{\varepsilon,l}_k} v^{\varepsilon,l}_{ix_k} + v^{\varepsilon,l}_k v^{\varepsilon,l}_{i\bar{x}_k} + v^{\varepsilon,l}_{k\bar{x}_k} v^{\varepsilon,l}_i, v^{\varepsilon,l}_{it}\right).$$

Furthermore, from (20) and (21) we obtain

$$\begin{aligned}(23)\qquad & 2\Delta t\left(p^{\varepsilon,l}, v^{\varepsilon,l+1}_{k\bar{x}_k}\right) = 2\Delta t h^3 \sum_{\Omega_h^l} p^{\varepsilon,l+1} v^{\varepsilon,l}_{k\bar{x}_k} + 2(\Delta t)^2\left(p^{\varepsilon,l}, v^{\varepsilon,l}_{k\bar{x}_k t}\right)\\ &- 2(\Delta t)^2\left(p^{\varepsilon,l}_t, v^{\varepsilon,l}_{k\bar{x}_k}\right) = -\varepsilon\|p^{\varepsilon,l+1}\|^2 + \varepsilon\|p^{\varepsilon,l}\|^2 - \varepsilon(\Delta t)^2\|p^{\varepsilon,l}_t\|^2\\ &- 2\varepsilon^{3/2}\Delta t\|p^{\varepsilon,l}_x\|^2 - 2\varepsilon^{3/2}(\Delta t)^2(p^{\varepsilon,l}_x, p^{\varepsilon,l}_{xt}) + 2(\Delta t)^2\left(p^{\varepsilon,l}, v^{\varepsilon,l}_{k\bar{x}_k t}\right)\\ &\qquad - 2(\Delta t)^2\left(p^{\varepsilon,l}_t, v^{\varepsilon,l}_{k\bar{x}_k}\right),\end{aligned}$$

and after this, from (19), (22) and (23), the following equation:

$$\begin{aligned}(24)\qquad & \|\mathbf{v}^{\varepsilon,l+1}\|^2 + \varepsilon\|p^{\varepsilon,l+1}\|^2 + (\Delta t)^2\|\mathbf{v}^{\varepsilon,l}_t\|^2 + \varepsilon(\Delta t)^2\|p^{\varepsilon,l}_t\|^2 + 2\nu\Delta t\|\mathbf{v}^{\varepsilon,l}_x\|^2\\ &+ 2\varepsilon^{3/2}\Delta t\|p^{\varepsilon,l}_x\|^2 = \|\mathbf{v}^{\varepsilon,l}\|^2 + \varepsilon\|p^{\varepsilon,l}\|^2 + 2\Delta t(\mathbf{f}^l, \mathbf{v}^{\varepsilon,l+1}) - 2\nu(\Delta t)^2(\mathbf{v}^{\varepsilon,l}_x, \mathbf{v}^{\varepsilon,l}_{xt})\\ &- 2\varepsilon^{3/2}(\Delta t)^2(p^{\varepsilon,l}_x, p^{\varepsilon,l}_{xt}) + 2(\Delta t)^2\left(p^{\varepsilon,l}, v^{\varepsilon,l}_{k\bar{x}_k t}\right) - 2(\Delta t)^2\left(p^{\varepsilon,l}_t, v^{\varepsilon,l}_{k\bar{x}_k}\right)\\ &+ (\Delta t)^2\left(\overset{+k}{v^{\varepsilon,l}_k} v^{\varepsilon,l}_{ix_k} + v^{\varepsilon,l}_k v^{\varepsilon,l}_{i\bar{x}_k} + v^{\varepsilon,l}_{k\bar{x}_k} v^{\varepsilon,l}_i, v^{\varepsilon,l}_{it}\right),\quad l = 0, 1, \ldots, N-1.\end{aligned}$$

We shall estimate the terms on the right-hand side of (24) with the aid of the Hölder and Cauchy inequalities (cf. [3], Chapter VI, §9]):

$$(25)\qquad 2\Delta t\,|(\mathbf{f}^l, \mathbf{v}^{\varepsilon,l+1})| \leqslant 2\Delta t\,\|\mathbf{f}^l\|\,\|\mathbf{v}^{\varepsilon,l+1}\|,$$

$$(26)\qquad 2\nu(\Delta t)^2|(\mathbf{v}^{\varepsilon,l}_x, \mathbf{v}^{\varepsilon,l}_{xt})| \leqslant \nu\Delta t\|\mathbf{v}^{\varepsilon,l}_x\|^2 + 12\nu\frac{(\Delta t)^3}{h^2}\|\mathbf{v}^{\varepsilon,l}_t\|^2,$$

$$(27)\qquad 2\varepsilon^{3/2}(\Delta t)^2|(p^{\varepsilon,l}_x, p^{\varepsilon,l}_{xt})| \leqslant \frac{\varepsilon^{3/2}}{2}\Delta t\|p^{\varepsilon,l}_x\|^2 + 24\varepsilon^{3/2}\frac{(\Delta t)^3}{h^2}\|p^{\varepsilon,l}_t\|^2,$$

$$\begin{aligned}(28)\qquad & 2(\Delta t)^2\left|\left(p^{\varepsilon,l}, v^{\varepsilon,l}_{k\bar{x}_k t}\right)\right| = 2(\Delta t)^2|(p^{\varepsilon,l}_x, \mathbf{v}^{\varepsilon,l}_t)| \leqslant\\ &\qquad \leqslant \frac{\varepsilon^{3/2}}{2}\Delta t\|p^{\varepsilon,l}_x\|^2 + 2\frac{(\Delta t)^3}{\varepsilon^{3/2}}\|\mathbf{v}^{\varepsilon,l}_t\|^2,\end{aligned}$$

$$(29)\qquad 2(\Delta t)^2|(p^{\varepsilon,l}_t, \mathbf{v}^{\varepsilon,l}_x)| \leqslant \frac{\varepsilon}{2}(\Delta t)^2\|p^{\varepsilon,l}_t\|^2 + 2\frac{(\Delta t)^2}{\varepsilon}\|\mathbf{v}^{\varepsilon,l}_x\|^2,$$

$$\begin{aligned}(30)\qquad & (\Delta t)^2\left|\left(\overset{+k}{v^{\varepsilon,l}_k} v^{\varepsilon,l}_{ix_k} + v^{\varepsilon,l}_k v^{\varepsilon,l}_{i\bar{x}_k} + v^{\varepsilon,l}_{kx_k} v^{\varepsilon,l}_i, v^{\varepsilon,l}_{it}\right)\right| \leqslant \frac{(\Delta t)^2}{2}\|\mathbf{v}^{\varepsilon,l}_t\|^2 +\\ &\qquad + \sqrt{3}\left(2\sqrt{6}\left(\frac{4}{3}\right)^{3/4}\right)^2\frac{(\Delta t)^2}{h^3}\|\mathbf{v}^{\varepsilon,l}\|^2\|\mathbf{v}^{\varepsilon,l}_x\|^2.\end{aligned}$$

Then from (24), using inequalities (25)–(30), conditions (15)–(17) and their consequence

$$1-24\varepsilon^{3/2}\frac{\Delta t}{h^2}\equiv\delta_3>0, \tag{31}$$

we obtain

$$\begin{gathered}\|\mathbf{v}^{\varepsilon,\,l+1}\|^2+\varepsilon\|p^{\varepsilon,\,l+1}\|^2+\delta_1(\Delta t)^2\|\mathbf{v}_t^{\varepsilon,\,l}\|^2+\delta_2\Delta t\|\mathbf{v}_x^{\varepsilon,\,l}\|^2+\delta_3\varepsilon(\Delta t)^2\|p_t^{\varepsilon,\,l}\|^2\\+\varepsilon^{3/2}\Delta t\|p_x^{\varepsilon,\,l}\|^2\leqslant 2\Delta t\|\mathbf{f}^l\|\cdot\|\mathbf{v}^{\varepsilon,\,l+1}\|+\|\mathbf{v}^{\varepsilon,\,l}\|^2+\varepsilon\|p^{\varepsilon,\,l}\|^2,\\ l=0,\,1,\ldots,\,N-1;\end{gathered} \tag{32}$$

from this inequality, on the basis of arguments like those in [**3**], Chapter VI, §9.3, we obtain for each $m=0,1,\cdots,N-1$ the following fundamental inequality—the difference analogue of the energy inequality for the solution of problem (1), (3), (4):

$$\begin{gathered}\|\mathbf{v}^{\varepsilon,\,m+1}\|^2+\varepsilon\|p^{\varepsilon,\,m+1}\|^2+\varepsilon^{3/2}\Delta t\sum_{l=0}^{m}\|p_x^{\varepsilon,\,l}\|^2+\delta_1(\Delta t)^2\sum_{l=0}^{m}\|\mathbf{v}_t^{\varepsilon,\,l}\|^2\\+\delta_2\Delta t\sum_{l=0}^{m}\|\mathbf{v}_x^{\varepsilon,\,l}\|^2+\delta_3\varepsilon(\Delta t)^2\sum_{l=0}^{m}\|p_t^{\varepsilon,\,l}\|^2\leqslant 2(\|\mathbf{a}\|^2+\varepsilon\|P\|^2)\\+5\left(\Delta t\sum_{l=0}^{m}\|\mathbf{f}^l\|\right)^2,\quad m=0,\,1,\,\ldots,\,N-1.\end{gathered} \tag{33}$$

Let $\mathbf{v}(x,t)$, $p(x,t)$ be the classical solution of the initial-boundary problem (5), and let $\mathbf{v}_h^l\equiv\mathbf{v}^l$, $p_h^l\equiv p^l$ be the values of this solution at the lattice points Ω_h^l, $l=0,1,\cdots,N$. Then for these values the following finite-difference equations of the type (10) with residuals r_i^l and ρ^l are valid:

$$\begin{gathered}v_{it}^l-\nu v_{ix_k\bar{x}_k}^l+\frac{1}{2}\overset{+k}{v_k^l}v_{ix_k}^l+\frac{1}{2}v_k^lv_{i\bar{x}_k}^l+p_{x_i}^l=f_i^l+r_i^l,\quad i=1,\,2,\,3,\\ v_{k\bar{x}_k}^l=\rho^l,\quad l=1,\,2,\,\ldots,\,N,\end{gathered} \tag{34}$$

and, in addition, $\mathbf{v}^l$ satisfies the following initial and boundary conditions:

$$\mathbf{v}^0=\mathbf{a}_h,\quad \mathbf{v}^l|_{\partial\Omega_h^l}=0,\ l=1,\,2,\,\ldots,\,N. \tag{35}$$

Let us set $\mathbf{u}^l\equiv\mathbf{v}^{\epsilon,l}-\mathbf{v}^l$ and $q^l\equiv p^{\epsilon,l}-p^l$, $l=0,1,\cdots,N$. Then $\mathbf{u}^l$ and q^l satisfy the following finite-difference equations and initial-boundary conditions:

$$\begin{gathered}u_{it}^l-\nu u_{ix_k\bar{x}_k}^l+\frac{1}{2}\overset{+k}{v_k^{\varepsilon,\,l}}u_{ix_k}^l+\frac{1}{2}\overset{+k}{u_k^l}v_{ix_k}^l+\frac{1}{2}v_k^{\varepsilon,\,l}u_{i\bar{x}_k}^l\\+\frac{1}{2}u_k^lv_{i\bar{x}_k}^l+\frac{1}{2}v_i^{\varepsilon,\,l}v_{k\bar{x}_k}^{\varepsilon,\,l}+q_{x_i}^l=r_i^l,\quad i=1,\,2,\,3,\\ \varepsilon q_t^l-\varepsilon^\alpha q_{x_k\bar{x}_k}^l+u_{k\bar{x}_k}^l=-\rho^l-\varepsilon p_t^l+\varepsilon^\alpha p_{x_k\bar{x}_k}^l,\quad l=1,\ldots,\,N,\end{gathered} \tag{36}$$

$$\mathbf{u}^0=0,\quad \mathbf{u}^l|_{\partial\Omega_h^l}=0,\quad l=1,\,\ldots,\,N. \tag{37}$$

Multiplying the ith equation of (36) by $2\Delta th^3u_i^{l+1}$ and the last by $2\Delta th^3q^{l+1}$, then summing over all the lattice points Ω_h^l, we obtain the inequalities

$$\begin{gathered}\|\mathbf{u}^{l+1}\|^2-\|\mathbf{u}^l\|^2+(\Delta t)^2\|\mathbf{u}_t^l\|^2+2\nu\Delta t\|\mathbf{u}_x^l\|^2=2\Delta t(\mathbf{r}^l,\,\mathbf{u}^{l+1})\\-2\nu(\Delta t)^2(\mathbf{u}_x^l,\,\mathbf{u}_{xt}^l)-2\Delta th^3\sum_{\Omega_h^l}q_{x_i}^lu_i^{l+1}+j_1^l,\end{gathered} \tag{38}$$

$$\varepsilon\|q^{l+1}\|^2-\varepsilon\|q^l\|^2+\varepsilon(\Delta t)^2\|q_t^l\|^2+2\varepsilon^{3/2}\Delta t\|q_x^l\|^2+2\varepsilon^{3/2}(\Delta t)^2(q_x^l,\ q_{xt}^l)$$
$$+2\Delta t h^3\sum_{\Omega_h^l} q^{l+1}u^l_{k\bar{x}_k}=-2\Delta t(\rho^l,\ q^{l+1})+j_2^l, \tag{39}$$

in which

$$j_1^l=-\Delta t\left(\overset{+k}{v_k^{\varepsilon,\,l}}u^l_{ix_k}+\overset{+k}{u_k^l}v^l_{ix_k}+v_k^{\varepsilon,\,l}u^l_{i\bar{x}_k}+u_k^lv^l_{i\bar{x}_k}+v_i^{\varepsilon,\,l}v^{\varepsilon,\,l}_{k\bar{x}_k},\ u_i^{l+1}\right), \tag{40}$$

$$j_2^l=2\Delta t\left(-\varepsilon p_t^l+\varepsilon^\alpha p^l_{x_k\bar{x}_k},\ q^{l+1}\right). \tag{41}$$

Arguing as we did in obtaining (21), we have

$$-2\Delta t h^3\sum_{\Omega_h^l} q^l_{x_i}u_i^{l+1}=2\Delta t h^3\sum_{\Omega_h^l} q^l u^{l+1}_{i\bar{x}_i}=2\Delta t(q^l,\ u^{l+1}_{i\bar{x}_i}), \tag{42}$$

$$2\Delta t h^3\sum_{\Omega_h^l} q^{l+1}u^l_{k\bar{x}_k}=2\Delta t\left(q^{l+1},\ u^l_{kx_k}\right)=2(\Delta t)^2\left(q_t^l,\ u^l_{k\bar{x}_k}\right)$$
$$-2(\Delta t)^2\left(q^l,\ u^l_{k\bar{x}_kt}\right)+2\Delta t\left(q^l,\ u^{l+1}_{k\bar{x}_k}\right), \tag{43}$$

and from (29) and (43) we obtain

$$2\Delta t\left(q^l,\ u^{l+1}_{k\bar{x}_k}\right)=-\varepsilon\|q^{l+1}\|^2+\varepsilon\|q^l\|^2-\varepsilon(\Delta t)^2\|q_t^l\|^2-2\varepsilon^{3/2}\Delta t\|q_x^l\|^2$$
$$-2\varepsilon^{3/2}(\Delta t)^2(q_x^l,\ q_{xt}^l)+2(\Delta t)^2\left(q^l,\ u^l_{k\bar{x}_kt}\right)-2(\Delta t)^2\left(q_t^l,\ u^l_{k\bar{x}_k}\right)$$
$$-2\Delta t(\rho^l,\ q^{l+1})+j_2^l. \tag{44}$$

From (38), (39) and (44) we next obtain the following fundamental equality:

$$\|\mathbf{u}^{l+1}\|^2+\varepsilon\|q^{l+1}\|^2+(\Delta t)^2\|\mathbf{u}_t^l\|^2+\varepsilon(\Delta t)^2\|q_t^l\|^2+2\nu\Delta t\|\mathbf{u}_x^l\|^2+2\varepsilon^{3/2}\Delta t\|q_x^l\|^2$$
$$=\|\mathbf{u}^l\|^2+\varepsilon\|q^l\|^2+2\Delta t(\mathbf{r}^l,\ \mathbf{u}^{l+1})-2\Delta t(\rho^l,\ q^{l+1})-2\nu(\Delta t)^2(\mathbf{u}_x^l,\ \mathbf{u}_{xt}^l)$$
$$-2\varepsilon^{3/2}(\Delta t)^2(q_x^l,\ q_{xt}^l)+2(\Delta t)^2\left(q^l,\ u^l_{k\bar{x}_kt}\right)-2(\Delta t)^2\left(q_t^l,\ u^l_{k\bar{x}_k}\right)+j_1^l+j_2^l. \tag{45}$$

Using identity transformations and the fact that

$$\left(\overset{+k}{u_k^l}u^l_{ix_k}+u_i^lu^l_{i\bar{x}_k}+u_i^lu^l_{k\bar{x}_k},\ u_i^l\right)=0 \tag{46}$$

(compare (21′)), we easily obtain

$$j_1^l=-(\Delta t)^2\left(\overset{+k}{u_k^l}u^l_{ix_k}+u_k^lu^l_{i\bar{x}_k}+u_i^lu^l_{i\bar{x}_k},\ u^l_{it}\right)-\Delta t\left(\overset{+k}{v_k^l}u^l_{ix_k}+\overset{+k}{u_k^l}v^l_{ix_k}\right.$$
$$\left.+v_k^lu^l_{i\bar{x}_k}+u_i^lv^l_{k\bar{x}_k}+v_i^lu^l_{k\bar{x}_k}+v_i^lv^l_{k\bar{x}_k},\ u_i^{l+1}\right), \tag{47}$$

and hence, applying the Hölder and Cauchy inequalities, we get the following estimate (see (30)):

$$|j_1^l|\leqslant\frac{(\Delta t)^2}{2}\|\mathbf{u}_t^l\|^2+\sqrt{3}\left[2\sqrt{6}\left(\frac{4}{3}\right)^{3/4}\right]^2\frac{(\Delta t)^2}{h^3}\|\mathbf{u}^l\|^2\|\mathbf{u}_x^l\|^2$$
$$+\frac{\delta_2}{2}\Delta t\|\mathbf{u}_x^l\|^2+C_1\Delta t(\|\mathbf{u}^l\|^2+\|\mathbf{u}^{l+1}\|^2), \tag{48}$$

in which the constant C_1 depends only on $\max_{Q_T}|\mathbf{v}(x,t)|$, $\max_{Q_T}|\mathbf{v}_x(x,t)|$, ν and meas Ω. Furthermore,

$$|j_2^l|\leqslant\varepsilon\Delta t\|q^{l+1}\|^2+C_2\varepsilon\Delta t, \tag{49}$$

where C_2 depends only on $\max_{Q_T}|p_t(x,t)|$, $\max_{Q_T}|p_{xx}(x,t)|$ and $\operatorname{meas}\Omega$,

(50) $$2\Delta t\,|(\mathbf{r}^l,\ \mathbf{u}^{l+1})| \leqslant \Delta t\,\|\mathbf{r}^l\|^2 + \Delta t\,\|\mathbf{u}^{l+1}\|^2,$$

(51) $$2\Delta t\,|(\rho^l,\ q^{l+1})| \leqslant \varepsilon\Delta t\,\|q^{l+1}\|^2 + \frac{\Delta t}{\varepsilon}\|\rho^l\|^2.$$

It is clear that

(52) $$\begin{gathered}\|\mathbf{u}^{l+1}\|^2 \leqslant 2\,\{(\Delta t)^2\,\|\mathbf{u}_t^l\|^2 + \|\mathbf{u}^l\|^2\},\\ \|q^{l+1}\|^2 \leqslant 2\,\{(\Delta t)^2\|q_t^l\|^2 + \|q^l\|^2\}.\end{gathered}$$

Estimating the remaining terms in the right member of (45) with the aid of inequalities of the type (26) – (29), and using conditions (16), (17), (31) as well as the fact that for small Δt

(53) $$\delta_4 \equiv \delta_1 - (2C_1 + 1)\,\Delta t > 0,$$

(54) $$\delta_5 \equiv \delta_3 - 4\Delta t > 0,$$

we obtain the following inequality from (45) (cf. (32)):

(55) $$\begin{aligned}\|\mathbf{u}^{l+1}\|^2 &+ \varepsilon\,\|q^{l+1}\|^2 + \delta_4\,(\Delta t)^2\,\|\mathbf{u}_t^l\|^2 + \frac{\delta_2}{2}\,\Delta t\,\|\mathbf{u}_x^l\|^2 + \delta_3\varepsilon\,(\Delta t)^2\,\|q_t^l\|^2\\ &+ \varepsilon^{3/2}\Delta t\,\|q_x^l\| \leqslant (1 + C_1\Delta t)\,(\|\mathbf{u}^l\|^2 + \varepsilon\,\|q^l\|^2) + \Delta t\,\|\mathbf{r}^l\|^2\\ &+ \frac{\Delta t}{\varepsilon}\|\rho^l\|^2 + C_2\varepsilon\Delta t, \quad l = 0,\ 1,\ \ldots,\ N-1,\end{aligned}$$

From this, summing over l from 1 to m $(m = 1, \cdots, N)$, we get (cf. [3], 1970 ed., p. 267)

(56) $$\begin{aligned}\|\mathbf{u}^{m+1}\|^2 &+ \varepsilon\,\|q^{m+1}\|^2 + \varepsilon^{3/2}\Delta t \sum_{l=0}^{m}\|q_x^l\|^2 + \delta_4\,(\Delta t)^2 \sum_{l=0}^{m}\|\mathbf{u}_t^l\|^2 + \frac{\delta_2}{2}\,\Delta t \sum_{l=0}^{m}\|\mathbf{u}_x^l\|^2\\ &+ \delta_5\varepsilon\,(\Delta t)^2 \sum_{l=0}^{m}\|q_t^l\|^2 \leqslant C_3\Big\{\|\mathbf{u}^0\|^2 + \varepsilon\,\|q^0\|^2 +\\ &+ \Delta t\Big[\varepsilon C_2 + \sum_{l=0}^{m}\Big(\|\mathbf{r}^l\|^2 + \frac{1}{\varepsilon}\|\rho^l\|^2\Big)\Big]\Big\},\end{aligned}$$

in which the constant C_3 depends only upon C_1 and T. From (56) in turn it follows that

(57) $$\begin{aligned}|\mathbf{u},\ q|_\varepsilon^2 &\equiv \max_{1,\ldots,N}(\|\mathbf{u}^m\|^2 + \varepsilon\,\|q^m\|^2) + (\Delta t)^2 \sum_{l=0}^{N}\|\mathbf{u}_t^l\|^2 + \Delta t \sum_{l=0}^{N}\|\mathbf{u}_x^l\|^2\\ &+ \varepsilon\,(\Delta t)^2 \sum_{l=0}^{N}\|q_t^l\|^2 + \varepsilon^{3/2}\Delta t \sum_{l=0}^{N}\|q_x^l\|^2\\ &\leqslant C_4\Big\{\|\mathbf{u}^0\|^2 + \varepsilon\,\|q^0\|^2 + \Delta t\Big[\varepsilon C_2 + \sum_{l=0}^{N}\Big(\|\mathbf{r}^l\|^2 + \frac{1}{\varepsilon}\|\rho^l\|^2\Big)\Big]\Big\},\end{aligned}$$

where C_4 depends upon C_3, δ_2, δ_4 and δ_5.

Suppose that the classical solution $\mathbf{v}(x,t)$, $p(x,t)$ of the initial-boundary problem (5) has continuous derivatives $\mathbf{v}_t$, $\mathbf{v}_x$, $\mathbf{v}_{xx}$ and p_t, p_x, p_{xx}. Then the residuals $\mathbf{r}^l$ and ρ^l in (34) have the following orders of smallness:

$$\mathbf{r}^l \sim \Delta t + h, \quad \rho^l \sim h, \tag{58}$$

and hence it follows from condition (15) that the right member of (57), and therefore $|\mathbf{u},q|_\epsilon$ also—the energy norm associated with the system (1)—vanishes as ϵ, Δt, $h \to 0$. Theorem 2 is proved.

The author is grateful to O. A. Ladyženskaja for her interest in this work and review of the results.

Bibliography

[1] A. P. Oskolkov, *A certain quasilinear parabolic system with small parameter that approximates a system of Navier-Stokes equations,* Zap. Naučn. Sem. Leningrad. Otdel. Mat. Inst. Steklov. (LOMI) **21** (1971), 79-103 = J. Soviet Math. **1** (1973), 452-470. MR **45** # 7314.

[2] ———, *On the asymptotic behavior of solutions of systems with a small parameter which approximate the system of Navier-Stokes equations,* Trudy Mat. Inst. Steklov. **125** (1973), 147-163 = Proc. Steklov Inst. Math. **125** (1973), 137-153.

[3] O. A. Ladyženskaja, *Mathematical problems in the dynamics of a viscous incompressible fluid,* 2nd rev. aug. ed., "Nauka", Moscow, 1970; English transl. of 1st ed., *The mathematical theory of viscous incompressible flow,* Gordon and Breach, New York, 1963; rev. 1969. MR **40** # 7610; **27** # 5034b; **42** # 6442.

[4] R. Temam, *Une méthode d'approximation de la solution des équations de Navier-Stokes,* Bull. Soc. Math. France **96** (1968), 115-152. MR **38** # 6249.

[5] ———, *Sur l'approximation de la solution des équations de Navier-Stokes par la méthode des pas fractionnaires.* I, II, Arch. Rational Mech. Anal. **32** (1969), 135-153; **33** (1970), 377-385. MR **38** # 6250; **39** # 5968.

Translated by
R. N. GOSS

Trudy Mat. Inst. Steklov.
125 (1973)

Proc. Steklov Inst. Math.
125 (1973)

THE GRID METHOD OF SOLVING PROBLEMS OF THE DYNAMICS OF A VISCOUS INCOMPRESSIBLE FLUID

V. JA. RIVKIND

Abstract. Various convergent difference schemes are proposed for the Navier-Stokes equations in which the approximation of the nonlinear terms differs from those proposed earlier (variable direction type schemes without the addition of the terms $\frac{1}{2}\mathbf{u}\operatorname{div}\mathbf{u}$ in the equation of motion). This is achieved at the expense of a "section" of the nonlinear terms, and is possible subject to a knowledge of a priori estimates for $\max_Q|\mathbf{u}|$ and $\max_Q|\partial\mathbf{u}/\partial x|$. Based on approximations of the Navier-Stokes equations by a system of quasilinear parabolic equations, explicit difference schemes are constructed for the step relation $\Delta t \sim h^2$.

Bibliography: 16 items.

In [**1**, **2**] various implicit linear difference schemes were given for the nonstationary Navier-Stokes equations, and their convergence was proved without making any assumptions concerning the smoothness of the exact solution. Moreover, special grid designs were formulated to approximate the nonlinear terms of the equation, the designs being such as to preserve energy estimates valid for the exact solutions. Further, in [**2**–**8**] and related papers, certain modified equations are formulated, which approximate the Navier-Stokes equations, making it possible to construct economical calculational schemes (of the variable directions type). In addition, the auxiliary term $\frac{1}{2}\mathbf{u}\operatorname{div}\mathbf{u}$ is introduced into the equation of motion; this term is different from zero for the solution of the approximating equation. The proof of convergence of these schemes requires sufficient smoothness of the exact solution. Explicit schemes for these equations which converge for the step relationship $\Delta t \sim h^3$ were studied in [**2**, **3**]. Many other schemes (see, for example, [**8**–**11**] and others) have been used in practice which are as yet without sufficient theoretical justification.

In this paper, to obtain smooth (bounded, together with their first derivatives) solutions we propose implicit linear, convergent difference schemes using various approximations for the nonlinear terms. In one of these schemes the nonlinear terms carry back to the previous (in time) level. Thanks to this the implicit scheme has a matrix of coefficients which is independent of the time, thereby reducing considerably the expense of calculating it. It has been possible to construct a scheme of this kind at the expense of a formal modification of the nonlinear terms in the Navier-Stokes equations and their reduction to terms with a bounded nonlinearity; this is justified in the presence of the necessary a priori estimates for the exact solution. At the expense of such a modification, it turns out to be possible to construct convergent difference schemes, both economical and explicit with respect to the pressure, which are similar to those given in [**2**, **6**] but without the introduction of the supplementary term $\frac{1}{2}\mathbf{u}\operatorname{div}\mathbf{u}$ in the equation of motion. Moreover, in these schemes, as in the implicit schemes, various methods are proposed for ap-

---

AMS (MOS) subject classifications (1970). Primary 76D05, 65N05, 65N10; Secondary 35D05.

proximating the nonlinear terms. Besides the implicit and economical difference schemes, we present in this paper explicit difference schemes converging for the step relationship $\Delta t \sim h^2$. To obtain such a difference scheme we use a system of equations close to that given in [12] and we present a special new approximation of the Navier-Stokes equations by a system of quasilinear degenerate parabolic equations. We also give estimates of the rate of convergence of the approximate solutions to the exact solution.

In the bounded cylinder $Q_T = \Omega \times [0, T]$ we consider the first boundary-value problem for the Navier-Stokes equations (our notation is the same as that in [3])

$$(1_1) \qquad L(\mathbf{u}, q) = \frac{\partial \mathbf{u}}{\partial t} - \nu \Delta \mathbf{u} + \mathbf{u}_k \frac{\partial \mathbf{u}}{\partial x_k} + \operatorname{grad} q = \mathbf{F}, \quad \operatorname{div} \mathbf{u} = 0,$$

$$(1_2) \qquad \mathbf{u}\,|_{S_T = S \times [0, T]} = 0, \quad \mathbf{u}(x, t)\,|_{t=0} = \mathbf{a}(x).$$

In obtaining an approximate solution we replace the problem (1) by one equivalent to it for smooth solutions. For this purpose we construct the vector-valued function

$$(2) \qquad \mathbf{f}_i\left(\mathbf{u}, \frac{\partial \mathbf{u}}{\partial x}\right) = \mathbf{u}_k \xi_A(|\mathbf{u}_k|) \frac{\partial \mathbf{u}_i}{\partial x_k} \xi_{A_1}\left(\left|\frac{\partial \mathbf{u}_i}{\partial x_k}\right|\right), \quad i = 1, \ldots, n.$$

Here $\xi_D(|\tau|)$ is a smooth nonnegative function, monotonic with respect to $|\tau|$, which is equal to one for $|\tau| \leqq D$ and to zero for $|\tau| \geqq 2D$. Further, we consider in Q_T the auxiliary problem (3):

$$(3_1) \qquad \tilde{L}(\mathbf{u}, q) = \frac{\partial \mathbf{u}}{\partial t} - \nu \Delta \mathbf{u} + \mathbf{f}\left(\mathbf{u}, \frac{\partial \mathbf{u}}{\partial x}\right) + \operatorname{grad} q = \mathbf{F},$$

$$(3_2) \qquad \operatorname{div} \mathbf{u} = 0,$$

$$(3_3) \qquad \mathbf{u}\,|_{S_T} = 0, \quad \mathbf{u}\,|_{t=0} = \mathbf{a}(x).$$

DEFINITION. By a *generalized solution of the problem* (3) *of the class* $\overset{\circ}{\mathbf{V}}{}_2^{1,0}(Q_T) \cap \overset{\circ}{\mathscr{J}}(Q_T)$ we mean a function $\mathbf{u}$ satisfying the integral identity

$$(4) \qquad \begin{aligned} &\int_{Q_T} \left(-\left\langle \mathbf{u}, \frac{\partial \eta}{\partial t}\right\rangle + \nu \left\langle \frac{\partial \mathbf{u}}{\partial x}, \frac{\partial \eta}{\partial x}\right\rangle + \left\langle \mathbf{f}\left(\mathbf{u}, \frac{\partial \mathbf{u}}{\partial x}\right), \eta\right\rangle\right) dx dt \\ &\qquad + \int_{\Omega_t} \langle \mathbf{u}, \eta\rangle\, dx\,|_{t=T} - \int_{\Omega} \langle \mathbf{a}, \eta\rangle\, dx\,|_{t=0} = \int_{Q_t} \langle \mathbf{F}, \eta\rangle\, dx dt \end{aligned}$$

for arbitrary $\eta(x,t) \in \overset{\circ}{\mathbf{W}}{}_2^{1,1}(Q_T) \cap \overset{\circ}{\mathscr{J}}(Q_T)$ and $\langle a, b\rangle = \sum_1^n a_i b_i$.

LEMMA. *Problem* (3) *has no more than one generalized solution of the class* $\overset{\circ}{\mathbf{V}}{}_2^{1,0}(Q_T) \cap \overset{\circ}{\mathscr{J}}(Q_T)$.

Indeed, suppose that there were two solutions $\mathbf{u}'$ and $\mathbf{u}''$. From the identity (4) for $\mathbf{u}'$ and $\mathbf{u}''$ we obtain the following identity for their difference:

$$\begin{aligned} &\int_{Q_T} \left(-\left\langle \mathbf{w}, \frac{\partial \eta}{\partial t}\right\rangle + \nu \left\langle \frac{\partial \mathbf{w}}{\partial x}, \frac{\partial \eta}{\partial x}\right\rangle + \left\langle \left(\mathbf{f}\left(\mathbf{u}', \frac{\partial \mathbf{u}'}{\partial x}\right) - \mathbf{f}\left(\mathbf{u}'', \frac{\partial \mathbf{u}''}{\partial x}\right)\right), \eta\right\rangle\right) dx dt \\ &\qquad + \int_{\Omega_t} \mathbf{w}(x, t)\, \eta(x, t)\, dx = 0. \end{aligned}$$

Considering in this identity the vector-valued function $\Phi = \mathbf{f}(\mathbf{u}', \partial\mathbf{u}'/\partial x) - \mathbf{f}(\mathbf{u}'', \partial\mathbf{u}''/\partial x)$ as a free term, we can show from the corresponding linear problem (see [**3**]) that

$$\frac{1}{2}\| \mathbf{w}(x, t) \|^2 + \nu \int_0^t \left\| \frac{\partial \mathbf{w}}{\partial x} \right\|^2 dt = \int_{Q_t} \langle \Phi, \mathbf{w} \rangle \, dx dt.$$

From Cauchy's inequality

$$\int_{Q_T} \langle \Phi, \mathbf{w} \rangle \, dx dt \leqslant \frac{1}{2\varepsilon} \int_{Q_t} |\mathbf{w}|^2 \, dx dt + \frac{\varepsilon}{2} \int_{Q_t} |\Phi|^2 \, dx dt.$$

From the form of the function Φ it follows that

$$\int_{Q_t} \Phi^2 dx dt \leqslant c \int_{Q_t} \left[|\mathbf{w}|^2 + \left| \frac{\partial \mathbf{w}}{\partial x} \right|^2 \right] dx dt,$$

where the constant c is a quantity bounding the quantities

$$\left| \frac{\partial \mathbf{f}_k(\mathbf{u}, \mathbf{p})}{\partial u_i} \right|; \quad \left| \frac{\partial \mathbf{f}_k(\mathbf{u}, \mathbf{p}))}{\partial p_{ij}} \right|; \quad k, i, j = 1, \ldots, n; \quad p_{ij} = \frac{\partial u_i}{\partial x_j}.$$

It is easy to see that

$$c \leqslant \max \left| \frac{\partial \mathbf{f}}{\partial u_i}, \frac{\partial \mathbf{f}}{\partial p_{ij}} \right| \leqslant 9AA_1. \tag{5}$$

Choosing ϵ so that $c\epsilon < \nu$, we obtain

$$\frac{1}{2} \int_{\Omega} |\mathbf{w}(x, t)|^2 \, dx \leqslant \int_{Q_t} |\mathbf{w}(x, t)|^2 \, dx dt. \tag{6}$$

Uniqueness of the solution of the problem follows from (6).

THEOREM 1. *Assume that we can find a solution of the boundary-value problem* (1) *for the Navier-Stokes equations such that* $|\mathbf{u}| \leqq M$ *and* $|\partial\mathbf{u}/\partial x| \leqq M_1$, *where* M_1, $M = \text{const} > 0$. *Then the solution of the problem* (3) *with* $A = M$ *and* $A_1 = M_1$ *is a solution of the problem* (1).

For values of $|\mathbf{u}| \leqq M$ and $|\partial\mathbf{u}/\partial x| \leqq M_1$ the two problems in question coincide, and, since the solution of the problem (3) is unique, it obviously coincides with the desired solution.

REMARK. Some sufficient conditions which guarantee the satisfaction of the premises of Theorem 1 are given in [**3**].

We now use the auxiliary problem (3) to obtain approximate solutions. We first recall some notation from [**3**]. In E_n let there be constructed the uniform grid $x_i = k_i h$, $h > 0$, $k_i = 0, \pm 1, \cdots$; $t_k = k\Delta t$, $k = 0, 1, \cdots, N$; $N = T/\Delta t$. Let Ω^k be a section of the cylinder by the plane $t = k\Delta t$, S_k its boundary, Ω_h^k the set of interior nodes of the grid, and S_h^k the set of boundary nodes, and, finally, let S_{ih}^k be the set of nodes of S_h^k which when displaced by h in the direction of the coordinate axis x_i go over into some point of Ω_k^h.

On this grid we can construct very diverse convergent approximations of the nonlinear terms $\mathbf{f}$ of the system (3). We pause to consider only the simplest of them (we denote grid approximations of $\mathbf{f}$ by $\mathbf{f}_h$):

a) explicit approximations, i.e. all the nonlinear terms relate back to the previous level:

$$f_{ih}^{(1)}(v_h, v_{hx}) = \xi_M\left(\left|\overset{-0}{v}_{kh}\right|\right) \overset{-0}{v}_{kh} \overset{-0}{v}_{ihx_k} \xi_{M_1}\left(\left|\overset{-0}{v}_{ihx_k}\right|\right);$$

b) approximations with second order accuracy (formally):

$$f_{ih}^{(2)}(v_h, v_{hx}) = \frac{1}{2} \xi_M\left(\left|\overset{-0}{v}_{kh}\right|\right)(v_{ihx_k} + v_{ih\bar{x}_k}) \xi_{M_1}\left(\left|\overset{-0}{v}_{ihx_k}\right|\right) \overset{-0}{v}_{kh};$$

c) the simplest one-sided approximation:

$$f_{ih}^{(3)}(v_h, v_{hx}) = \xi_M\left(\left|\overset{-0}{v}_{kh}\right|\right) \overset{-0}{v}_{kh} v_{ihx_k} \xi_{M_1}\left(\left|\overset{-0}{v}_{ihx_k}\right|\right);$$

d) approximation depending on the sign of v_{kh}:

$$f_{ih}^{(4)}(v_h, v_{hx}) = \frac{1}{2} \xi_M\left(\left|\overset{-0}{v}_{kh}\right|\right)\left(1 - \operatorname{sign} \overset{-0}{v}_{kh}\right) \xi_{M_1}\left(\left|\overset{-0}{v}_{ih\bar{x}_k}\right|\right) v_{h\bar{x}_k} \overset{-0}{v}_{kh}$$
$$+ \frac{1}{2} \xi_M\left(\left|\overset{-0}{v}_{kh}\right|\right)\left(1 + \operatorname{sign} \overset{-0}{v}_{kh}\right) \xi_{M_1}\left(\left|\overset{-0}{v}_{ihx_k}\right|\right) v_{ihx_k} \overset{-0}{v}_{kh}.$$

We present convergence theorems for one of the implicit schemes using the explicit approximation of the nonlinear terms $\mathbf{f}_h^1$. For implicit schemes involving the other approximation data, and also for a set of other schemes, the derivation of the estimates and the proof of convergence is carried out completely analogously.

The difference scheme being studied here has in Ω_h^k $(k = 1, \cdots, N)$ the form

$$(7_1) \qquad v_{ih\bar{t}} - \nu v_{ihx_k\bar{x}_k} + f_{ih}^{(1)}(v_h, v_{hx}) + p_{h\bar{x}_i} = F_{ih}, \quad i = 1, \ldots, n,$$

$$(7_2) \qquad v_{ihx_i} = 0,$$

$$(7_3) \qquad v_{ih}|_{S_h} = 0, \quad v_{ih}|_{t=0} = a_h.$$

The equations (7_1) are assumed to be satisfied at the points Ω_h^k $(k = 1, \cdots, N)$, the equations (7_2) at the points $\Omega_h^k \cup \sum_{i=1}^n S_{ih}^k = \Omega_h^{k'}$, and to the equations (7) it is necessary to add in a condition on p_k, for example,

$$(7_4) \qquad \sum_{\Omega_h^{k'}} p_h = 0, \quad k = 1, \ldots, N,$$

(henceforth we omit the subscript h). As in [1], we select an energy estimate for the equations (7). For this we multiply both sides of (7_1) on the lth level by v_{ih}^l and sum the resulting equations with respect to i and the points Ω_h^l. Carrying out these operations, we obtain

$$(8) \qquad \begin{aligned} &\|\mathbf{v}^l\|^2 - \|\mathbf{v}^{l-1}\|^2 + 2\nu(\Delta t)\|\mathbf{v}_x\|^2 + (\Delta t)^2\|\mathbf{v}_{\bar{t}}^l\|^2 \\ &= 2(\Delta t)\int_\Omega \langle F_i, v_i\rangle\, dx - 2(\Delta t)\int_\Omega \langle \mathbf{f}_h^{(1)}(v_h, v_{hx}), \mathbf{v}^l\rangle\, dx, \quad l = 1, \ldots, M. \end{aligned}$$

From the condition

$$(9) \qquad \left|f\left(\mathbf{u}, \frac{\partial \mathbf{u}}{\partial x}\right)\right| \leqslant \frac{9MM_1\sqrt{3}}{2}; \quad |f_h^1(v_h, v_{hx})| \leqslant \frac{9MM_1\sqrt{3}}{2}$$

we obtain, as in [3], the result

$$(10) \qquad \begin{aligned} \|\mathbf{v}^m\|^2 + 2\nu(\Delta t)\sum_{l=1}^m \|\mathbf{v}_x^l\|^2 + (\Delta t)^2 \sum_{l=1}^m \|\mathbf{v}_{\bar{t}}\|^2 \leqslant 2\|\mathbf{a}_h\|^2 \\ + 5(\Delta t)\sum_{l=1}^m \|\mathbf{F}^l\|^2 + T\cdot 81M^2M_1^2. \end{aligned}$$

The uniqueness of the solution of the system of linear algebraic equations (7) on each level l follows from the derivation of (8) and (9). One of the exact methods for solving the system (7) (the matrix sweep method) can be formulated as in [6]. We note that the formulation of the basic matrix coefficients in the direct operation of the matrix sweep method can be carried out once for all levels with respect to the time l.

THEOREM 2. *Let the conditions of Theorem 1 be satisfied. Then the grid solution of problem (7) converges to a solution bounded in $C_1(Q_T)$ of problem (1) so that*

$$\sup_{t\in[0,T]}\int_\Omega |\mathbf{u}(x,t)-\mathbf{v}'_h(x,t)|^2\,dx+\int_{Q_t}\left|\frac{\partial \mathbf{u}}{\partial x}-\frac{\partial \mathbf{v}'_h}{\partial x}\right|^2 dxdt$$

$$\leqslant\int_0^T\int_\Omega\left\{\left|\frac{\partial \mathbf{u}(x,t)}{\partial x}-\frac{\partial \mathbf{u}(x-\Delta x,t)}{\partial x}\right|^2+\left|\frac{\partial \mathbf{u}(x,t)}{\partial t}-\frac{\partial \mathbf{u}(x,t-\Delta t)}{\partial t}\right|^2\right. \tag{11}$$

$$\left.+|\mathbf{F}(x,t)-\mathbf{F}(x-\Delta x,t)|^2+|\mathbf{F}(x,t)-\mathbf{F}(x,t-\Delta t)|^2\right\}dxdt,$$

where $\mathbf{u}$ *and* $\mathbf{F}$ *are assumed to be continued by zero outside* Ω, *and* $\mathbf{v}'_h$ *is the multilinear complement of* $\mathbf{v}_h$.

The proof of this theorem proceeds according to the following scheme. With the exact solution $\mathbf{u}(x,t)$ of problem (1) we form the grid function $\mathbf{u}_h$:

$$u_{ih}(k_1h,\ldots,k_nh)=\frac{1}{h^{n-1}}\int_{\bar\sigma_i}u_i(x)\,d\sigma, \tag{12}$$

where $\bar\sigma_i$ is the boundary of the cube $k_jh\leqq x_j\leqq(k_j+1)h,\ j=1,\cdots,n$, lying in the plane $x_i=k_ih$. Further, as in [13,3], we show that the u_{ih} satisfy the equations (7) with a residual expressible in terms of the approximation error of the integral identity (4) which corresponds to it in the differences by summation. An estimate of this error of approximation leads to (11).

Along with the implicit difference schemes similar to [3] we can also consider economical difference schemes based on the approximations of the Navier-Stokes equations given in [1–5], and also on other approximations, however, without the term $\frac{1}{2}\mathbf{u}\operatorname{div}\mathbf{u}$ in the equations of motion. By way of example we give here, as in the case of the implicit schemes, only the simplest grid scheme approximating the system of equations (see [3])

$$\tilde L(\mathbf{u}_\varepsilon,q_\varepsilon)=\mathbf{F}, \tag{13_1}$$

$$\varepsilon\frac{\partial q_\varepsilon}{\partial t}+\operatorname{div}\mathbf{u}_\varepsilon=0, \tag{13_2}$$

$$\mathbf{u}_\varepsilon|_S=0,\quad \mathbf{u}_\varepsilon|_{t=0}=\mathbf{a}(x),\quad q|_{t=0}=q_{\varepsilon0}(x). \tag{13_3}$$

It has the form

$$\frac{1}{\Delta t}\left(v_i^{m-\frac{k-1}{3}}-v_i^{m-\frac{k}{3}}\right)+L_k^m\left(v_i^{m-\frac{k-1}{3}}\right)+\delta_k^i p_{x_k}^{m-\frac{k-1}{3}}=\frac{1}{3}F_i^{m-\frac{k-1}{3}}, \tag{14_1}$$

$$\frac{\varepsilon}{\Delta t}\left(p^{m-\frac{k-1}{3}}-p^{m-\frac{k}{3}}\right)+v_{k\bar x_k}^{m-\frac{k-1}{3}}=0, \tag{14_2}$$

$$i,\ k=1,\ 2,\ \ldots;\quad m=1,\ \ldots,\ N.$$

The equations (14_1) and (14_2) are considered at the same points used for the equations (7_1) and (7_2). To them we adjoin the conditions

$$(14_3)\qquad \mathbf{v}_{S_h^{m-\frac{k}{3}}}=0,\quad m=1,\ 2,\ldots,\ N;\quad \mathbf{v}|_{t=0}=\mathbf{a}_h,\ \ p|_{t=0}=q_{\varepsilon 0h}.$$

The operator $L_k^m(v_i^{m-(k-1)/3})$ can be chosen in a different way depending on the nature of the difference replacement of the term $f_k(\mathbf{u},\partial\mathbf{u}/\partial x)$ (similar to the substitutions a) – d) in the implicit schemes). We consider only the simplest case

$$(15)\qquad L_k^m(w)=-\nu w_{x_k\bar{x}_k}+\frac{1}{2}v_k^{m-\frac{k}{3}}\xi_M\left(\left|v_k^{m-\frac{k}{3}}\right|\right)\overset{-0}{w}_{x_k}\xi_{M_1}\left(\left|\overset{-0}{w}_{x_k}\right|\right),$$

where by $\overset{-0}{w}$ we mean the value of the grid function w on the previous level, closest to the given level, on which the function w has been calculated. To solve the problem (14) – (15) we multiply (14_1) by $v_i^{m-(k-1)/3}$ and (14_2) by $p^{m-(k-1)/3}$, and then by summation by parts we establish the energy inequality

$$(16)\qquad \begin{aligned}&\|\mathbf{v}^m\|^2+2\nu(\Delta t)\sum_{l=1,\,k=1}^{m;\,3}\left\|\mathbf{v}_{x_k}^{l-\frac{k-1}{3}}\right\|^2+\varepsilon\|p^m\|^2+\sum_{l=1,\,k=1}^{m;\,3}\left\|\mathbf{v}^{l-\frac{k-2}{3}}-\mathbf{v}^{l-\frac{k-1}{3}}\right\|^2\\ &\qquad+\varepsilon\sum_{l=1,\,k=1}^{m;\,3}\left\|p^{l-\frac{k-2}{3}}-p^{l-\frac{k-1}{3}}\right\|^2\\ &\qquad\leqslant c_1e^{cT}\left\{\|\mathbf{a}_h\|^2+(\Delta t)\sum_{l=1,\,k=1}^{m;\,3}\left\|\mathbf{F}^{l-\frac{k-1}{3}}\right\|^2+\varepsilon\|p^0\|^2\right\},\\ &\qquad c_1,\ c=\text{const},\end{aligned}$$

where c and c_1 are defined as the constants appearing on the right sides of (5) and (9). From the derivation of this inequality we obtain the uniqueness of the solution of the problem (14), (15) on each level l. For a solution of (14), (15), analogously to [**6, 13**] we establish

THEOREM 3. *Let the conditions of Theorem* 1 *be satisfied and, in addition, suppose that*

$$(17)\qquad \|\mathbf{u}\|_{W_2^2(Q_T)}+\left\|\frac{\partial^2 q}{\partial t^2}\right\|_{L_2(Q_T)}+\max\left(\left|\frac{\partial q}{\partial t}\right|+|q|\right)\leqslant c,\qquad c=\text{const}>0.$$

Then the solution of (14), (15) *converges to the solution of the boundary-value problem* (1) *in such a way that the following estimate is valid for the difference* $\mathbf{w}^l=\mathbf{u}_h^l-\mathbf{v}_h^l$:

$$(18)\qquad \begin{aligned}&\sup_{l=0,\,\frac{1}{3},\,\ldots,\,m-\frac{2}{3},\,m}\|\mathbf{w}^l\|^2+(\Delta t)\sum_{l=1,\,k=1}^{m;\,3}\left\|\mathbf{w}_{x_k}^{m-\frac{k-1}{3}}\right\|^2\leqslant e^{cT}\Big(\Delta t+\varepsilon+\frac{\Delta t}{\varepsilon}+h\\ &\quad+\int_{Q_t}\left\{\sum_{i=1}^{3}(F_i(x+\Delta x,\ t)-F_i(x,\ t))^2+(F_i(x,\ t+\Delta t)-F_i(x,\ t))^2\right\}dxdt,\end{aligned}$$

where the constant c is determined by the constants in (16) *and* $\mathbf{u}_k$ *is the Steklov mean of the exact solution of* (1) *calculated at the nodes of the grid.*

We remark that in addition to the scheme (14), (15) we can also consider other difference schemes, similar to those in [**3, 6**], including even the explicit-implicit schemes given in [**6**] in which the replacements for the nonlinear terms are similar to the substitutions (15) and to others corresponding to the substitutions a) – d) in the implicit schemes.

We present two explicit schemes, which are convergent for the step relation $\Delta t \sim h^2$. For this purpose we consider the following regularization of the Navier-Stokes equations. To ease the exposition, we assume that Ω is the unit cube with boundaries parallel to the coordinate axes. The regularized equations have the following form:

$$(19_1) \qquad \tilde{L}(\mathbf{u}_\varepsilon, q_\varepsilon) = \mathbf{F},$$

$$(19_2) \qquad \tilde{L}_1(\mathbf{u}_\varepsilon, q_\varepsilon) \equiv \varepsilon \frac{\partial q_\varepsilon}{\partial t} - \varepsilon \frac{\partial}{\partial x_i} x_i (1 - x_i) \frac{\partial q_\varepsilon}{\partial x_i} + \operatorname{div} \mathbf{u}_\varepsilon = 0;$$

the equations (19_1) and (19_2) are supplemented with the addition of the boundary and initial conditions (13_3) and a condition for the boundedness of q_ϵ on S:

$$(19_3) \qquad |q_\epsilon|\,|_S < \infty.$$

To determine the generalized solution of the problem (19) and to prove existence and uniqueness theorems we introduce into the cylinder Q_T the space of functions $V_{2,r_i}(Q_T)$, obtained from functions in Q_T with compact support by closure in the norm

$$(20) \qquad \|\psi\|_{V_{2,r_i}(Q_T)} = \sup_{t \in [0,T]} \int_\Omega \psi^2(x, t)\, dx + \int_0^T \int_\Omega r_i \left| \frac{\partial \psi}{\partial x_i} \right|^2 dx dt, \qquad r_i = x_i(1 - x_i).$$

Definition. By a *generalized solution of the problem* (19) we shall mean the totality of functions $\mathbf{u}_\epsilon(x,t)$, $q_\epsilon(x,t)$, $u_{i\epsilon}(x,t) \in V_2^{1,0}(Q_T)$, $q_\epsilon(x,t) \in V_{2,r_i}(x,t)$, $i = 1, \cdots, n$, such that the integral identities

$$(21_1) \qquad \int_{Q_T} \left(\left\langle -\mathbf{u}_\varepsilon, \frac{\partial \eta}{\partial t} \right\rangle + \nu \left\langle \frac{\partial \mathbf{u}_\varepsilon}{\partial x_i}, \frac{\partial \eta}{\partial x_i} \right\rangle + \left\langle \mathbf{f}\left(\mathbf{u}_\varepsilon, \frac{\partial \mathbf{u}_\varepsilon}{\partial x}\right), \eta \right\rangle + q_\varepsilon \operatorname{div} \eta \right) dx dt$$
$$+ \int_\Omega \langle \mathbf{u}_\epsilon, \eta \rangle \, dx \,|_{t=t} - \int_\Omega \langle \mathbf{a}, \eta(x, 0) \rangle \, dx = \int_{Q_t} \langle \mathbf{F}, \eta \rangle \, dx dt; \quad 0 \leqslant t \leqslant T,$$

$$(21_2) \qquad \int_{Q_t} \left(-\varepsilon q_\varepsilon \frac{\partial \psi}{\partial t} + \varepsilon r_i \frac{\partial q_\varepsilon}{\partial x_i} \frac{\partial \psi}{\partial x_i} + \psi \operatorname{div} \mathbf{u}_\varepsilon \right) dx dt$$
$$+ \varepsilon \int_\Omega q_\varepsilon(x, t) \psi(x, t)\, dx - \varepsilon \int_\Omega q_{\varepsilon 0}(x) \psi(x, 0)\, dx = 0$$

are satisfied for arbitrary $\eta(x,t) \in \mathring{\mathbf{W}}_2^{1,1}(Q_T)$ and $\psi(x,t) \in V_{2,r_i}(Q_T)$, which in Q_T have first order derivatives with respect to t that are bounded in $L_2(Q)$.

Theorem 4. *The initial-boundary-value problem* (19) *has no more than one generalized solution in* Q_T.

The proof of the theorem is similar to that in [**3**], 1970 ed., pp. 180–181, with those essential simplifications taken into account in the proof which relate to the assumptions of boundedness of $|\mathbf{f}|$, $|\partial f_k(\mathbf{u},p)/\partial u_i|$, $|\partial f_k(\mathbf{u},p)/\partial p_{ij}|$, $k, i, j = 1, \cdots, n$ (see relations (5) and (9)).

Theorem 5. *The generalized solution of problem* (19) *exists and has for*

$\mathbf{F}(x,t) \in L_2(Q_T)$ *and* $q_{\epsilon 0}(x) \in W_2^2(\Omega)$ *the derivatives* $\partial \mathbf{u}_\epsilon / \partial t \in L_2(Q)$, $\sqrt{\epsilon}\, \partial p_\epsilon / \partial t \in L_2(Q_T)$.

The existence in Theorem 5 can be proved by Galerkin's method similarly to the procedure followed in [3, 14, 15].

THEOREM 6. *Let the solution of problem* (1) *satisfy the conditions of Theorem* 1 *and the solution of* (19) *the conditions of Theorem* 5. *Then the generalized solution of problem* (19) *converges to the solution of problem* (1) *and, in addition,*

$$\sup \| \mathbf{u}_\varepsilon - \mathbf{u} \|_{L_2(\Omega_t)} + \int_{Q_T} \left| \frac{\partial \mathbf{u}_\varepsilon}{\partial x} - \frac{\partial \mathbf{u}}{\partial x} \right| dxdt + \varepsilon \| q_\varepsilon - q \|_{V_{2,r_i}(Q_T)} \leqslant e^{cT} \varepsilon \left(\int_{Q_T} \left(\frac{\partial q}{\partial t}\right)^2 + r_i \left(\frac{\partial q}{\partial x_i}\right)^2 \right) dxdt + \int_{\Omega} q^2_{\varepsilon_0}(x)\, dx. \tag{22}$$

To prove this theorem we subtract equation (4) from (21_1) and put $\boldsymbol{\eta} = \mathbf{u}_\epsilon - \mathbf{u}$ (this can be done by virtue of the assumptions of Theorem 1 and the assertions of Theorem 5). We then obtain

$$\sup_{t \in [0, T]} \int_{\Omega} | \mathbf{w}_\varepsilon (x, t) |^2 dx + \nu \int_{Q_T} \left| \frac{\partial \mathbf{w}_\varepsilon}{\partial x} \right|^2 dxdt + \int_{Q_T} z_\varepsilon \operatorname{div} \mathbf{w}_\varepsilon dxdt = \int_{Q_t} \left\langle \mathbf{f}\left(\mathbf{u}, \frac{\partial \mathbf{u}}{\partial x}\right) - \mathbf{f}\left(\mathbf{u}_\varepsilon, \frac{\partial \mathbf{u}_\varepsilon}{\partial x}\right), \mathbf{w}_\varepsilon \right\rangle dxdt, \tag{23}$$

where $\mathbf{w}_\epsilon(x,t) = \mathbf{u}_\epsilon(x,t) - \mathbf{u}(x,t)$ and $z_\epsilon(x,t) = q_\epsilon(x,t) - q(x,t)$.

Further, taking smoothness into account, we write down the obvious equation

$$\int_{Q_t} \left(-\varepsilon q \frac{\partial \psi}{\partial t} + \varepsilon r_i \frac{\partial q}{\partial x_i} \frac{\partial \psi}{\partial x_i} + \psi \operatorname{div} \mathbf{u} \right) dxdt + \varepsilon \int_{\Omega} q(x, t)\psi(x, t)\, dx - \varepsilon \int_{\Omega} q(x, 0)\psi(x, 0)\, dx$$
$$= -\int_{Q} \left(\varepsilon q \frac{\partial \psi}{\partial t} + \varepsilon r_i \frac{\partial q}{\partial x_i} \frac{\partial \psi}{\partial x_i} \right) dxdt + \varepsilon \int_{\Omega} q(x, t)\psi(x, t) - \varepsilon \int_{\Omega} q(x, 0)\psi(x, 0)\, dx, \tag{24}$$

where $\psi(x,t)$ is an arbitrary function from $V_{2,r_i}(Q_T)$ having a derivative $\partial\psi/\partial t$ which is bounded in $L_2(Q_T)$.

Putting $\psi(x,t) = z_\epsilon(x,t)$ into equations (21_2) and (24) and then subtracting (24) from (21_2), we find from (23), (21_2), (24), (5), (9) and Cauchy's inequality that

$$\int_{\Omega} (| \mathbf{w}_\varepsilon (x, t) |^2 + \varepsilon z^2 (x, t))\, dx + \int_{Q_t} \left(\nu \left| \frac{\partial \mathbf{w}_\varepsilon (x, t)}{\partial x} \right|^2 + \varepsilon r_i \left(\frac{\partial z_\varepsilon}{\partial x_i} \right)^2 \right) dxdt$$
$$\leqslant \frac{\varepsilon}{2} \left(\int_{\Omega_t} r_i \left(\frac{\partial q}{\partial x_i} \right)^2 dxdt + \int_{\Omega} q^2_{\varepsilon_0}(x)\, dx \right) + \int_{Q_t} \left(\frac{\varepsilon}{2} r_i \left(\frac{\partial z_\varepsilon}{\partial x_i} \right)^2 + \frac{\nu}{2} \left| \frac{\partial \mathbf{w}_\varepsilon}{\partial x} \right|^2 \right) dxdt + c \int_{Q_t} \frac{2}{\nu} | \mathbf{w}_\varepsilon (x, t) |^2 dxdt, \tag{25}$$

where the constant c is defined by the constants in (5) and (9). The validity of the theorem in question then follows from (25) (see [3]).

We now form an explicit difference scheme for the problem (19_1), (19_2), (13_3) (we omit the subscript ϵ for now). For $k=1,\cdots,N-1$ we can write this problem in the form

$$(26_1)\qquad v_{it}-\nu v_{ix_k\bar{x}_k}+f_{ih}(v,\ v_x)+p_{\bar{x}_i}=F_i,$$

$$(26_2)\qquad \begin{gathered}\varepsilon p_t-\varepsilon\,(r_{ih}p_{\bar{x}_i})_{x_i}+v^{+0}_{ix_i}=0,\\ f_{ih}(v_i,\ v_x)=v_{kh}\zeta_M(|\,v_k\,|)\,v_{ix_k}\zeta_{M_1}(|\,v_{ix_k}\,|).\end{gathered}$$

The equations (26_1) and (26_2) can be written down at the same points as the equations (7_1) and (7_2), respectively. To the equations (26_1) and (26_2) we add the initial and boundary conditions

$$(26_3)\qquad v_i\,|_{t=0}=a_{ih},\ \ v_i\,|_S=0,\ \ p\,|_{t=0}=q_{\varepsilon 0h}.$$

There is no need to write an additional boundary condition for p on S. The equation (26_2) is constructed so as to be automatically satisfied (see [7, 16]). The role of the boundary condition is played here by the equation

$$(26_4)\qquad r_{ih}\,|_{S_h}\equiv(x_i)\,(1-x_i)\,|_{S_h}=0.$$

To obtain an a priori estimate the difference equation (26_1), taken at the kth level, is multiplied by v_i^{k+1} and summed over Ω_h with respect to i from 1 to n and with respect to k from 1 to m. After a summation by parts we obtain

$$(27)\qquad \begin{aligned} h^n\sum_{\Omega_h}|\,\mathbf{v}^{m+1}\,|^2&+h^n(\Delta t)^2\sum_{k=1}^{m+1}\sum_{\Omega_h}|\,\mathbf{v}_{\bar{t}}^k\,|^2+2\nu\,(\Delta t)\,h^n\sum_{l=1}^{m+1}\sum_{\Omega_h}|\,\mathbf{v}_x^l\,|^2\\ &-2\nu\,h^n(\Delta t)^2\sum_{l=1}^{m+1}\sum_{\Omega_h}v^l_{ix_k}v^l_{ix_k\bar{t}}=2\,(\Delta t)\,h^n\sum_{l=1}^{m}\sum_{\Omega_h}(-1)\,(p^l_{\bar{x}_i}v_i^{l+1})\\ &+2(\Delta t)\,h^n\sum_{l=1}^{m}\sum_{\Omega_h}\langle-\mathbf{f}\,(v^l,\ v^l_x)+\mathbf{F}^l_h,\ \mathbf{v}^{l+1}\rangle+h^n\sum_{\Omega_h}|\,\mathbf{a}_h\,|^2.\end{aligned}$$

We rewrite the first term on the right side of (27) in the form

$$(28)\qquad (\Delta t)\,h^h\sum_{l=1}^{m}\sum_{\Omega_h}-p^l_{\bar{x}_i}v_i^{l+1}=(\Delta t)\,h^n\sum_{l=1}^{m}\sum_{\Omega_h}p^lv^l_{ix_i}+(\Delta t)^2h^n\sum_{l=1}^{m}\sum_{\Omega_h}(-p^l_{x_i})\,(v^{l+1}_{i\bar{t}}).$$

To estimate the first of the terms on the right side of equation (28) we multiply both sides of equation (26_2), taken at the $(l-1)$th level, by p^l and carry out a summation by parts. Then after simplifications we obtain

$$(29)\qquad \begin{aligned} h^n\sum_{l=1}^{m}\sum_{\Omega_h}p^lv^l_{ix_i}&=-\frac{1}{2}\,\varepsilon h^n\sum_{\Omega_h}(p^m)^2-\varepsilon\frac{(\Delta t)}{2}h^n(\Delta t)\sum_{l=1}^{m}\sum_{\Omega_h}(p^l_{\bar{t}})^2\\ &-\varepsilon h^n(\Delta t)\sum_{l=1}^{m}\sum_{\Omega_h}r_{ih}(p^l_{\bar{x}_i})^2-\varepsilon h^n(\Delta t)\sum_{l=1}^{m}\sum_{\Omega_h}r_{ih}p^l_{\bar{x}_i\bar{t}}p^l_{x_i}+\frac{1}{2}\,\varepsilon h^n\sum_{\Omega_h}|\,q_{\varepsilon 0h}\,|^2.\end{aligned}$$

Choosing (Δt) in (29) so as to satisfy

$$4n(\Delta t)/h^2 \leqq 1, \tag{30}$$

we obtain

$$h^n(\Delta t)\sum_{l=1}^{m}\sum_{\Omega_h} p_h^l v_{ix_i}^l \leqslant -\frac{\varepsilon}{2}h^n\sum_{\Omega_h}(p_h^m)^2 - \frac{\varepsilon}{2}h^n(\Delta t)\sum_{l=1}^{m}\sum_{\Omega_h} r_{ih}(p_{\bar{x}_i}^l)^2 + \varepsilon h^n\sum_{\Omega_h}(q_{\varepsilon 0h})^2. \tag{31}$$

We substitute (28) and (31) into the left side of (27). Then noting the condition (30) and, additionally, putting

$$4n\nu(\Delta t)/h^2 < 1, \tag{32}$$

we have

$$\begin{aligned} h^n\sum_{\Omega_h}|\mathbf{v}^{m+1}|^2 &+ \frac{\nu}{2}(\Delta t)h^n\sum_{l=1}^{m+1}\sum_{\Omega_h}|\mathbf{v}_x^l|^2 + \varepsilon h^n\sum_{\Omega_h}(p^m)^2 \\ &+ \varepsilon h^n\frac{(\Delta t)}{2}\sum_{l=1}^{m}\sum_{\Omega_h} r_{ih}(p_{\bar{x}_i}^l)^2 + \frac{\Delta t}{4}h^n(\Delta t)\sum_{l=1}^{m+1}\sum_{\Omega_h}|\mathbf{v}_{\bar{t}}^l|^2 \\ &\leqslant \varepsilon\frac{h}{2}(\Delta t)h^n\sum_{l=1}^{m}\sum_{\Omega_h}\sum_{i=1}^{3}(p_{\bar{x}_i})^2 + \frac{2n(\Delta t)}{\varepsilon h}\frac{(\Delta t)}{4}(\Delta t)h^n\sum_{l=1}^{m}\sum_{\Omega_h}|\mathbf{v}_{\bar{t}}^{l+1}|^2 \\ &+ \frac{1}{2}h^n(\Delta t)\sum_{l=1}^{m+1}(\mathbf{v}^l)^2 + h^n\sum_{\Omega_h}|\mathbf{a}_h|^2 + h^n\sum_{\Omega_h}(q_{\varepsilon 0h})^2 \\ &+ \frac{1}{2}h^n(\Delta t)\sum_{l=0}^{m}\sum_{\Omega_h}\{|\mathbf{f}_h^l(\mathbf{v},\mathbf{v}_x)|^2 + |\mathbf{F}_h^l|^2\}. \end{aligned} \tag{33}$$

From the obvious inequality

$$\frac{\varepsilon h}{2}(\Delta t)h^n\sum_{l=1}^{n}\sum_{\Omega_h}(p_{x_i}^l)^2 \leqslant \varepsilon\frac{h^n\Delta t}{2}\sum_{l=1}^{n}\sum_{\Omega_h} r_{ih}(p_{x_i}^l)^2,$$

assuming, in addition to (31) and (32), that

$$2n(\Delta t)/\epsilon h < 1, \tag{34}$$

we obtain the estimate

$$h^n\sum_{\Omega_h}|\mathbf{v}^{m+1}|^2 \leqslant h^n(\Delta t)\sum_{l=1}^{m+1}\sum_{\Omega_h}|\mathbf{v}^l|^2 + c_1, \tag{35}$$

where the constant is determined by the constants from (5) and (9) and the value of the norms $\|\mathbf{F}_h\|_{L_2(Q_T)}$, $\|\mathbf{a}_h\|_{L_2(\Omega)}$ and $\|p_h^0\|_{L_2(\Omega)}$. The stability of the difference scheme (26) follows from (35). If now we consider the equations for the difference and assume that $\mathbf{u} \in W_2^2(Q_T)$ and $q(x,t) \in W_2^2(Q_T)$, then as in [3] we obtain an estimate of the rate of convergence.

From the equations (26_1) and (26_2) we construct summator identities similar to the integral identities (21_1) and (21_2). They appear as follows:

$$\begin{aligned} I_1(\mathbf{v}_h, \mathbf{p}_h, \eta_h) \equiv h^n(\Delta t)\sum_{\Omega_h}\sum_{l=0}^{m}(\langle\mathbf{v}_{ht}^l, \eta_h^l\rangle + \nu\langle\mathbf{v}_{hx_i}^l, \eta_{hx_i}^l\rangle \\ + \langle\mathbf{f}_h^l(\mathbf{v}_h, \mathbf{v}_{hx_i}), \eta_{hl}\rangle + p_{\varepsilon h\bar{x}_i}\eta_{ih} - F_{ih}\eta_{ih}) = 0, \end{aligned} \tag{36_1}$$

$$(36_2) \qquad I_2(v_h,\ p_{\varepsilon h},\ \psi_h) \equiv h^n (\Delta t) \sum_{\Omega_h} \sum_{l=0}^{m} (\varepsilon p^l_{ht}\psi^l_h + \varepsilon r_{ih} p^l_{hx_i}\psi_{hx_i} + \psi^l_h v^l_{ihx_i}) = 0.$$

We consider functionals I_1 and I_2 on the grid functions $\mathbf{u}_h$ and q_h, which are obtained as Steklov averages of the exact solution $\mathbf{u}$ and q of the exact solution of equations (1) with step h taken at the grid nodes. Then as in [**13**] we obtain

$$(37_1) \qquad I_1(\mathbf{u}_h,\ q_h,\ '\eta_h) = R_1(\mathbf{u},\ q,\ \eta,\ \mathbf{u}_h,\ q_h,\ \eta_h),$$

$$(37_2) \qquad I_2(\mathbf{u}_h,\ q_h,\ \eta_h) = R_2(\mathbf{u},\ q,\ \psi,\ \mathbf{u}_h,\ q_h,\ \psi_h),$$

where R_1 and R_2 can be obtained as the errors of approximation of the integral identities (21_1) and (24) as summators considered on the functions $\mathbf{u}$ and q, the exact solutions of (1). Next, subtracting from (36_1) and (36_2) the equations (37_1) and (37_2), and putting $\eta_h = \mathbf{v}_h - \mathbf{u}_h$ and $\psi_h = q_h - p_h$, similarly to the manner in which we obtained (35) we can derive an estimate for the differences $\mathbf{v}_h - \mathbf{u}_h$ and $p_h - q_h$ in terms of estimates for $|R_1|$ and $|R_2|$. We estimate $|R_1|$ and $|R_2|$ in turn through the norm of the differences $\|\mathbf{u} - \mathbf{u}'_h\|_{W_2^1(Q)}$ and the quantities

$$\varepsilon\left\{\left\|\frac{\partial q}{\partial t}\right\|_{L_2(Q_T)} + \left\|\sqrt{r_i}\,\frac{\partial q}{\partial x_i}\right\|_{L_2(Q_T)}\right\} + \sum_{i=1}^{3}\left\|\frac{\partial q}{\partial x_i} - \frac{\partial q'_h}{\partial x_i}\right\|_{L_2(Q_T)},$$

where $\mathbf{u}'_h$ and q'_h are the multilinear complements of $\mathbf{u}_h$ and q_h. A method of obtaining such estimates was given in [**13, 6**]. But if $\|\mathbf{u}\|_{W_2^2(Q_T)}$ and $\|q\|_{W_2^2(Q_T)}$ are bounded, such estimates lead to the following result.

THEOREM 7. *Suppose that the exact solution of the boundary-value problem* (1) *satisfies the conditions of Theorem* 3 *and that the steps* (Δt) *and* h *tend to zero in such a way that*

$$(38) \qquad \frac{4n(\Delta t)}{h^2} \leqslant 1; \quad \frac{4n\nu(\Delta t)}{h^2} \leqslant 1; \quad \frac{2n(\Delta t)}{\varepsilon h} \leqslant 1.$$

Then the approximate solution $\mathbf{v}_h$ *of the problem* (19), (20), (14) *converges to the exact solution of the problem* (1), *namely* $\mathbf{u}(x,t)$, *and, in addition,*

$$\sup_{t\in[0,T]} \int_{\Omega} (\mathbf{u}(x,\ t) - \mathbf{v}'_{h\varepsilon}(x,\ t))^2\,dx + \int_{Q_t} \left|\frac{\partial \mathbf{u}}{\partial x} - \frac{\partial \mathbf{v}'_{h\varepsilon}}{\partial x}\right|^2 dx\,dt$$

$$(39) \qquad \leqslant c\left\{\varepsilon + h + (\Delta t) + \int_{Q_T} (|\mathbf{F}(x,\ t) - \mathbf{F}(x - \Delta x,\ t)|^2 + |\mathbf{F}(x,\ t) - \mathbf{F}(x,\ t - \Delta t)|^2)\,dx\,dt\right\},$$

where the constant c is determined by the values of the constants from (35), (5), (9) *and the estimates of* $\|\mathbf{u}\|_{W_2^2(Q_T)}$ *and* $\|q\|_{W_2^1(Q_T)}$.

REMARK. The fact that Ω is a cube is not essential. In the case of an arbitrary smooth domain Ω it is necessary to consider, instead of the operator $(\partial/\partial x_i)x_i(1 - x_i)\partial q_\varepsilon/\partial x_i$ in (19_2), the operator $(\partial/\partial x_i)\zeta^2(x)\partial q_\varepsilon/\partial x_i$, where $\zeta^2(x)$ is a cutoff function in Ω which vanishes with an approach to the boundary proportional to the first power of the distance to the boundary.

We consider yet one more explicit difference scheme, starting from the approximation (one close to it was considered in [**12**])

(38₁) $$\tilde{L}(\mathbf{u}_\epsilon,\ q_\varepsilon)=\mathbf{F},$$

(38₂) $$\varepsilon\frac{\partial q_\varepsilon}{\partial t}-\varepsilon\Delta q_\varepsilon+\operatorname{div}\mathbf{u}_\varepsilon=0.$$

To the initial and boundary conditions (13_3) it is necessary here to add a boundary condition on q_ϵ:

(38₃) $$q_\varepsilon|_S=0 \quad \text{or} \quad \left.\frac{\partial q_\varepsilon}{\partial \mathbf{n}}\right|_S=0;$$

$\mathbf{n}$ is the normal to (S).

As in the case of equations (19_1) and (19_2), the solution of the equations (38_1), (38_2), and (38_3) converges for $\epsilon\to 0$ to the exact solution of the problem (1). For this solution we can also establish an estimate of the rate of convergence for the pressure q_ϵ in a norm with a weight vanishing on the boundary S of Ω.

The explicit difference scheme constructed on the approximations (38_1), (38_2), (38_3), (13_3) preserves the equation (26_1); equation (26_2) is replaced by the difference equation

(39₁) $$\varepsilon p_t-\varepsilon p_{x_i\bar{x}_i}+v_{ih\varepsilon x_i}^{+0}=0.$$

On the boundary S_h we place the condition

(39₂) $$q_{\epsilon h}|_{S_h}=0.$$

The results we obtain for the difference scheme (26_1), (39_1), (39_2), (26_3) are similar to those established for the explicit difference scheme (26_1), (26_2), (26_3); however the condition (34) is replaced by the requirement

(40) $$\Delta t/\epsilon<1.$$

The latter follows from the fact that the expression $\epsilon/2(\Delta t)h^n\sum_{i=1}^m\sum_{\Omega_h}\sum_{i=1}^n p_{x_i}^2$ may be estimated in terms of an analogous term appearing in the right member of the result obtained by replacing (26_2) by the equation (38_2) corresponding to it in the given approximation. Thus for $\epsilon\sim h^2$ we obtain a difference scheme converging to the exact solution for the step relation $(\Delta t)\sim h^2$. For its solution a theorem analogous to Theorem 7 is valid, the difference being that the condition (38) is replaced by the requirement

(41) $$\frac{4n\,(\Delta t)}{h^2}\leqslant 1;\quad \frac{4n\nu\,(\Delta t)}{h^2}\leqslant 1;\quad \frac{2n\,(\Delta t)}{\varepsilon}\leqslant 1$$

(the condition $2n\Delta t/\epsilon\leq 1$ could also be replaced by the requirement $n\Delta t/\sqrt{\epsilon}h\leq 1$ and, correspondingly, $2n(\Delta t)/h^2\leq 1$).

In the present paper we have considered how a modification of the nonlinear terms (2) – (3) furthers the construction of grid schemes for nonstationary problems. However, it is entirely clear that the use of a given modification makes it easy to obtain estimates of the rate of convergence in other projected methods also: Galerkin's method, the method of lines, and others. Moreover, if it is known that the stationary equations have several solutions which, together with their first order derivatives, are bounded, then all these solutions will also be solutions of the modified equations (3) for corresponding values of M and M_1. By the same token, the search for solutions of this kind can be reduced to a search of the corresponding solutions of equation (3). We remark that the modification (2) – (3) makes it possible to justify a number of iterational processes, for example, the method of establishing a solution for sufficiently small Re numbers (values of M and M_1). A modification similar to (2) – (3) can also be used for solving equations with a nonlinear viscosity. Convergent difference schemes of a similar nature can also be constructed for quasilinear systems of equations of the type proposed by O. A. Ladyženskaja as modifications of the Navier-Stokes equations. It is obvious, in the case of two-dimensional flows, that instead of the nonlinear terms $\mathbf{f}(\mathbf{u}, \partial\mathbf{u}/\partial x)$ given in (2) – (3), we can construct nonlinear terms of the form

$$\varphi\left(\mathbf{u}, \frac{\partial\mathbf{u}}{\partial x}\right) = \mathbf{u}_k \zeta_M(|\mathbf{u}_k|) \frac{\partial\mathbf{u}}{\partial x_k}. \tag{42}$$

For the equations obtained from the Navier-Stokes equations through replacements of the nonlinear terms of this kind, the results obtained are valid even for the modifications considered in this paper. Moreover, it need only be required of the exact solution of equations (1) that solutions exist of bounded absolute value. This is guaranteed, for example, by showing that a solution belongs to $W_2^2(\Omega)$ for $t \in [0, T]$. Replacements of this kind can also be considered in the case of three variables; however, then all the proofs presented in this paper would become substantially more complicated while, at the same time, with the assumptions made here that the solution of problem (1) be a function $\mathbf{u} \in W_2^2(Q)$, one could, for suitable smoothness of the boundary S and of the function $\mathbf{F}$, even prove the smoothness of the solution, which is a premise of Theorem 1.

In conclusion we should remark that along with the linear regularizations investigated in [**2**, **8**, **12**], and also proposed here, we could suggest also a set of other linear and nonlinear regularizations of the Navier-Stokes equations. However we have tended here to limit ourselves to the simplest of the possible regularizations.

I express my thanks to O. A. Ladyženskaja for her interest in my work and a discussion of the results.

Bibliography

[1] A. Kživickiĭ and O. A. Ladyženskaja, *The method of nets for the nonstationary Navier-Stokes equations,* Trudy Mat. Inst. Steklov. **92** (1966), 93-99 = Proc. Steklov Inst. Math. **92** (1966), 105-112. MR **34** #4651.

[2] O. A. Ladyženskaja, *Stable difference schemes for the Navier-Stokes equations,* Zap. Naučn. Sem. Leningrad. Otdel. Mat. Inst. Steklov. (LOMI) **14** (1969), 92-126 = Sem. Math. V. A. Steklov Math. Inst. Leningrad **14** (1969), 46-62. MR **44** #1250.

[3] ———, *Mathematical problems in the dynamics of a viscous incompressible fluid,* 2nd rev. aug. ed., "Nauka", Moscow, 1970; English transl. of 1st ed., *The mathematical theory of viscous incompressible flow,* Gordon and Breach, New York, 1963; rev. 1969. MR **40** #7610; **27** #5034b; **42** #6442.

[4] R. Temam, *Une méthode d'approximation de la solution des équations de Navier-Stokes,* Bull. Soc. Math. France **96** (1968), 115-152. MR **38** #6249.

[5] ———, *Sur l'approximation de la solution des équations de Navier-Stokes par la méthode des pas fractionnaires.* I, Arch. Rational Mech. Anal. **32** (1969), 135-153. MR **38** #6250.

[6] O. A. Ladyženskaja and V. Ja. Rivkind, *On convergent difference schemes for the Navier-Stokes equations,* Čisl. Metody Meh. Splošnoĭ Sredy **2** (1971), no. 1, 55-73. (Russian) MR **48** #3365.

[7] ———, *On the alternating directions method for the calculation in cylindrical coordinates of the flow of a viscous incompressible fluid,* Izv. Akad. Nauk SSSR Ser. Mat. **35** (1971), 259-268 = Math. USSR Izv. **5** (1971), 267-277. MR **44** #1306.

[8] N. N. Janenko, *The method of fractional steps for solving multidimensional problems of mathematical physics,* "Nauka", Novosibirsk, 1967; English transl., Springer-Verlag, Berlin and New York, 1971. MR **36** #4815; **46** #6613.

[9] I. Ju. Brailovskaja, T. V. Kuskova and L. A. Čudov, *Difference methods of solution of the Navier-Stokes equations (a survey),* Computing Methods and Programming, XI: Numerical Methods in Gas Dynamics, Izdat. Moskov. Univ., Moscow, 1968, pp. 3-18. (Russian) MR **44** #1229.

[10] L. M. Simuni, *Motion of a viscous incompressible fluid in a plane pipe,* Ž. Vyčisl. Mat. i Mat. Fiz. **5** (1965), 768-772 = U.S.S.R. Comput. Math. and Math. Phys. **5** (1965), no. 6, 229-234.

[11] I. E. Fromm, *A method for computing nonsteady incompressible, viscous fluid flows,* Los Alamos Sci. Labor., Rep. LA-2910, 1963.

[12] A. P. Oskolkov, *A certain quasilinear parabolic system with small parameter that approximates a system of Navier-Stokes equations,* Zap. Naučn. Sem. Leningrad. Otdel. Mat. Inst. Steklov. (LOMI) **21** (1971), 79-103 = J. Soviet Math. **1** (1973), 452-470. MR **45** #7314.

[13] V. Ja. Rivkind, *On an estimate of the rapidity of convergence of homogeneous difference schemes for elliptic and parabolic equations with discontinuous coefficients,* Problemy Mat. Anal., vyp. 1, Izdat. Leningrad. Univ., Leningrad, 1966, pp. 110-119 = Problems in Math. Anal., no. 2, Plenum Press, New York, 1968, pp. 102-110. MR **36** #4819.

[14] Ju. A. Dubinskiĭ, *Quasilinear elliptic and parabolic equations of arbitrary order,* Uspehi Mat. Nauk **23** (1968), no. 1 (139), 45-90 = Russian Math. Surveys **23** (1968), no. 1, 45-91. MR **37** #4405.

[15] V. Ja. Rivkind and N. N. Ural′ceva, *Classical solvability and linear schemes for the approximate solution of diffraction problems for quasilinear elliptic and parabolic equations,* Problemy Mat. Anal., vyp. 3, Izdat. Leningrad. Gos. Univ., Leningrad, 1972, pp. 69-110 = J. Soviet Math. **1** (1973), 235-264.

[16] V. Ja. Rivkind and R. Merc, *The finite difference method for solving degenerate elliptic and parabolic equations,* Vestnik Leningrad Univ. **22** (1967), no. 13, 64-75. (Russian) MR **36** #2326.

Translated by
J. F. HEYDA

Trudy Mat. Inst. Steklov.
125 (1973)

Proc. Steklov Inst. Math.
125 (1973)

ON THE SPECTRUM OF THE SCHRÖDINGER OPERATOR WITH A RAPIDLY OSCILLATING POTENTIAL

M. M. SKRIGANOV

Abstract. A study is made of the spectrum of the many-dimensional Schrödinger operator with a potential rapidly oscillating and unbounded at infinity. Criteria for selfadjointness and semiboundedness of such an operator are obtained; criteria are also obtained for the finiteness of the negative part of the spectrum and for the absence of positive eigenvalues. For such operators a theorem on the expansion in eigenfunctions is proved.

Bibliography: 10 items.

A proof was given in [1] of the existence of the complete wave operators

$$W_{\pm}(H, H_0) = s - \lim_{t \to \pm\infty} e^{itH} e^{-itH_0}, \tag{0.1}$$

for the Schrödinger operators

$$H = -\Delta + q(x), \quad \operatorname{Im} q = 0, \quad H_0 = -\Delta, \quad x \in R^n, \tag{0.2}$$

with a potential $q(x)$ oscillating sufficiently rapidly, and possibly even unbounded for $|x| \to \infty$. For example, the complete wave operators (0.1) exist for $q(x) = \varphi(\hat{x})(1 - |x|)^{\alpha} \sin \exp |x|$, where $\varphi(\hat{x})$ is an arbitrary smooth function of the angular coordinates $\hat{x}$ of the vector x and α is an arbitrary real number. Scattering by such potentials is a purely quantum phenomenon and may be explained by the tunnel effect: the "thickness" of the potential barrier is finite, although the integral describing it converges nonabsolutely.

From the existence of the complete wave operators (0.1) the only conclusion that can be drawn is that the absolutely continuous part of the operator H is unitarily equivalent to the operator H_0. In the present paper we study the spectrum of H in more detail. In §1 we carry over the known results of qualitative spectral theory to operators H with rapidly oscillating potentials. In §2 for Schrödinger operators with a broad class of rapidly oscillating potentials we prove a theorem on expansion in eigenfunctions, giving therewith a complete description of such operators.

The author thanks V. S. Buslaev for his interest in the work, and V. B. Matveev and L. D. Faddeev for a discussion of the results.

§1. Let $x \in R^n$; we put $r = |x|$ and $\hat{x} = r^{-1}x$. We denote by $\Delta_{\hat{x}}$ the Laplace operator on the unit sphere S^{n-1} in R^n.

The basic assumption we make concerning the potential is the following.

CONDITION A. *The potential* $q(x)$ *is a real continuous function and the limit*

$$\Phi(x) = \lim_{N \to \infty} \int_r^N d\xi \lim_{M \to \infty} \int_{\xi}^M q(\eta \hat{x})\, d\eta \tag{1.1}$$

exists, where the function $\Phi(x)$ *is twice continuously differentiable.*

AMS (MOS) subject classifications (1970). Primary 35J10, 35P25, 47A40; Secondary 81A45.

We note that $\Phi(x)$ satisfies the equation

$$\frac{\partial^2 \Phi(x)}{\partial r^2} = q(x) \tag{1.2}$$

and that $\Phi(x) = o(1)$ for $r \to \infty$.

In $L_2(R^n)$ we introduce the operator consisting of multiplication by a function:

$$(Tf)(x) = T(x) f(x), \quad T(x) = e^{\Phi_\eta(x)}, \quad \Phi_\eta(x) = \Phi(x) \eta(x), \tag{1.3}$$

where $\eta(x)$ is a smooth real-valued function, equal to 0 for $r < 1$ and to 1 for $r > 2$. It is obvious that T is selfadjoint and that T and T^{-1} are bounded.

On the set $C_0^2(R^n)$ of finite twice continuously differentiable functions we assign the symmetric operator

$$\tilde{H} f(x) = -\Delta f(x) + q(x) f(x), \quad f(x) \in C_0^2(R^n). \tag{1.4}$$

We denote the operator $-\Delta$ by H_0.

Theorem 1.1. *Let the potential $q(x)$ satisfy Condition* A *and, in addition, let the following estimate hold for $r \to \infty$:*

$$|\nabla \Phi(x)| + r^{-2} |\Delta_{\hat{x}} \Phi(x)| = O(1). \tag{1.5}$$

Then on the set $C_0^2(R^n)$ the operator H is semibounded and essentially selfadjoint. Its closure is a selfadjoint operator H with the domain of definition

$$D(H) = TD(H_0), \quad D(H_0) = W_2^2(R^n). \tag{1.6}$$

Proof. We note that the operator T maps the set $C_0^2(R^n) \subset L_2(R^n)$ onto itself. We use the obvious relation

$$\begin{gathered}(\tilde{H} - z) T = T [I + Q (H_0 - z)^{-1}] (H_0 - z), \\ Q = T^{-1} (\tilde{H} T - T H_0),\end{gathered} \tag{1.7}$$

where I is the identity operator in $L_2(R^n)$ and Q is a differential operator of first order, namely,

$$Qf(x) = \langle Q_1(x), \nabla f(x) \rangle + Q_2(x) f(x); \tag{1.8}$$

here $\langle \cdot, \cdot \rangle$ is the scalar product in R^n and

$$Q_1(x) = -2 \nabla \Phi_\eta(x), \tag{1.9}$$

$$Q_2(x) = q(x) - \Delta \Phi_\eta(x) - (\nabla \Phi_\eta(x))^2. \tag{1.10}$$

For $r > 2$, writing the Laplace operator in polar coordinates

$$\Delta = \frac{\partial^2}{\partial r^2} + \frac{n-1}{r} \frac{\partial}{\partial r} + \frac{1}{r^2} \Delta_{\hat{x}} \tag{1.11}$$

and using (1.2), we obtain

$$Q_2(x) = -r^{-2} \Delta_{\hat{x}} \Phi - (\nabla \Phi)^2 - \frac{n-1}{r} \frac{\partial \Phi}{\partial r}. \tag{1.12}$$

Thanks to the estimate (1.5) the coefficients Q_1 and Q_2 are bounded. From this it follows that for all real $z < -a < 0$ and sufficiently large $a > 0$ the following estimate holds:

$$\| Q (H_0 - z)^{-1} \|_{L_2(R^n)} < 1. \tag{1.13}$$

Consequently we transform the second factor in (1.7). Using the essential selfadjointness of H_0 on the domain $C_0^2(R^n)$, we see from (1.7) that for $z < -a$ the operator $\widetilde{H} - z$ maps the set $TC_0^2(R^n)$ onto a set dense in $L_2(R^n)$. Thus $\widetilde{H}$ is essentially selfadjoint on the set $TC_0^2(R^n) = C_0^2(R^n)$, and the points of the real semi-axis $z < -a$ are nonspectral points for this operator, i.e. $\widetilde{H}$ is semibounded. From equation (1.7) we also obtain the equation (1.6). This completes the proof of Theorem 1.1.

We now prove several assertions concerning the spectrum of H.

THEOREM 1.2. *Let the potential satisfy Condition* A *and, in addition, let the following estimate hold for* $r \to \infty$:

$$|\nabla\Phi(x)| + r^{-2}|\Delta_{\hat{x}}\Phi(x)| = o(1). \tag{1.14}$$

Then the condensation spectrum of the operator H coincides with the semi-axis $[0, \infty)$.

PROOF. For real λ we consider the family of operators, selfadjoint on $W_2^2(R^n)$,

$$L_\lambda = THT - \lambda(I - T^2); \tag{1.15}$$

for $f(x) \in W_2^2(R^n)$ it is easy to verify that

$$\begin{gathered} L_\lambda f(x) = -\nabla(a(x)\nabla f(x)) + p_\lambda(x) f(x), \\ a(x) = T^2(x), \end{gathered} \tag{1.16}$$

where

$$p_\lambda(x) = T^2(x)[q(x) - \Delta\Phi_\eta(x) - (\nabla\Phi_\eta(x))^2] + \lambda[1 - T^2(x)]; \tag{1.17}$$

for $r > 2$, by virtue of (1.2) and (1.11),

$$p_\lambda(x) = -T^2(x)\left[r^{-2}\Delta_{\hat{x}}\Phi + (\nabla\Phi)^2 + \frac{n-1}{r}\frac{\partial\Phi}{\partial r}\right] + \lambda[1 - T^2(x)]. \tag{1.18}$$

The functions $p_\lambda(x)$, $a(x)$ and $(a(x))^{-1}$ are bounded; in addition, from the condition (1.14) there follow for $r \to \infty$ the estimates

$$p_\lambda(x) = o(1) \quad \text{and} \quad a(x) = 1 + o(1). \tag{1.19}$$

Therefore it follows from Theorem 3.5 in M. Š. Birman's paper [2] that the condensation spectrum of the operator L_λ coincides for arbitrary real λ with the semi-axis $[0, \infty)$. By definition, this means that for arbitrary $\lambda_0 \geqq 0$ there exists a sequence of functions $f_n = f_n(\lambda, \lambda_0)$, bounded and noncompact in $L_2(R^n)$, belonging to $W_2^2(R^n)$, such that

$$\lim_{n\to\infty} \|L_\lambda f_n - \lambda_0 f_n\|_{L_2(R^n)} = 0. \tag{1.20}$$

Consider the sequence $g_n = Tf_n(\lambda_0, \lambda_0)$. It is bounded, noncompact in $L_2(R^n)$, and belongs to $D(H)$; moreover,

$$\begin{aligned} Hg_n - \lambda_0 g_n &= T^{-1}[THTf_n(\lambda_0, \lambda_0) - \lambda_0 T^2 f_n(\lambda_0, \lambda_0)] \\ &= T^{-1}[L_{\lambda_0} f_n(\lambda_0, \lambda_0) - \lambda_0 f_n(\lambda_0, \lambda_0)]. \end{aligned} \tag{1.21}$$

Consequently from (1.20) we obtain

$$\lim_{n\to\infty} \|Hg_n - \lambda_0 g_n\|_{L_2(R^n)} = 0, \tag{1.22}$$

i.e. an arbitrary point $\lambda_0 \geqq 0$ is a point of the condensation spectrum of H. This completes the proof of Theorem 1.2.

THEOREM 1.3. *Let the potential satisfy Condition* A *and, in addition, let the following estimate*[1] *be valid for* $r \to \infty$:

$$|\Phi(x)| + |\nabla\Phi(x)| + r^{-2}|\Delta_{\hat{x}}\Phi(x)| = o(r^{-1}). \tag{1.23}$$

Then the operator H has no positive eigenvalues.

PROOF. As in the proof of Theorem 1.2, we consider the family L_λ (see (1.15)) for real λ. Using (1.23), we obtain for $r \to \infty$

$$|\nabla a(x)| + |p_\lambda(x)| = o(r^{-1}) \tag{1.24}$$

for arbitrary λ and $\lim_{r\to\infty} a(x) = 1$.

It has been proved by S. N. Roze [3] that in this case the operator L_λ has no positive eigenvalues.

Assume now that the operator H has a positive eigenvalue $\lambda_0 > 0$. Let $f \in D(H)$ be a corresponding eigenfunction

$$Hf = \lambda_0 f. \tag{1.25}$$

We put $g = T^{-1}f$, $g \in W_2^2(R^n)$; then

$$THTg = \lambda_0 T^2 g \tag{1.26}$$

or

$$[THT - \lambda_0(I - T^2)]g = L_{\lambda_0} g = \lambda_0 g, \tag{1.27}$$

i.e. $\lambda_0 > 0$ is a positive eigenvalue of L_{λ_0}. This contradiction completes the proof of Theorem 1.3.

THEOREM 1.4. *Let the potential satisfy Condition* A *and, in addition, let the following estimate be valid for* $r \to \infty$:

$$|\nabla\Phi(x)| + r^{-2}|\Delta_{\hat{x}}\Phi(x)| = \begin{cases} o(r^{-2}), & n \geqslant 3, \\ o(r^{-2}\ln^{-2} r), & n = 2. \end{cases} \tag{1.28}$$

Then the discrete spectrum of the operator H can consist only of a finite number of nonpositive eigenvalues of finite multiplicity.

PROOF. We note that the following identity holds for the quadratic forms of the operators H and $L = THT$:

$$(Hg, g) = (Lf, f), \quad g = Tf. \tag{1.29}$$

According to Theorem 1.3, H has no positive discrete spectrum. Let us assume that the common multiplicity of the nonpositive part of the spectrum of H is infinite. Then on some infinite-dimensional linear set $\mathfrak{M}$ of $D(H)$ the quadratic form of H is nonpositive. It follows from (1.29) that the quadratic form of L is also nonpositive on an infinite-dimensional linear set $\mathfrak{N} = T^{-1}\mathfrak{M}$ of $W_2^2(R^n)$. By the relations (1.15)–(1.19) with $\lambda = 0$ the operator L is

[1] For $n = 1$ it is sufficient for the validity of Theorem 1.3 to require merely that $\int_x^{\pm\infty} q(t)\,dt = O(|x|^{-1-\epsilon})$, $\epsilon > 0$, for $x \to \pm\infty$.

strongly elliptic, and its quadratic form can be estimated from below by the quadratic form of the Schrödinger operator $C = -\Delta + c(x)$, with a potential $c(x)$ satisfying, by virtue of (1.29), the relations

$$c(x) = \begin{cases} o(r^{-2}), & n \geqslant 3, \\ o(r^{-2}\ln^{-2} r), & n = 2. \end{cases} \tag{1.30}$$

However, as M. Š. Birman [4] showed, the common multiplicity of the nonpositive part of the spectrum of the operator C with the condition (1.30) cannot be infinite. This completes the proof of Theorem 1.4.

§2. In this section we prove a theorem concerning an expansion in eigenfunctions of the Schrödinger operator with a rapidly oscillating potential $q(x)$. To simplify the presentation we restrict the discussion to the case $n = 3$; however, our proof carries over immediately to the case of arbitrary dimension.

In this section we assume that $q(x)$ satisfies

CONDITION B. *The potential $q(x)$ is a real once continuously differentiable function and the limit* (1.1) *exists, wherein the function $\Phi(x)$ is twice continuously differentiable. In addition, the following estimate*[2] *holds for $r \to \infty$:*

$$\nabla\Phi(x) = O(r^{-2-\varepsilon}), \quad r^{-2}\Delta_{\hat{x}}\Phi(x) = O(r^{-2-\varepsilon}), \quad \varepsilon > 0. \tag{2.1}$$

Suppose that the function $\psi(x)$ satisfies the equation

$$-\Delta\psi(x) + q(x)\psi(x) = k^2\psi(x); \tag{2.2}$$

then the function

$$\varphi(x) = T^{-1}(x)\psi(x) \tag{2.3}$$

satisfies the equation

$$-\Delta\varphi(x) + Q\varphi(x) = k^2\varphi(x), \tag{2.4}$$

where Q is the first-order differential operator (1.8).

From Condition B it follows that the coefficients $Q_1(x)$ and $Q_2(x)$ of Q are once continuously differentiable and for $r \to \infty$ satisfy the estimates

$$Q_{1,2}(x) = O(r^{-2-\epsilon}), \qquad \epsilon > 0. \tag{2.5}$$

We study the following integral equation:

$$\varphi(x, k) = e^{i\langle x, k\rangle} - \int \frac{e^{i|k||x-y|}}{4\pi|x-y|} Q\varphi(y, k)\, dy \tag{2.6}$$

for $k \in R^3$ and $|k| > 0$.

Equation (2.6) is a generalization of the Lippmann-Schwinger equation to the case of the Schrödinger equation (2.2) with a rapidly oscillating potential.

It is natural to consider the equation (2.6) in the Banach space $\mathfrak{B}$ of once continuously differentiable functions on R^3 which, together with their first derivatives, vanish for $r \to \infty$.

---

[2] For arbitrary $n \geq 3$ the right sides of (2.1) are to be replaced by $O(r^{-(n+1)/2-\epsilon})$, $\epsilon > 0$.

We define the norm in $\mathfrak{B}$ as follows:

$$|u|_{\mathfrak{B}} = \max_{x \in R^3} (|u(x)| + |\nabla u(x)|). \tag{2.7}$$

We shall need the following lemmas concerning the behavior of the integral

$$u(x) = -\int \frac{e^{i|k||x-y|}}{4\pi |x-y|} v(y)\, dy. \tag{2.8}$$

LEMMA 2.1. *If $v(x)$ is a continuous function satisfying the estimate*

$$|v(x)| \leqq C(1+r)^{-2-\delta}, \qquad \delta > 0, \tag{2.9}$$

then for $r \to \infty$

$$|u(x)| + |\nabla u(x)| = O(r^{-1}) + O(r^{-\delta}). \tag{2.10}$$

Lemma 2.1 is proved in a manner similar to that used in proving Lemma 3.1 in the paper [**5**] by T. Ikebe.

LEMMA 2.2. *Under the assumptions of Lemma* 2.1, *the following estimate is valid when $|x - x'| \to 0$:*

$$\nabla u(x) - \nabla u(x') = O(|x - x'| \ln |x - x'|), \tag{2.11}$$

where the right side of (2.11) *depends only on the constants C and δ in* (2.9).

The proof of (2.11) is readily obtained by an obvious modification of the proof of the analogous estimate (4.3) in [**5**].

LEMMA 2.3. *Let $v(x)$ be a continuous function which satisfies for $r \to \infty$ the estimate $v(x) = O(r^{-3-\delta})$, $\delta > 0$. Then $u(x)$ satisfies the radiation condition*

$$\frac{\partial u(x)}{\partial r} - i|k| u(x) = o(r^{-1}). \tag{2.12}$$

Lemma 2.3 was proved in [**6**] by A. Ja. Povzner.

In the space $\mathfrak{B}$ we consider the operator

$$\Theta_k u(x) = -\int \frac{e^{i|k||x-y|}}{4\pi |x-y|} Qu(y)\, dy. \tag{2.13}$$

By Lemma 2.1 the operator Θ_k is bounded in $\mathfrak{B}$. By Lemma 2.2 and the Arzelà-Ascoli theorem it is completely continuous in $\mathfrak{B}$. Let $\varphi_0(x,k)$ denote the first iteration of the equation (2.6):

$$\varphi_0(x, k) = -\int \frac{e^{i|k||x-y|}}{4\pi |x-y|} e^{i\langle y, k\rangle} \{i \langle Q_1(y), k\rangle + Q_2(y)\}\, dy. \tag{2.14}$$

From Lemma 2.1 it is clear that $\varphi_0(x,k) \in \mathfrak{B}$ for all $k \in R^3$. We now put

$$\varphi(x,k) = e^{i\langle x,k\rangle} + \varphi_1(k,k). \tag{2.15}$$

For $\varphi_1(x,k)$ we obtain from (2.6) the equation

$$\varphi_1(x, k) = \varphi_0(x, k) - \int \frac{e^{i|k||x-y|}}{4\pi |x-y|} Q\varphi_1(y, k)\, dy, \tag{2.16}$$

which is a Fredholm equation of the second kind in the space $\mathfrak{B}$,

$$\varphi_1 = \varphi_0 + \Theta_k \varphi_1. \tag{2.17}$$

We consider the corresponding homogeneous equation

$$f = \Theta_k f \tag{2.18}$$

or

$$f(x) = -\int \frac{e^{i|k||x-y|}}{4\pi |x-y|} Qf(y)\, dy. \tag{2.19}$$

Lemma 2.4. *If $f(x) \in \mathfrak{B}$ and satisfies* (2.19) *for $|k| > 0$, then $f(x) \equiv 0$.*

Proof. If $f(x) \in \mathfrak{B}$ and satisfies (2.19), then $f(x)$ is twice continuously differentiable. Therefore it is readily verified that $f(x)$ satisfies the differential equation

$$-\Delta f(x) + Qf(x) = k^2 f(x). \tag{2.20}$$

Multiplying (2.20) by $T^2(x)$, we obtain

$$-\nabla (a(x) \nabla f(x)) + p_{k^2}(x) f(x) = k^2 f(x), \tag{2.21}$$

where the real coefficients $a(x)$ and $p_{k^2}(x)$ are defined by (1.17) for $\lambda = k^2$.
We note that the following equations are also valid for $f(x)$:

$$f = \Theta_k f = \Theta_k^2 f = \ldots = \Theta_k^m f, \quad m \geqslant 1. \tag{2.22}$$

If now we apply Lemma 2.1 to equation (2.19) several times, we obtain the estimate $f(x) = O(r^{-1})$ for $r \to \infty$. This estimate makes it possible to apply Lemma 2.3 to equation (2.19) and thus to affirm that $f(x)$ satisfies the radiation condition

$$\partial f(x)/\partial r - i|k| f(x) = o(r^{-1}) \tag{2.23}$$

for $r \to \infty$.

It is now easy to verify that $f(x)$ satisfies the conditions of Theorem 2.2 in the paper [7] by D. M. Èĭdus; from this it follows that $f(x) \equiv 0$. This completes the proof of Lemma 2.4.

Applying the Fredholm alternative to equation (2.17) and using the relations (2.3) and (2.15), we obtain the following result.

Theorem 2.1. *For arbitrary $k \in R^3$, $|k| > 0$, equation* (2.16) *has a unique solution $\varphi_1(x, k) \in \mathfrak{B}$. The function*

$$\psi(x, k) = T(x)\, [e^{i(x,k)} + \varphi_1(x, k)] \tag{2.24}$$

satisfies the Schrödinger equation (2.2).

The proof we give of the theorem on an expansion with respect to the solutions $\psi(x, k)$ follows that given by D. Thoe in [8]. Let H_ρ be Schrödinger operators, selfadjoint on $W_2^2(R^3)$:

$$H_\rho = -\Delta + q_\rho(x) \tag{2.25}$$

with finite potentials $q_\rho(x) = q(x)\eta_\rho(x)$, where $\eta_\rho(x) \in C_0^\infty(R^3)$ and $\eta_\rho(x) = 1$ for $r < \rho$ and 0 for $r > \rho + 1$. We denote the function $T(x)$ and the operators T and Θ_k, constructed according to the equations (1.3) and (2.1) for the operators H_ρ with the potential $q_\rho(x)$ in place of $q(x)$, by $T_\rho(x)$, T_ρ and $\Theta_{\rho,k}$, respectively. If $\psi_\rho(x, k)$ are the eigenfunctions of the continuous spectrum of H_ρ (their existence being known from [8]), it is then easy to verify that the functions $\varphi_{1,\rho}(x,k)$, defined by the relation

$$\psi_\rho(x, k) = T_\rho(x)[e^{i\langle x, k\rangle} + \varphi_{1,\rho}(x, k)], \tag{2.26}$$

satisfy a Fredholm equation of the second kind in the space $\mathfrak{B}$:

$$\varphi_{1,\rho} = \varphi_{0,\rho} + \Theta_{\rho,k}\varphi_{1,\rho}, \tag{2.27}$$

where $\varphi_{0,\rho}(x,k) = \Theta_{\rho,k}e^{i\langle x,k\rangle}$. It is easily shown that $\| \Theta_{\rho,k} - \Theta_k \|_{\mathfrak{B}} \to 0$ and that $\| \varphi_{0,\rho} - \varphi_0 \|_{\mathfrak{B}} \to 0$ for $\rho \to \infty$. It follows from this that the solutions $\varphi_{1,\rho}$ of (2.27) converge in $\mathfrak{B}$ for $\rho \to \infty$ to the solution φ_1 of (2.17). Using (2.26) and (2.24), we obtain

$$\lim_{\rho \to \infty} \max_{x \in R^3} | \psi_\rho(x, k) - \psi(x, k) | = 0. \tag{2.28}$$

In $L_2(R^3)$ we define an operator U on finite smooth functions:

$$\hat{f}(k) = (Uf)(k) = (2\pi)^{-\frac{3}{2}} \int \overline{\psi(x, k)} f(x)\, dx. \tag{2.29}$$

The theorem on expansion in eigenfunctions of H can be formulated in the following way.

Theorem 2.2. *The operator U is isometric in $L_2(R^3)$, and the following relations are valid for it*:

$$U^*U = I - P, \tag{2.30}$$

$$UU^* = I, \tag{2.31}$$

where P is the projector on the discrete spectrum of H.

Proof. We prove (2.30). This relation is known (see [**5**]) to be equivalent to the following assertion: for an arbitrary smooth finite function $f(x)$ and an arbitrary interval $\delta = (a,b)$, $0 < a < b < \infty$, the equation

$$(E(\delta)f, f)_{L_2(R^3)} = \int_{a<k^2<b} |\hat{f}(k)|^2\, dk \tag{2.32}$$

is valid. Here we denote the spectral measure of H by $E(\cdot)$. A theorem on the expansion in eigenfunctions of Schrödinger operators with finite potentials $q_\rho(x)$ is well known [**8**]. Hence equations analogous to (2.32) hold for such operators:

$$(E_\rho(\delta)f, f)_{L_2(R^3)} = \int_{a<k^2<b} |\hat{f}_\rho(k)|^2\, dk, \tag{2.33}$$

where $E_\rho(\cdot)$ is the spectral measure of H_ρ and

$$\hat{f}_\rho(k) = (2\pi)^{-\frac{3}{2}} \int \overline{\psi_\rho(x, k)} f(x)\, dx. \tag{2.34}$$

To prove (2.32), and hence (2.30), it is sufficient to assure ourselves of the possibility of passage to the limit for $\rho \to \infty$ in (2.33).

Passage to the limit on the right side of (2.33) is obviously possible by virtue of (2.28). Since on the set $C_0^2(R^3)$, which is dense in $L_2(R^3)$, the operators H converge strongly for $\rho \to \infty$ to the operator H, it follows, in accord with Theorem 2, §79 of [**9**], that the spectral measures $E_\rho(\delta)$ converge strongly to $E(\delta)$ for arbitrary δ containing no points of the discrete spectrum of H.

According to Theorem 1.3 the operator H has no positive eigenvalues. Thus passage to the limit for $\rho \to \infty$ is also possible on the left side of (2.33).

T. Ikebe [**10**] proved (2.31) for the case of a sufficiently rapidly decreasing potential. A simple modification in his reasoning makes it possible to prove (2.31) for our case also.

Bibliography

[1] V. B. Matveev and M. M. Skriganov, *Wave operators for the Schrödinger equation with rapidly oscillating potential,* Dokl. Akad. Nauk SSSR **202** (1972), 755-757 = Soviet Math. Dokl. **13** (1972), 185-188. MR **45** #9183.

[2] M. Š. Birman, *On the spectrum of singular boundary problems,* Mat. Sb. **55** **(97)** (1961), 125-174; English transl., Amer. Math. Soc. Transl. (2) **53** (1966), 23-80. MR **26** #463.

[3] S. N. Roze, *On the spectrum of an elliptic operator of second order,* Mat. Sb. **80** **(122)** (1969), 195-209 = Math. USSR Sb. **9** (1969), 183-198. MR **40** #6090.

[4] M. Š. Birman, *On the number of eigenvalues in a quantum scattering problem,* Vestnik Leningrad. Univ. **16** (1961), no. 13, 163-166. (Russian) MR **25** #2803.

[5] T. Ikebe, *Eigenfunction expansions associated with the Schroedinger operators and their applications to scattering theory,* Arch. Rational Mech. Anal. **5** (1960), 1-34. MR **23** #B1398.

[6] A. Ja. Povzner, *The expansion of arbitrary functions in terms of eigenfunctions of the operator* $-\Delta u + cu$, Mat. Sb. **32** **(74)** (1953), 109-156; English transl., Amer. Math. Soc. Transl. (2) **60** (1967), 1-49. MR **14**, 755.

[7] D. M. Èĭdus, *The principle of limiting amplitude,* Uspehi Mat. Nauk **24** (1969), no. 3 (147), 91-156 = Russian Math. Surveys **24** (1969), no. 3, 97-167.

[8] D. W. Thoe, *Eigenfunction expansions associated with Schroedinger operators in* R^n, $n \geqq 4$, Arch. Rational Mech. Anal. **26** (1967), 335-356. MR **36** #1856.

[9] N. I. Ahiezer and I. M. Glazman, *The theory of linear operators in Hilbert space,* 2nd rev. ed., "Nauka", Moscow, 1966; English transl. of 1st ed., Ungar, New York, 1961. MR **34** #6527; **41** #9015a.

[10] T. Ikebe, *Remarks on the orthogonality of eigenfunctions for the Schrödinger operator in* R^n, J. Fac. Sci. Univ. Tokyo Sect. I **17** (1970), 355-361. MR **43** #6777.

Translated by
J. F. HEYDA

Trudy Mat. Inst. Steklov.
125 (1973)

Proc. Steklov Inst. Math.
125 (1973)

ON A BOUNDARY VALUE PROBLEM FOR A STATIONARY SYSTEM OF NAVIER-STOKES EQUATIONS

V. A. SOLONNIKOV AND V. E. ŠČADILOV

Abstract. We consider a boundary value problem for which in a region Ω with boundary S we seek a solution of a linearized stationary system of Navier-Stokes equations which on one part S_1 of the boundary S satisfies the usual adhesion conditions and on the other part S_2 satisfies the conditions $vn = 0$ and $Tn - n(nTn) = 0$, where v is the velocity vector, n is the normal to the boundary, and T is the stress tensor. We establish the existence of a generalized solution and show that it is smooth everywhere except on $S_1 \cap S_2$.

Bibliography: 8 items.

§1. Introduction

In this article we will consider the following boundary value problem for a linearized stationary system of Navier-Stokes equations in a region $\Omega \subset R^3$ with boundary $S = S_1 \cup S_2$:

(1) $$\nu \Delta v - \operatorname{grad} p = f, \qquad \operatorname{div} v = 0,$$

(2) $$vn|_{S_1} = \alpha(x), \qquad t - n(tn)|_{S_1} = b(x), \qquad v|_{S_1} = a(x).$$

Here v is the velocity vector, p is the pressure, n is the unit outer normal to S, t is a vector with components $t_i = \sum_{k=1}^3 t_{ik}(x) n_k(x)$,

$$tn = \sum_{i=1}^{3} t_i(x) n_i(x),$$

and

$$t_{ik} = -\delta_{ik} p + \nu\left(\frac{\partial v_i}{\partial x_k} + \frac{\partial v_k}{\partial x_i}\right).$$

The given vector $b(x)$ on S_1 must satisfy the compatibility condition $bn = 0$.

The region Ω can be finite or infinite, and its boundary S is assumed to be a bounded piecewise-smooth surface which can consist of a finite number of connected components. If Ω is infinite, the conditions in (2) must be supplemented by a condition at infinity which we will write in the form

(3) $$v_{|x| = \infty} = v^\infty = \text{const.}$$

The problem (1)–(3) can be regarded as a model of the problem of fluid motion around a free surface. Another variation of the problem with the boundary condition $t|_S = d(x)$ was investigated in [1].

For the sake of subsequent study of a nonlinear and nonstationary system of Navier-Stokes equations satisfying (2) and (3), the authors set themselves the goal of determining to what extent the facts established for the first boundary value problem in [3] remain true for problem (1)–(3).

As we will show in §3, it is easy to prove the existence of a generalized solution of (1)–(3) with square-integrable first derivatives v_{ix_k}.

AMS (MOS) subject classifications (1970). Primary 35Q10; Secondary 76D05, 35D05, 35D10.

Furthermore, we will show in §4 that with zero boundary conditions it is the case that if $f \in L_2(\Omega')$, where Ω' is an interior subregion of Ω or a subregion whose boundary intersects S_1 or S_2 but not $S_1 \cap S_2$ and this part of the boundary is of class C^3, then in an arbitrary subregion $\Omega'' \subset \Omega'$ such that $\bar{\Omega}'' \cap (\Omega \setminus \Omega') = 0$ the generalized solution $v(x)$ has square-integrable generalized second derivatives and (1) is satisfied almost everywhere. The bounds given in this article for $\partial^2 v_i / \partial x_k \partial x_l$, in contrast to those in [**2**] and [**3**], do not depend on potential theory; they are derived from Lemma 3 in §2 concerning orthogonal decomposition of the space of vectors in $W_2^1(\Omega)$ which satisfy homogeneous boundary conditions (2). Besides this lemma, §2 also contains an elementary proof of Korn's inequality (see [**4**] and also [**5**]) for vectors in the same space for either a bounded or an unbounded region Ω.

§2. Auxiliary propositions

Let $\Omega \subset R^3$ be a region with a compact piecewise-smooth (twice continuously differentiable) boundary S. Depending on whether the region is interior or exterior with respect to S we will write Ω_i or Ω_e. First we recall the definition of the norms in the spaces $L_p(\Omega)$ and $W_p^l(\Omega)$ with integers $l > 0$ and $p \geq 1$:

$$\| u \|_{L_p(\Omega)} = \| u \|_{p,\Omega} \equiv \left(\int_\Omega | u(x) |^p \, dx \right)^{\frac{1}{p}},$$

$$\| u \|_{W_p^l(\Omega)} = \sum_{|\alpha| \leq l} \| D^\alpha u(x) \|_{p,\Omega}.$$

For $\| u \|_{2,\Omega}$ we will write $\| u \|_\Omega$ or even more simply just $\| u \|$. These symbols will also be used in the case where $u(x)$ is a vector-valued function with components $u_k(x)$, $k = 1, 2, 3$.

We now define the space $H(\Omega)$ which is of basic importance for this work.

Definition 1. The space $H(\Omega)$ is the closure in the norm

$$\| u \|_H = \left(\sum_{i,j=1}^{3} \left\| \frac{\partial u_i}{\partial x_j} \right\|^2 \right)^{\frac{1}{2}} = \| u_x \|$$

of the linear set $\dot{H}$ of real continuously differentiable vectors satisfying the boundary conditions

$$un|_{S_1} = 0, \qquad u|_{S_2} = 0. \tag{4}$$

The space H is a Hilbert space with the scalar product

$$(u, v)_H = \sum_{i,j=1}^{3} \int_\Omega \frac{\partial u_i}{\partial x_j} \frac{\partial v_i}{\partial x_j} \, dx. \tag{5}$$

If $\Omega = \Omega_i$, then for all $u \in H$ we have $\| u \| \leq c \| u \|_H$ with the constant c given in [**5**]. Therefore $H(\Omega_i)$ is a subspace of the vector space $W_2^1(\Omega_i)$, and by Sobolev's imbedding theorem for $q \leq 6$ we have

$$\| u \|_{q,\Omega_i} \leq c \| u \|_H. \tag{6}$$

If $\Omega = \Omega_e$, equation (6) is true for $q = 6$. This is implied by the following lemma.

Lemma 1. *For each continuously differentiable function $v(x)$ of compact support we have*

$$\| v \|_{6,\Omega_e} \leqq c \| v_x \| . \tag{7}$$

Proof. We put Ω_i inside a ball $K_M: |x - x_0| \leqq M$, $x_0 \in \Omega_i$, and we introduce a function $\zeta(x) \in C_0^\infty(R^3)$ equal to one in K_M and zero outside the ball K_{M+1} which is concentric with K_M. We have $v = v_1 + v_2$, $v_1 = v\zeta$, $v_2 = v(1 - \zeta)$,

$$\| v_1 \|_{6,\Omega_e} \leqslant c \| v_{1x} \| \leqslant c (\| v_x \| + \| v \|_{K_{M+1} \setminus K_M}),$$

$$\| v_2 \|_{6,\Omega_e} = \| v_2 \|_{6,R^3} \leqslant c \| v_{2x} \| \leqslant c (\| v_x \| + \| v \|_{K_{M+1} \setminus K_M}),$$

since $v_1 = 0$ on the sphere $|x - x_0| = M + 1$ and $v_2 = 0$ in $K_M \cap \Omega_e$, and therefore v_2 can be extended to all of R^3 with preservation of class by setting $v_2(x) = 0$ for $x \in \Omega_i$. The region $K_{M+1} \setminus K_M$ can be covered by a finite number of parallelepipeds of the form $a_i \leqq x \leqq b_i$ which do not intersect S. If we replace a_i by $-\infty$ or b_i by $+\infty$, we can imbed Q_i in an infinite region R_i which also does not intersect S and is obtained from the octant R^+: $x_j \geqq 0$ by a rotation and a shift. Since (7) holds for all such regions, we have

$$\| v \|^2_{K_{M+1} \setminus K_M} \leqslant \sum_i \| v \|^2_{Q_i} \leqslant c^2 \sum_i \| v_x \|^2_{R_i} \leqslant c_1^2 \| v_x \|^2$$

and

$$\| v \|_{6,\Omega_e} \leqq \| v_1 \|_{6,\Omega_e} + \| v_2 \|_{6,\Omega_e} \leqq c \| v_x \|_{\Omega_e}.$$

The lemma is proved.

Corollary. *Inequality* (7) *holds for all vectors* $u(x) \in H(\Omega_e)$, *and we have*

$$\| u \|_{2,S} \leqq c \| u \|_H, \qquad \| u \|_{q,\omega} \leqq c \| u \|_H,$$

where ω *is a bounded subregion of* Ω_e *and* $q \leqq 6$.

Remark. Lemma 1 can be generalized as follows. For each function $v(x) \in C^l(\Omega_e)$ of compact support we have

$$\| D^\beta u \|_{q,\Omega_e} \leqslant c \sum_{|\alpha| = l} \| D^\alpha u \|_{p,\Omega_e},$$

where $p \geqq 1$, $n > (l - |\beta|)p > 0$, $1/q = 1/p - (l - |\beta|)/n$.

The proof is the same as the proof of (7).

We let $D(\Omega_i)$ denote the subspace of functions in $L_2(\Omega_i)$ which are orthogonal to a constant function, and we let $\overset{\circ}{H}(\Omega)$ denote the space $H(\Omega)$ in the case where $S_2 = S$; we consider the auxiliary problem of representing an arbitrary function $p \in D(\Omega_i)$ or $p \in L_2(\Omega_e)$ in the form

$$p(x) = \operatorname{div} v(x), \tag{8}$$

where $v \in H(\Omega)$ and

$$\| v_x \| \leqq c \| p \| \tag{9}$$

with a constant c which does not depend on $p(x)$.

This problem has an infinite number of solutions. For us it is important to prove the existence of at least one.

LEMMA 2. *Problem* (8), (9) *always has a solution if one of the following conditions is satisfied*:

a) *The boundary* S *of the region* Ω *is of class* C^2.

b) Ω *is the half-space* $x_3 > 0$ *or the cube* $0 \leq x_i \leq 1$, $i = 1, 2, 3$.

PROOF. We look for v in the form $v = \operatorname{grad} \psi + u$, where

$$\Delta \psi = p(x), \qquad \partial \psi / \partial n \,|_{S=0}, \tag{10}$$

$\psi \to 0$ as $|x| \to \infty$ in the case $\Omega = \Omega_e$, and

$$\operatorname{div} u(x) = 0, \quad u|_S = -\operatorname{grad} \psi|_S. \tag{11}$$

Problem (11) was considered in [3], where it was shown that a solution existed which satisfied the inequality

$$\|u_x\| \leqslant c \|\operatorname{grad} \psi\|_{W_2^{1/2}(S)} \leqslant c \|\operatorname{grad} \psi_x\|. \tag{12}$$

Here $W_2^{1/2}(S)$ is the space of Slobodeckiĭ, one of the equivalent norms of which is

$$\|u\|_{W_2^{1/2}(S)} = \inf_{\substack{v \in W_2^1(\Omega) \\ v|_S = u}} \|v\|_{W_2^1(\Omega)}.$$

Inequality (9) follows from (12) and the known bound $\|\operatorname{grad} \psi_x\| \leq c \|p\|$ for a solution of the Neumann problem (10).

The same arguments are valid for the half-space $x_3 > 0$ if we allow for the fact that the solution can be written explicitly in terms of potentials with difference kernels with respect to the variables x_1 and x_2. Thus instead of (9) we have

$$\|D^\beta v_x\| \leq c \|D^\beta p\|, \quad \beta_3 = 0, \tag{13}$$

if $D^\beta p \in L_2(R_+^3)$.

Now suppose Ω is the unit cube $0 \leq x_i \leq 1$, $i = 1, 2, 3$. Each function $p \in D(\Omega)$ can be written as a Fourier series

$$p(x) = \sum_{k_1+k_2+k_3>0} p_{k_1k_2k_3} \cos k_1\pi x_1 \cos k_2\pi x_2 \cos k_3\pi x_3, \quad \sum_k p_{k_1k_2k_3}^2 < \infty,$$

and the solution of (10) is of the form

$$\psi(x) = -\sum_{\substack{k_1+k_2+k_3>0 \\ k_i \geqslant 0}} \frac{p_{k_1k_2k_3}}{k_1^2 + k_2^2 + k_3^2} \cos k_1\pi x_1 \cos k_2 \pi x_2 \cos k_3\pi x_3$$

and satisfies the same inequality $\|\operatorname{grad} \psi_x\| \leq c \|p\|$.

We seek a solution of (11) in the form

$$u = \operatorname{curl} \varphi. \tag{14}$$

The boundary conditions for φ are of the form

$$\frac{\partial \varphi_3}{\partial x_2} - \frac{\partial \varphi_2}{\partial x_3} = -\frac{\partial \psi}{\partial x_1} \equiv \psi_1, \tag{15$_1$}$$

$$\frac{\partial \varphi_1}{\partial x_3} - \frac{\partial \varphi_3}{\partial x_1} = -\frac{\partial \psi}{\partial x_2} \equiv \psi_2, \tag{15$_2$}$$

$$\frac{\partial \varphi_2}{\partial x_1} - \frac{\partial \varphi_1}{\partial x_2} = -\frac{\partial \psi}{\partial x_3} \equiv \psi_3. \tag{15$_3$}$$

In Ω we consider functions $\Phi_i \in W_2^2(\Omega)$ which satisfy the following boundary conditions: if i, k, l is a cyclic permutation of the subscripts 1, 2, 3, then

$$\Phi_i|_{\substack{x_k=0\\x_k=1}} = \Phi_i|_{\substack{x_l=0\\x_l=1}} = 0, \qquad (16)$$
$$\left.\frac{\partial\Phi_i}{\partial x_k}\right|_{\substack{x_k=0\\x_k=1}} = 0, \quad \left.\frac{\partial\Phi_i}{\partial x_l}\right|_{\substack{x_l=0\\x_l=1}} = \psi_k.$$

Since $\psi_k = 0$ for $x_k = 0$ and $x_k = 1$, these conditions do not conflict; that is, compatibility conditions are satisfied on the edges of the cube Ω parallel to the x_i axis. Therefore by methods in the theory of functions (see, for example, [**6**]) we can construct in Ω functions $\Phi_i \in W_2^2(\Omega)$ which satisfy (16) and the inequalities

$$\|\Phi_i\|_{W_2^2(\Omega)} \leqslant c\left(\|\psi_k\|_{W_2^{1/2}(\Omega_l^0)} + \|\psi_k\|_{W_2^{1/2}(\Omega_l^1)}\right) \leqslant c\|\psi\|_{W_2^2(\Omega)}, \qquad (17)$$

where Ω_l^0 and Ω_l^1 are the faces $x_l = 0$ and $x_l = 1$ of the cube Ω.

The vector $\chi = \varphi - \Phi$ has to satisfy the boundary conditions

$$\frac{\partial\chi_3}{\partial x_2} - \frac{\partial\chi_2}{\partial x_3} = \psi_1 - \left(\frac{\partial\Phi_3}{\partial x_2} - \frac{\partial\Phi_2}{\partial x_3}\right) \equiv \omega_1,$$
$$\frac{\partial\chi_1}{\partial x_3} - \frac{\partial\chi_3}{\partial x_2} = \psi_2 - \left(\frac{\partial\Phi_1}{\partial x_3} - \frac{\partial\Phi_3}{\partial x_1}\right) \equiv \omega_2, \qquad (18)$$
$$\frac{\partial\chi_2}{\partial x_1} - \frac{\partial\chi_1}{\partial x_2} = \psi_3 - \left(\frac{\partial\Phi_2}{\partial x_1} - \frac{\partial\Phi_1}{\partial x_2}\right) \equiv \omega_3.$$

We define the χ_i and their derivatives on the surface of the cube as follows: $\chi = 0$, and the normal derivatives on each face are found by means of (18). Since $\omega = 0$ on all edges of the cube, compatibility conditions are satisfied as above for the functions (16), and we can extend χ_i inside Ω so that $\chi_i \in W_2^2(\Omega)$ and

$$\sum_{i=0}^{3}\|\chi_i\|_{W_2^2(\Omega)} \leqslant c\sum_{i,l=1}^{3}\left(\|\omega_i\|_{W_2^{1/2}(\Omega_l^0)} + \|\omega_i\|_{W_2^{1/2}(\Omega_l^1)}\right) \leqslant c\|\psi\|_{W_2^{1/2}(\Omega)}.$$

Therefore it follows from (17) that

$$\|u\|_{W_2^1(\Omega)} \leqslant \|\varphi\|_{W_2^2(\Omega)} \leqslant c\|\psi\|_{W_2^2(\Omega)} \leqslant c\|p\|$$

and

$$\|v\|_{W_2^1(\Omega)} \leqslant \|\operatorname{grad}\psi\|_{W_2^1(\Omega)} + \|u\|_{W_2^1(\Omega)} \leqslant c\|p\|.$$

The lemma is proved.

We continue to investigate the space $H(\Omega)$. Let $[u, v]$ be any scalar product which is equivalent to $(u, v)_H$; that is, such that

$$\mu\|u\|_H^2 \leqslant [u, u] \leqslant \nu\|u\|_H^2, \quad \mu, \nu > 0.$$

We consider a linear functional $v \in H$ of the form $(p, \operatorname{div} v) = \int p \operatorname{div} v\,dx$, where $p \in D(\Omega_i)$ or $p \in L_2(\Omega_e)$. By the Riesz representation theorem we have

$$(p, \operatorname{div} v) - [u, v], \qquad (19)$$

where $u = Ap$; here A is a bounded linear operator defined on $D(\Omega_i)$ or $L_2(\Omega_e)$. Boundedness is easy to prove by letting $v = Ap$ in (19). If problem (8), (9) is solvable, the inverse operator is also bounded, since if in (19) we substitute the solution for v we obtain

$$\|p\|^2=(p,\ \operatorname{div} v)\leqslant \nu\|u\|_H\|v\|_H\leqslant c\|Ap\|_H\|p\|;$$

that is, $\|p\|\leq c\|Ap\|_H$.

We introduce the notation $H_1=\overline{R(A)}$ and the following definition.

DEFINITION 2. The space $J(\Omega)$ is the subspace of solenoidal vectors in $H(\Omega)$.

LEMMA 3. *$H=H_1\oplus J$, and if problem* (8), (9) *is solvable, then $H_1=R(A)$.*

PROOF. If $v\in H\ominus H_1$, then for all $p\in D(\Omega_i)$ or $p\in L_2(\Omega_e)$ we have

$$[Ap,v]=(p,\operatorname{div} v)=0,$$

and thus $\operatorname{div} v=\text{const}=0$, $v\in J$.

If problem (8), (9) is solvable, then, as we showed above,

$$c_1\|p\|\leqq\|Ap\|_H\leqq c_2\|p\|, \tag{20}$$

which implies that the range of A is closed. The lemma is proved.

We proceed to the proof of Korn's inequality for vectors in $H(\Omega)$. In $H(\Omega)$ we introduce a new scalar product

$$E(u,\ v)=\int_\Omega\sum_{i,k=1}^{3}\left(\frac{\partial u_i}{\partial x_k}+\frac{\partial u_k}{\partial x_i}\right)\left(\frac{\partial v_i}{\partial x_k}+\frac{\partial v_k}{\partial x_i}\right)dx.$$

We recall that the vectors for which $E(u,u)=0$ form a finite-dimensional affine space of vectors of the form

$$u=A+B\times x,$$

which intersects H only if the following conditions are satisfied:

1) $\Omega=\Omega_i$ is a region obtained by revolution around the vector B.

2) $S_1=S$.

Under these conditions $u_0=B\times x\in J\subset H$.

DEFINITION 3. We let $\mathbf{H}$ and $\mathbf{J}$ denote the subspaces of vectors in H and J which are orthogonal to $u_0(x)$ in $L_2(\Omega)$ if conditions 1) and 2) are satisfied; otherwise we set $\mathbf{H}=H$ and $\mathbf{J}=J$.

LEMMA 4. *For each $u(x)\in\mathbf{H}(\Omega)$ we have*

$$\|u\|_H^2\leqq cE(u), \tag{21}$$

where $E(u)=E(u,u)$.

PROOF. We will prove the lemma under the more general assumption that $S_1=S$. First suppose that $u\in C^2(\Omega)$, $un|_S=0$ and $(u,u_0)=0$ (such vectors are dense in $\mathbf{H}(\Omega)$). By integration by parts we obtain

$$\begin{aligned}E(u)=2\sum_{i,k=1}^{3}\int_\Omega\left[\left(\frac{\partial u_i}{\partial x_k}\right)^2+\frac{\partial u_i}{\partial x_k}\frac{\partial u_k}{\partial x_i}\right]dx=2\|u\|_H^2\\ +2\int_\Omega(\operatorname{div} u)^2\,dx+2\sum_{i,k=1}^{3}\int_S u_i n_k\frac{\partial u_k}{\partial x_i}\,dS.\end{aligned}\tag{22}$$

We consider a part σ of the surface S which in local coordinates at some point $\xi\in S$ is given by an equation

$$y_3=F(y_1,y_2)\in C^2, \tag{23}$$

where y_k and x_l are connected by the relation

(24) $$y_k = \sum_{l=1}^{3} \alpha_{kl} (x_l - \xi_e), \quad \alpha_{3l} = n_l(\xi).$$

We have

$$\sum_{i,k=1}^{3} u_i n_k \frac{\partial u_k}{\partial x_i} = \sum_{l=1}^{3} v_l \frac{\partial v_3}{\partial y_l} \quad \left(v_l = \sum_{i=1}^{3} \alpha_{li} u_i \right),$$

and since

$$v_3 - \sum_{\alpha=1}^{2} \frac{\partial F}{\partial y_\alpha} v_\alpha = 0 \text{ on } \sigma, \quad \left. \frac{\partial F}{\partial y_\alpha} \right|_{y=0} = 0,$$

we have

(25) $$\left. \sum_{i,k=1}^{3} u_i n_k \frac{\partial u_k}{\partial x_i} \right|_{x=\xi} = \sum_{\alpha,\beta=1}^{2} v_\alpha(0) v_\beta(0) \left. \frac{\partial^2 F(y)}{\partial y_\alpha \partial y_\beta} \right|_{y=0}.$$

Thus the surface integral in (22) can be bounded by means of $\int_S u^2 dS$, and

(26) $$\| u \|_H^2 \leqslant \frac{1}{2} E(u) + c_0 \| u \|_{2,S}^2.$$

We now show that there exists a positive number M such that

(27) $$\| u \|_{2,S}^2 \leqslant \delta \| u \|_H^2 + ME(u), \quad \delta = \frac{1}{2c_0}.$$

If such a number did not exist, then for each natural number m there would exists a vector $u^m \in \mathbf{H}(\Omega)$ such that

$$\| u^m \|_{2,S}^2 \geqslant \delta \| u^m \|_H^2 + mE(u^m) \equiv G_m(u^m),$$

and for $v^m = u^m / \| u^m \|_{2,S}$ we would have

$$\| v^m \|_{2,S} = 1, \quad G_m(v^m) = \frac{G_m(u^m)}{\| u^m \|_{2,S}^2} \leqslant 1.$$

Thus from the sequence v^m we could choose a subsequence v^{m_k} which would converge weakly in $H(\Omega)$ and strongly in $L_2(S)$ to a vector $v \in \mathbf{H}(\Omega)$; also, $E(v^{m_k}) \leq 1/m_k \to 0$. Then it follows that $E(v) = 0$; consequently $v = cu_0 = 0$. This contradicts the fact that $\| v \|_{2,S} = \lim \| v^{m_k} \|_{2,S} = 1$, and (27) has been proved by contradiction.

From (26) and (27) we obtain (21).

REMARK 1. If $\Omega = \Omega_e$ and S is convex, then as we see from (25) the last integral in (22) is positive and $\| u \|_H^2 \leq \frac{1}{2} E(u)$.

REMARK 2. As is clear from the proof, inequality (21) (with a different constant) is valid not only for the space $\mathbf{H}(\Omega)$ but also for any subspace of $H(\Omega)$ which does not contain the vector $u_0(x)$.

§3. A generalized solution

As in [1], for the definition of a generalized solution we start from the following Green's formula which holds for smooth solenoidal vectors v and φ and functions $p \in C^1(\Omega)$:

(28) $$\int_\Omega (\nu \Delta v - \operatorname{grad} p) \varphi dx = -\frac{\nu}{2} E(v, \varphi) + \int_S t \varphi dS.$$

First we consider the case where Ω is bounded and the boundary conditions are homogeneous.

Definition 4. By a *generalized solution of the problem*

$$\begin{gathered}\nu\Delta v-\operatorname{grad}p=f,\quad \operatorname{div}v=0,\\ vn\,|_{S_1}=0,\quad t-n\,(tn)\,|_{S_1}=0,\quad v\,|_{S_2}=0\end{gathered}\tag{29}$$

in the class $\mathbf{J}(\Omega)$ we mean a vector $v\in\mathbf{J}(\Omega)$ which satisfies the identity

$$-\frac{\nu}{2}E(v,\varphi)=(f,\varphi)=\int_\Omega f\varphi dx\tag{30}$$

for all $\varphi\in\mathbf{J}(\Omega)$.

The basis for this definition is formula (28) which implies that each classical solution of (29) is a generalized solution. From this formula it is also clear that in the case where $S_1=S$ and Ω is a bounded region obtained by revolution around a vector B the vector f should satisfy the condition

$$(f,u_0)=0,\tag{31}$$

since this is a necessary condition for existence of a classical solution of (29). But if (31) is satisfied, (30) will be true not only for $\varphi\in\mathbf{J}(\Omega)$ but also for $\varphi\in J(\Omega)$.

We now show that if a generalized solution $v\in W_2^2(\Omega')$ for an arbitrary subregion $\Omega'\subset\Omega$ and $f\in L_2(\Omega')$, then the system of Navier-Stokes equations is satisfied almost everywhere in Ω' and $t-n(tn)=0$ for $x\in S_1'\equiv S_1\cap\partial\Omega'$. In fact, if we substitute into the identity smooth solenoidal vectors φ which are finite in Ω' and integrate by parts we obtain

$$\int_{\Omega'}(\nu\Delta v-f)\,\varphi dx=0;$$

and consequently $\nu\Delta v-f=\operatorname{grad}p$ in Ω'.

To verify the boundary conditions we pick an arbitrary smooth submanifold $\sigma\subset S_1'$ and define a linear set $J_\sigma(\Omega')$ of vectors $\varphi\in J(\Omega)$ which equal zero outside Ω' and which have boundary values on S outside σ. If we again integrate by parts in (30), from (28) we obtain

$$\int_\sigma t\varphi dS=0,\quad \varphi\in J_\sigma(\Omega').\tag{32}$$

It now follows that $t_\tau|_\sigma\equiv t-n(tn)|_\sigma=0$. In fact, it was shown in [**3**], Chapter I, §2, that we can construct a vector $\varphi\in J_\sigma(\Omega')$ which on the boundary takes the value $t_\tau\Phi$, where Φ is a smooth nonnegative function equal to zero outside σ, and for this vector (32) reduces to $\int_\sigma t_\tau^2\Phi dS=0$.

Theorem 1. *Problem* (29) *has a unique generalized solution in the class* $\mathbf{J}(\Omega)$ *for each vector* f *for which the expression* (f,φ) *is a linear functional on the space* $J(\Omega)$ *and* $(f,u_0)=0$. *Here* $\|u\|_H\leq c\langle f\rangle$, *where* $\langle f\rangle$ *is the norm of the functional* (f,φ).

The proof is the same as for the first boundary value problem in [**3**]. It depends on the Riesz representation theorem and the fact that the expression $E(v,\varphi)$ is a scalar product in $\mathbf{J}(\Omega)$ and the metric generated by it is equivalent to the original metric by virtue of Korn's inequality (21).

By analogy we can treat the case of an unbounded region Ω.

In the case of nonhomogeneous boundary conditions we define a generalized solution as follows. We assume that

$$\tilde{\alpha}(x) = \begin{cases} an, & x \in S_2 \\ \alpha, & x \in S_1 \end{cases} \in W_2^{1/2}(S), \quad a(x) \in W_2^{1/2}(S_2);$$

then there exists a solenoidal vector $w(x)$ which in the case of a bounded region Ω belongs to $W_2^1(\Omega)$ and satisfies the boundary conditions

$$wn|_{S_1} = \alpha(x), \qquad w|_{S_2} = a(x), \tag{33}$$

and in the case of an unbounded region Ω has locally square-integrable generalized first derivatives, satisfies (33), and for $|x| \geqq M$ equals a constant vector v^∞. It is easy to see that the expression $E(w, \varphi)$ is a linear functional on $\varphi \in J(\Omega)$.

DEFINITION 5. By a *generalized solution of problem* (1)–(3) we mean a vector $v = w + u$, where $u \in J(\Omega)$ satisfies the integral identity

$$\frac{\nu}{2} E(u, \varphi) = -\frac{\nu}{2} E(w, \varphi) - (f, \varphi) + \int_{S_1} b\varphi dS. \tag{34}$$

By analogy condition (31) for the case where Ω is a region obtained by revolution around a vector β will be

$$(f, u_0) - \int_{S_1} \beta u_0 dS = 0.$$

THEOREM 1′. *If the right side of* (34) *defines a linear functional on* $\varphi \in J(\Omega)$, *problem* (1)–(3) *has a unique generalized solution.*

The pressure p was defined above as a function in $W_2^1(\Omega')$ under the assumption that $v \in W_2^2(\Omega)$, but it is possible to define the pressure directly for a generalized solution. We will do this for the problem with nonhomogeneous boundary conditions, assuming that the right side of (34) defines a linear functional $L(\varphi)$ on $\varphi \in H(\Omega)$. We let $l(\varphi)$ denote the functional corresponding to the difference of the left and right sides of (34). It is known that on vectors $\varphi \in J(\Omega)$ we have $l(\varphi) = 0$. By means of Lemma 3 we write each vector $\varphi \in H(\Omega)$ as a sum $\varphi = \varphi_1 + \varphi_2$, $\varphi_1 \in J(\Omega)$, $\varphi_2 \in H_1(\Omega)$, and we obtain

$$l(\varphi) = l(\varphi_2) = (\psi, \varphi_2)_H = (\psi, \varphi); \quad \psi \in H_1.$$

We assume that Ω is a region in which problem (8), (9) is solvable; then $\psi = Ap$, where $p \in D(\Omega_i)$ or $p \in L_2(\Omega_e)$ and

$$l(\varphi) = (p, \operatorname{div} \varphi)$$

or

$$\frac{\nu}{2} E(u, \varphi) = -\frac{\nu}{2} E(w, \varphi) - (f, \varphi) + \int_{S_1} b\varphi dS + (p, \operatorname{div} \varphi), \quad \varphi \in H(\Omega). \tag{35}$$

Here $\|p\| \leqq c\|Ap\|_H = c\langle l\rangle \leqq c_1(\|u\|_H + \langle L\rangle)$.

It is natural to call the square-integrable function $p(x)$ the pressure corresponding to the generalized solution $v = u + w$.

If Ω is a region in which we cannot show that problem (8), (9) is solvable, but if in a subregion Ω' a solution does exist, then we can construct a pressure

p' in Ω'. To do this it is necessary in (34) to consider only vectors in $J(\Omega') \subset J(\Omega)$ which equal zero outside Ω'. The function p' will satisfy (35) for each $\varphi \in H(\Omega')$. It is easy to show that such functions which correspond to different regions Ω' differ only by a constant term.

§4. Differential properties of a generalized solution

For the sake of simplicity we consider problem (29) and we begin with bounds for the second derivatives of its generalized solution in the norm of $L_2(\Omega'), \Omega' \subseteq \Omega$.

THEOREM 2. *If* $f \in L_2(\Omega')$, $\Omega' \subseteq \Omega$, *then* $v \in W_2^2(\Omega'')$ *in* $\Omega'' \subset \Omega'$, *equation (29) is satisfied almost everywhere in* Ω'' *for some* $p \in W_2^1(\Omega'')$, *and we have the inequality*

$$\|v\|_{W_2^2(\Omega'')} + \|p_x\|_{2,\Omega''} \leqslant c\|f\|_{2,\Omega'} + c_1\|v\|_{W_2^1(\Omega')}. \tag{36}$$

Here either Ω'' *is a strictly interior subregion of* Ω' *or the following conditions are satisfied*: $S' = \partial\Omega' \cap S$ *is a surface of class* C^3 *lying entirely in* S_1 *or in* S_2, *and* $S'' = \partial\Omega'' \cap \partial\Omega'$ *is a strictly interior subregion of* S'.

PROOF. We will treat only the more complicated second case since the first case (with Ω'' inside Ω) can be treated as in [3], Chapter II, Theorem 3.

We pass to local coordinates at a point $\beta \in S''$ by means of (24) and simultaneously we introduce a new unknown vector $u = (u_1, u_2, u_3)$ by means of the formula $u_k = \sum_1^3 \alpha_{kl} v_l$, which represents a generalized solution of problem (29) in the variables $\{y\}$ with right side f'; $f'_k = \sum_1^3 \alpha_{ki} f_i$.

We let σ denote the part of the surface S'', given by (23), in which $|y_1|$, $|y_2| \leq d$, where d is chosen such that the region $\omega: \{|y_1| \leq d, |y_2| \leq d, F(y_1, y_2) \leq y_3 \leq F + 2d\}$ is contained in Ω''.

We introduce the change of variable

$$z_1 = y_1, \quad z_2 = y_2, \quad z_3 = y_3 - F(y_1, y_2), \tag{37}$$

which transforms ω into a cube Q with sides of length $2d$, and we have a new unknown vector w with components

$$w_1 = u_1, \quad w_2 = u_2, \quad w_3 = u_3 - \frac{\partial F}{\partial y_1} u_1 - \frac{\partial F}{\partial y_2} u_2. \tag{38}$$

Since $u(y)$ is a solenoidal vector, $w(z)$ is also solenoidal. In the identity (30) for $u(y)$ we pass to the coordinates $\{z\}$, replacing $u(y)$ by $w(z)$ and $\varphi(y)$ by

$$\psi(z) = \left(\varphi_1, \varphi_2, \varphi_3 - \frac{\partial F}{\partial y_1}\varphi_1 - \frac{\partial F}{\partial y_2}\varphi_2\right),$$

where for $\varphi(y)$ we choose vectors in $J(\Omega)$ which vanish outside ω. We let $J(\omega)$ denote the space of restrictions of these vectors to ω, and we let $J(Q)$ denote the space of the corresponding vectors $\psi(z)$; they vanish on all faces of the cube Q except the face $z_3 = 0$, and on this face they satisfy the condition $\psi_3 = 0$ or $\psi = 0$ depending on whether S'' is contained in S_1 or S_2. As a result of this transformation we have changed (30) to

$$-\frac{\nu}{2} E(w, \psi) + \varepsilon(w, \psi) = (f', \psi), \quad \psi \in J(Q), \tag{39}$$

where ϵ is a bilinear form of the type

$$\varepsilon(w, \psi) = \int_Q \Big[\sum_{i,k,l,m=1}^{3} p_{lm}^{ik}(z) \frac{\partial w_i}{\partial z_l}\frac{\partial \psi_k}{\partial z_m} + \sum_{i,k,l=1}^{3} p_l^{ik}(z)\frac{\partial w_i}{\partial z_l}\psi_k + \sum_{i,k,m=1}^{3} \tilde{p}_m^{ik}(z) w_i \frac{\partial \psi_k}{\partial z_m} + \sum_{i,k=1}^{3} p^{ik}(z) w_i \psi_k \Big] dz,$$

where the p_{lm}^{ik}, p_l^{ik}, $\tilde{p}_m^{ik}$, p^{ik} are continuously differentiable and the p_{lm}^{ik} are bounded by $c_1|\operatorname{grad} F(z_1, z_2)|$ with a constant c_1 that does not depend on d, and this quantity is small near the origin; that is, near the point $\xi \in S''$, since $\partial F/\partial z_\alpha|_{z=0} = 0$.

If we argue as in the previous section and make use of Lemma 3, it is not hard to show that (39) is equivalent to

$$-\frac{\nu}{2}E(w, \psi) + \varepsilon(w, \psi) = (f, \psi) - (p, \operatorname{div}\psi), \quad \psi \in H(Q), \tag{40}$$

where $p \in D(Q)$ and

$$\|p\|_{2,Q} \leqslant c(\|w\|_{W_2^1(Q)} + \langle f \rangle_{H(Q)}), \tag{41}$$

and by means of a similarity transformation we can guarantee that the constant in (41) remains bounded when the length of the edge of Q goes to zero.

Now we will find a bound for the norm of the second derivatives of the functions w_i just as we would for generalized solutions of second-order elliptic equations. In (40) we let

$$\psi = [\zeta\Delta'(\zeta w_\rho)]_\rho, \tag{42}$$

where φ_ρ is the average of the function φ with respect to the variables $z' = (z_1, z_2)$ with radius ρ, and Δ' is the Laplace operator with respect to these variables, while $\zeta(z)$ is an infinitely differentiable function which equals zero in the half-space $z_3 > 0$ outside Q and equals one in the cube $Q' : |z_\alpha| \leqq 3d/4$, $\alpha = 1, 2$, $0 \leqq z_3 \leqq 3d/2$. If we substitute (42) into (40), after some simple calculations we obtain

$$\sum_{\alpha=1}^{2} \frac{\nu}{2} E\left(\frac{\partial \zeta w_\rho}{\partial z_\alpha}, \frac{\partial \zeta w_\rho}{\partial z_\alpha}\right) + \frac{\nu}{2}E'(w_\rho, \zeta w_\rho) + \varepsilon_\rho(w, \zeta w_\rho) = (f_\rho, \zeta\Delta'(\zeta w_\rho)) + (p_\rho, \operatorname{div}[\zeta\Delta'(\zeta w_\rho)]), \tag{43}$$

where

$$\begin{aligned} E' = &-\sum_{i,k=1}^{3}\int_Q \left(\frac{\partial w_{i\rho}}{\partial z_k} + \frac{\partial w_{k\rho}}{\partial z_i}\right)\left[\frac{\partial \zeta}{\partial z_k}\Delta'(\zeta w_{i\rho}) + \frac{\partial \zeta}{\partial z_i}\Delta'(\zeta w_{k\rho})\right] dz \\ &- \sum_{i,k=1}^{3}\sum_{\alpha=1}^{2}\int_Q \frac{\partial}{\partial z_\alpha}\left(w_{i\rho}\frac{\partial \zeta}{\partial z_k} + w_{k\rho}\frac{\partial \zeta}{\partial z_i}\right)\left(\frac{\partial^2 \zeta w_{i\rho}}{\partial z_k \partial z_\alpha} + \frac{\partial^2 \zeta w_{k\rho}}{\partial z_i \partial z_\alpha}\right) dz, \end{aligned} \tag{44}$$

$$\begin{aligned} \varepsilon_\rho = \varepsilon(w, [\zeta\Delta'(\zeta w_\rho)]_\rho) = &-\sum_{i,k,l,m=1}^{3}\Big\{\sum_{\alpha=1}^{2}\int_Q p_{lm}^{ik}(z)\frac{\partial^2 w_{i\rho}\zeta}{\partial z_l \partial z_\alpha}\frac{\partial^2 w_{k\rho}\zeta}{\partial z_m \partial z_\alpha} dz \\ &- \sum_{\alpha=1}^{2}\int_Q \frac{\partial}{\partial z_\alpha}\zeta\left[p_{lm}^{ik}(z)\frac{\partial w_{i\rho}}{\partial z_l} - \left(p_{lm}^{ik}(z)\frac{\partial w_i}{\partial z_l}\right)_\rho\right]\frac{\partial^2 w_{k\rho}\zeta}{\partial z_m \partial z_\alpha} dz + \end{aligned}$$

$$+\sum_{\alpha=1}^{2}\int_Q\left[\frac{\partial}{\partial z_\alpha}\zeta p_{lm}^{ik}(z)\frac{\partial w_{i\rho}}{\partial z_l}-p_{lm}^{ik}(z)\frac{\partial^2 w_{i\rho}\zeta}{\partial z_l\partial z_\alpha}\right]\frac{\partial^2 w_{k\rho}\zeta}{\partial z_m\partial z_\alpha}dz \tag{45}$$

$$\left.-\int_Q\left(p_{lm}^{ik}(z)\frac{\partial w_{i\rho}}{\partial z_l}\right)\frac{\partial\zeta}{\partial z_m}\Delta'(\zeta w_{k\rho})\,dz\right\}+\sum_{i,k,l=1}^{3}\int_Q\left(p_l^{ik}\frac{\partial w_i}{\partial z_l}\right)_\rho\zeta\Delta'(\zeta w_{k\rho})\,dz$$

$$-\sum_{i,k,m=1}^{3}\sum_{\alpha=1}^{2}\left\{\int_Q(\tilde p_m^{ik}w_i)_\rho\frac{\partial}{\partial z_m}\left(\frac{\partial\zeta}{\partial z_\alpha}\frac{\partial\zeta w_{k\rho}}{\partial z_\alpha}\right)dz+\int_Q\frac{\partial}{\partial z_\alpha}(\tilde p_m^{ik}w_i)_\rho\frac{\partial}{\partial z_m}\left(\zeta\frac{\partial\zeta w_{k\rho}}{\partial z_\alpha}\right)dz\right\}$$

$$+\sum_{i,k=1}^{3}\int_Q(p^{ik}(z)\,w_i)_\rho\,\zeta\Delta'(\zeta w_{k\rho})\,dz.$$

If we estimate each term in E' and ϵ_ρ by means of Hölder's inequality and if for the second term on the right side of (45) we also apply the Friedrichs lemma in [7], we obtain

$$|E'|\leqslant c(d)\,|\zeta\omega_\rho|\,\|w\|_{W_2^1(Q)}, \tag{46}$$

$$|\varepsilon_\rho|\leqslant c\max|\operatorname{grad}F|\,|\zeta w_\rho|^2+c(d)\,\|w\|_{W_2^1(Q)}\left(|\zeta w_\rho|+\|w\|_{W_2^1(Q)}\right), \tag{47}$$

where $|\varphi|^2=\sum_{i,l=1}^3\sum_{\alpha=1}^2\|\partial^2\varphi_l/\partial z_i\partial z_\alpha\|_{2,Q}^2$.

Since

$$\operatorname{div}\zeta\Delta'(\zeta w_\rho)=\operatorname{grad}\zeta\Delta'(\zeta w_\rho)+\Delta'(\zeta\operatorname{grad}\zeta w_\rho)$$
$$-\Delta'\zeta w_\rho\operatorname{grad}\zeta-2\sum_{\alpha=1}^{2}\frac{\partial\zeta}{\partial z_\alpha}\frac{\partial}{\partial z_\alpha}(w_\rho\operatorname{grad}\zeta),$$

we have

$$|(p_\rho,\ \operatorname{div}\zeta\Delta'(\zeta w_\rho))|\leqslant c(d)\,\|p\|_{2,Q}\left(|\zeta w_\rho|+\|w\|_{W_2^1(Q)}\right)\leqslant$$
$$\leqslant c(d)\left(\|w\|_{W_2^1(Q)}+\langle f\rangle_{H(Q)}\right)\left(|\zeta w_\rho|+\|w\|_{W_2^1(Q)}\right).$$

Therefore from (44), (46), and (47) it follows that

$$|\zeta w_\rho|^2\leqslant c(d)\left(|\zeta w_\rho|+\|w\|_{W_2^1(Q)}\right)\left(\|w\|_{W_2^1(Q)}+\|f\|_{2,Q}\right)+c\max|\operatorname{grad}F|\,|\zeta w_\rho|^2,$$

where c does not depend on d. If we fix d sufficiently small, we obtain

$$|\zeta w_\rho|^2\leqslant c\left(\|w\|_{W_2^1(Q)}+\|f\|_{2,Q}\right)^2$$

and after we let $\rho\to 0$ we obtain

$$\sum_{i,k=1}^{3}\sum_{\alpha=1}^{2}\left\|\frac{\partial^2 w_i}{\partial z_k\partial z_\alpha}\right\|_{2,Q'}^2\leqslant c\left(\|w\|_{W_2^1(Q')}+\|f\|_{2,Q}\right)^2. \tag{48}$$

By analogy we can obtain a bound for $\|\partial p/\partial z_\alpha\|_{2,Q'}$, $\alpha<3$. In (40) we set $\psi=(\zeta'\Delta'\Phi^\rho)$, where ζ' is a smooth function equal to one for $z\in Q'':|z_\alpha|\leqq d/2$, $\alpha=1,2$, $0\leqq z_3\leqq d$, and equal to zero outside Q', and Φ^ρ is a solution of the problem

$$\operatorname{div}\Phi^\rho=\zeta p_\rho,\quad \Phi^\rho|_{z_3=0}=0$$

in the half-space $z_3>0$ satisfying inequalities of the form (9) and (13).

If we perform the above transformations once again, we obtain

$$\sum_{\alpha=1}^{2}\frac{\nu}{2}E\left(\frac{\partial\zeta' w_\rho}{\partial z_\alpha},\ \frac{\partial\Phi^\rho}{\partial z_\alpha}\right)+\frac{\nu}{2}E'(w_\rho,\ \Phi^\rho)+\varepsilon_\rho(w,\ \Phi^\rho) \tag{49}$$
$$=(f_\rho,\ \zeta'\Delta'\Phi^\rho)+(p_\rho,\ \operatorname{div}\zeta'\Delta'\Phi^\rho).$$

The terms on the left side are bounded by

$$c\,(\|\zeta' w_\rho\|+\|w\|_{W_2^1(Q)})\,|\Phi^\rho|\leqslant c\,(\|w\|_{W_2^1(Q)}+\|f\|_Q)\,|\Phi^\rho|.$$

Furthermore,

$$|(f_\rho,\ \zeta'\Delta'\Phi^\rho)|\leqslant\|f\|_Q\,|\Phi^\rho|.$$

Finally,

$$(p_\rho,\ \operatorname{div}\zeta'\Delta'\Phi^\rho)=(p_\rho,\ \operatorname{grad}\zeta'\Delta'\Phi^\rho)-\sum_{\alpha=1}^{2}\left(\frac{\partial\zeta' p_\rho}{\partial z_\alpha},\ \frac{\partial\zeta' p_\rho}{\partial z_\alpha}\right) \tag{50}$$
$$\leqslant-\sum_{\alpha=1}^{2}\left\|\frac{\partial\zeta' p_\rho}{\partial z_\alpha}\right\|^2+c\,\|p_\rho\|\,|\Phi^\rho|.$$

Since

$$|\Phi^\rho|\leqslant c\left(\sum_{\alpha=1}^{2}\left\|\frac{\partial\zeta' p_\rho}{\partial z_\alpha}\right\|_Q^2\right)^{1/2},$$

it follows from (49) and (50) that

$$\sum_{\alpha=1}^{2}\left\|\frac{\partial\zeta' p_\rho}{\partial z_\alpha}\right\|^2\leqslant c\,[(\|w\|_{W_2^1(Q)}+\|f\|_Q)^2+\|p\|_Q^2]$$

and also

$$\left(\sum_{\alpha=1}^{2}\left\|\frac{\partial p_\rho}{\partial z_\alpha}\right\|_{2,\,Q''}^2\right)^{1/2}\leqslant c\,(\|w\|_{W_2^1(Q)}+\|f\|_R). \tag{51}$$

Thus we have found bounds for the derivatives $\partial p/\partial z_\alpha$ and $\partial^2 w_i/\partial z_k\partial z_\alpha$, $\alpha<3$; that is, we have found bounds for $\partial^2 v_i/\partial z_k\partial z_\alpha$ in the cube Q''. We will find bounds for the other derivatives by means of a Navier-Stokes system for v and p which is satisfied in each case inside Q and which in the coordinates $\{z\}$ is of the form

$$\nu\left(\Delta v_\beta+\operatorname{grad}^2 F\frac{\partial^2 v_\beta}{\partial z_3^2}-2\sum_{\alpha=1}^{2}F_\alpha\frac{\partial^2 v_\beta}{\partial z_\alpha\partial z_3}-\Delta' F\frac{\partial v_\beta}{\partial z_3}\right)=\frac{\partial p}{\partial z_\beta}-F_\beta\frac{\partial p}{\partial z_3}+f_\beta,$$
$$\beta=1,\ 2,\quad F_\beta=\frac{\partial F}{\partial z_\beta},$$
$$\nu\left(\Delta v_3+\operatorname{grad}^2 F\frac{\partial^2 v_3}{\partial z_3^2}-2\sum_{\alpha=1}^{2}F_\alpha\frac{\partial^2 v_3}{\partial z_\alpha\partial z_3}-\Delta' F\frac{\partial v_3}{\partial z_3}\right)=\frac{\partial p}{\partial z_3}+f_3, \tag{52}$$
$$\operatorname{div} v-\sum_{\alpha=1}^{2}F_\alpha\frac{\partial v_\alpha}{\partial z_3}=0.$$

We will use the letter h, with affixes, to denote functions which are bounded by the right side of (51). From the first two equations it follows that

$$\nu\,(1+\operatorname{grad}^2 F)\frac{\partial^2 v_\beta}{\partial z_3^2}=h_\beta^{(1)}-F_\beta\frac{\partial p}{\partial z_3}=h_\beta^{(2)}-\nu F_\beta(1+\operatorname{grad}^2 F)\frac{\partial^2 v_3}{\partial z_3^2};$$

that is,

$$\frac{\partial^2 v_\beta}{\partial z_3^2} = h_\beta^{(3)} - F_\beta \frac{\partial^2 v_3}{\partial z_3^2}.$$

If we differentiate (52) with respect to z_3, we obtain

$$\frac{\partial^2 v_3}{\partial z_3^2} = h + \sum_{\alpha=1}^{2} F_\alpha \frac{\partial^2 v_\alpha}{\partial z_3^2} = h + \sum_{\alpha=1}^{2} F_\alpha h_\alpha^{(3)} - \operatorname{grad}^2 F \frac{\partial^2 v_3}{\partial z_3^2}.$$

Consequently

$$\frac{\partial^2 v_3}{\partial z_3^2} = \frac{h + \sum_{\alpha=1}^{2} F_\alpha h_\alpha^{(3)}}{1 + \operatorname{grad}^2 F} = h_3, \quad \frac{\partial^2 v_\beta}{\partial z_3^2} = h_\beta^{(3)} - F_\beta h_3 = h_\beta,$$

and therefore $\partial p/\partial z_3 = h'$.

Therefore $\partial^2 v_i/\partial x_k \partial x_l$ and $\partial p/\partial x_k$ are bounded in Ω'':

$$|y_\alpha| \leqslant \frac{d}{2}, \quad \alpha = 1, 2, \quad F(y_1, y_2) \leqslant y_3 \leqslant F(y_1, y_2) + d.$$

Thus the assertion in the theorem is valid for more general regions Ω''.

Further research into the smoothness of a generalized solution can be carried out on the basis of the general theory of systems which are elliptic in the sense of Douglis-Nirenberg, since both boundary conditions are such that on S_1 and S_2 a complementarity condition is satisfied. From the results in [8] it follows that in regions which satisfy the hypothesis of Theorem 2 we can find local bounds of the type (36) in the norms of W_p^l and in Hölder norms.

Bibliography

[1] S. G. Kreĭn and G. I. Laptev, *On the problem of the motion of a viscous fluid in an open vessel*, Funkcional. Anal. i Priložen. **2** (1968), no. 1, 40-50 = Functional Anal. Appl. **2** (1968), 38-47. MR **40** # 1714.

[2] V. A. Solonnikov, *On estimates of Green's tensors for certain boundary problems*, Dokl. Akad. Nauk SSSR **130** (1960), 988-991 = Soviet Math. Dokl. **1** (1960), 128-131. MR **22** # 12300.

[3] O. A. Ladyženskaja, *Mathematical problems in the dynamics of a viscous incompressible fluid*, Fizmatgiz, Moscow, 1961; English transl., Gordon and Breach, New York, 1963. MR **27** # 5034a,b.

[4] A. Korn, *Über einige Ungleichungen, welche in der Theorie der elastischen und elektrischen Schwingungen eine Rolle spielen*, Bull. Internat. Acad. Sci. Cracovie. Cl. Sci. Math. Nat. **1909**, no. 9, 705-724. Jbuch. Fortschritte Math. **40**, 884.

[5] S. G. Mihlin, *The problem of the minimum of a quadratic functional*, GITTL, Moscow, 1952; English transl., Holden-Day Series in Math. Phys., Holden-Day, San Francisco, Calif., 1965. MR **16**, 41; **30** # 1427.

[6] V. P. Il'in and V. A. Solonnikov, *On some properties of differentiable functions of several variables*, Trudy Mat. Inst. Steklov. **66** (1962), 205-226; English transl., Amer. Math. Soc. Transl. (2) **81** (1969), 67-90. MR **27** # 2768.

[7] K. O. Friedrichs, *On the differentiability of the solutions of linear elliptic differential equations*, Comm. Pure Appl. Math. **6** (1953), 299-326. MR **15**, 430.

[8] V. A. Solonnikov, *General boundary value problems for Douglis-Nirenberg elliptic systems*. II, Trudy Mat. Inst. Steklov. **92** (1966), 233-297 = Proc. Steklov Inst. Math. **92** (1966), 269-339. MR **35** # 1953.

Translated by
F. A. CEZUS

Trudy Mat. Inst. Steklov.
125 (1973)

Proc. Steklov Inst. Math.
125 (1973)

BOUNDARY VALUE PROBLEMS FOR ELLIPTICO-HYPERBOLIC EQUATIONS

L. STUPJALIS

Abstract. This paper is devoted to the study of boundary value problems for equations which are of elliptic type in one of the parts of the region Ω of the variable x and of hyperbolic type in the other, where conditions of conjugacy of diffraction type are posed on the boundary separating the two parts of Ω. Unique solvability of initial boundary value problems is proved for these equations under wider conditions than in the previous work of O. A. Ladyženskaja and the author.

Bibliography: 7 items.

Let a bounded region Ω of n-dimensional Euclidean space be the union of two regions Ω_1 and Ω_2 and an $(n-1)$-dimensional surface Γ which separates them, so that $\Omega=\Omega_1\cup\Gamma\cup\Omega_2$. Let us denote by S the boundary of Ω, and by S_1 and S_2 the boundaries of Ω_1 and Ω_2. For such a region let $Q_T^{(k)}=\Omega_k\times[0,T]$, $k=1,2$; for definiteness in $Q_T^{(1)}$ let there be given an elliptic equation

$$(1)\qquad L_1^{(1)}u\equiv-\frac{\partial}{\partial x_i}\left[a_{ij}^{(1)}(x,\ t)\frac{\partial u}{\partial x_j}\right]-a^{(1)}(x,\ t)\,u=f_1^{(1)}(x,\ t),$$

and in $Q_T^{(2)}$ a hyperbolic equation

$$(2)\qquad L_3^{(2)}u\equiv\frac{\partial^2u}{\partial t^2}-\frac{\partial}{\partial x_i}\left[a_{ij}^{(2)}(x,\ t)\frac{\partial u}{\partial x_j}\right]-a_i^{(2)}(x,\ t)\frac{\partial u}{\partial x_i}-a^{(2)}(x,\ t)\,u=f_3^{(2)}(x,\ t).$$

For these equations we shall investigate the following boundary value problem: to determine a function $u(x,t)$ which satisfies equation (1) in $Q_T^{(1)}$, equation (2) in $Q_T^{(2)}$, the initial conditions

$$(3)\qquad u|_{t=0}=\varphi_0^{(2)}(x),\qquad u_t|_{t=0}=\varphi_1^{(2)}(x)$$

in Ω_2 for $t=0$, the boundary condition

$$(4)\qquad u|_{S_T}=0$$

on the boundary $S_T=S\times[0,T]$, and the conjugacy conditions

$$(5)\qquad [u]|_{\Gamma_T}=0,\quad \left[\frac{\partial u}{\partial N}\right]\bigg|_{\Gamma_T}=0$$

on $\Gamma_T=\Gamma\times[0,T]$, where the symbol $[v]$ denotes the jump experienced by a function v during the passage through Γ_T.

Without mentioning it, we shall assume throughout the paper that the following conditions are satisfied:

a) Ω_1 is a strictly interior subregion of Ω.

b) The surfaces S and Γ are twice continuously differentiable.

c) The coefficients of the operators

$$L_1^{(k)},\quad k=1,\ 2;\quad L_1^{(2)}=-\frac{\partial}{\partial x_i}\left(a_{ij}^{(2)}\frac{\partial}{\partial x_j}\right)-a_i^{(2)}\frac{\partial}{\partial x_i}-a^{(2)}$$

AMS (MOS) subject classifications (1970). Primary 35M05, 35A05.

satisfy the conditions

$$\nu \sum_{i=1}^{n} \xi_i^2 \leqslant a_{ij}^{(k)} \xi_i \xi_j \leqslant M \sum_{i=1}^{n} \xi_i^2, \quad \nu > 0, \quad a_{ij}^{(k)} = a_{ji}^{(k)},$$

$$\left| \frac{\partial a_{ij}^{(k)}}{\partial t}, \quad a_i^{(2)}, \quad \frac{\partial a_i^{(2)}}{\partial t}, \quad a^{(k)}, \quad \frac{\partial a^{(k)}}{\partial t} \right| \leqslant M.$$

d) $\operatorname{ess\,sup}_{Q_T^{(1)}} a^{(1)}(x,t) = a_0^{(1)} < 0$.

Without loss of generality we shall assume that $\varphi_0^{(2)} = \varphi_1^{(2)} = 0$.

Boundary value problems for equations of mixed type were considered in [1–4] under various restrictions on the known functions. During the proof of the solvability of the boundary value problem (1)–(5) the following restriction was imposed: it was required that the free term of the elliptic equation have a square integrable generalized derivative with respect to the time t, and that the elliptic operator appearing in it be formally selfadjoint. The example cited in [4] justifies this restriction, i.e. if the function $f_1^{(1)}(x,t)$ from equation (1) does not have a square integrable generalized derivative with respect to t, then the problem (1)–(5) does not have solutions in the class of functions which have a finite norm

$$\max_{0 \leqslant t \leqslant T} \left(\| u_t \|_{L_2(\Omega_2)} + \| u \|_{W_2^1(\Omega)} \right).$$

Consequently, for the solvability of the boundary value problem (1)–(5) for an arbitrary square integrable free term of the elliptic equation it is necessary to widen the class of generalized solutions. The question arises: is such an extension of the class of generalized solutions of this problem admissible, i.e. does there exist a class of functions in which the problem (1)–(5) has generalized solutions, and, in addition, is the uniqueness theorem preserved? In this paper a positive answer is given to this question in the case when the elliptic operator appearing in (1)–(5) is formally selfadjoint.

The proof of the unique solvability of the boundary value problem (1)–(5) is constructed according to the following scheme. First, the operator A corresponding to (1)–(5) is defined in the usual way on the set of smooth functions $D(A)$. For this operator the closure $\bar{A}$ is constructed, and the operator adjoint to $\bar{A}$ is formed. Moreover, using a theorem on the normal solvability of the closed operator $\bar{A}$, one proves that (1)–(5) has a unique generalized solution from $D(\bar{A})$ for any $f_1^{(1)}(x,t) \in L_2(Q_T^{(1)})$ and $f_3^{(2)}(x,t) \in L_2(Q_T^{(2)})$ (even for a wider class of right sides of equations (1) and (2)).

The existence of generalized solutions of (1)–(5) in $D(\bar{A})$ immediately follows from Lemma 4′ of [4] and an "energy inequality", i.e. an inequality of the form

$$\| u \| \leqq C \| Au \|,$$

which will be proved in the following section.

§1. Existence theorems

Let M be the set of all elements $u(x,t)$ from $L_2(Q_T)$, where $Q_T = \Omega \times [0,T]$, which have generalized derivatives of the form $\partial u / \partial x_i$ from $L_2(Q_T)$ and which satisfy the condition $u|_{S_T} = 0$. Let us define a scalar product $[u,v]_0$ on the set M by putting

$$[u,\ v]_0=\int\limits_{Q_T}\left(\sigma\frac{\partial k^*u}{\partial t}\frac{\partial k^*v}{\partial t}+\frac{\partial k^*u}{\partial x_i}\frac{\partial k^*v}{\partial x_i}+k^*uk^*v\right)dxdt,$$

where k^* is the operator defined on M according to the rule $k^*u=-\int_0^t u(x,\tau)d\tau$, and

$$\sigma(x,\ t)=\begin{cases}0, & \text{if } (x,\ t)\in Q_T^{(1)},\\ 1, & \text{if } (x,\ t)\in Q_T^{(2)}.\end{cases}$$

After the introduction of this scalar product and the completion of M, we convert it into a Hilbert space, which we shall denote by $\widetilde{W}_0(Q_T)$.

Let A be the operator defined by the equality

$$Au=(L_1^{(1)}u,(x,t)\in Q_T^{(1)};\ L_3^{(2)}u,(x,t)\in Q_T^{(2)}).$$

It acts from $\widetilde{W}_0(Q_T)$ into the space $W=L_2(Q_T^{(1)})\times L_2(Q_T^{(2)})$.[1] The domain of definition $D(A)$ consists of all elements $u(x,t)$ from $\widetilde{W}_0(Q_T)$ which in the regions $Q_T^{(1)}$ and $Q_T^{(2)}$ respectively belong to the spaces $W_2(Q_T^{(1)})$ and $W_2(Q_T^{(2)})$ and which satisfy the conditions

$$u\,|_{t=0}=0,\quad u_t\,|_{t=0}=0,\quad x\in\Omega_2,$$

$$u\,|_{S_T}=0,\quad [u]\,|_{\Gamma_T}=0,\quad \left[\frac{\partial u}{\partial N}\right]\bigg|_{\Gamma_T}=0.$$

The operator A is obviously not bounded. The range $R(A)$ of this operator forms a linear set in W.

The following lemma is proved in [**4**].

LEMMA 1. *If the functions $a_{ij}^{(k)}$, $a^{(k)}$ and $\partial a_{ij}^{(k)}/\partial x_i$ are bounded and measurable and have bounded derivatives with respect to t up to third order, and if $a_i^{(2)}$ has bounded derivatives with respect to t up to second order, then the range $R(A)$ of A is dense in $W=L_2(Q_T^{(1)})\times L_2(Q_T^{(2)})$.*

Let us prove the following lemma.

LEMMA 2. *If the functions $a_{ij}^{(k)}$, $a_i^{(2)}$ and $a^{(k)}$ are bounded and measurable and have a bounded first derivative with respect to t, and if $\operatorname{ess\,sup}_{Q_T^{(1)}}a^{(1)}(x,t)=a_0^{(1)}<0$, then for all $u\in D(A)$ the inequality*

$$\max_{0\le t\le T}(\|w_t\|_{L_2(\Omega_2)}+\|w\|_{W_2^{(1)}(\Omega)})\le C_1\|Au\|_W \tag{6}$$

is valid, where $w=k^*u$.

PROOF. We have

$$\int\limits_{Q_{t_0}}Au\Phi dxdt=\int\limits_{Q_{t_0}}\left[\sigma\frac{\partial^2u}{\partial t^2}-\frac{\partial}{\partial x_i}\left(a_{ij}\frac{\partial u}{\partial x_j}\right)-\sigma a_i\frac{\partial u}{\partial x_i}-au\right]\Phi dxdt$$

$$=\frac{1}{2}\int\limits_{\Omega_2}\left(\frac{\partial w}{\partial t}\right)^2dx\bigg|_{t=t_0}+\frac{1}{2}\int\limits_{\Omega}\left(a_{ij}\frac{\partial w}{\partial x_i}\frac{\partial w}{\partial x_j}-aw^2\right)dx\bigg|_{t=t_0}+$$

---

[1] The scalar product in W is defined as usual:

$$((f^{(1)};f^{(2)}),(f_1^{(1)};f_1^{(2)}))_W=(f^{(1)},f_1^{(1)})_{L_2(Q_T^{(1)})}+(f^{(2)},f_1^{(2)})_{L_2(Q_T^{(2)})}.$$

$$+\int_0^{t_0}\int_\Omega\left[\frac{1}{2}\frac{\partial a_{ij}}{\partial t}\frac{\partial w}{\partial x_i}\frac{\partial w}{\partial x_j}-\sigma\left(\frac{\partial a_i}{\partial t}\frac{\partial w}{\partial x_i}w+a_i\frac{\partial w}{\partial x_i}\frac{\partial w}{\partial t}\right)-\frac{1}{2}\frac{\partial a}{\partial t}w^2\right]dxdt$$

$$+\int_0^{t_0}\int_\Omega\Big[-\frac{\partial a_{ij}}{\partial t}\frac{\partial w(x,t)}{\partial x_i}\frac{\partial w(x,t_0)}{\partial x_j}+\sigma\frac{\partial a_i}{\partial t}\frac{\partial w(x,t)}{\partial x_i}w(x,t_0)$$

$$+\frac{\partial a}{\partial t}w(x,t)\,w(x,t_0)\Big]dxdt,$$

where $Q_{t_0}=\Omega\times[0,t_0]$, $0\leqq t_0\leqq T$, $\Phi(x,t)=-\int_{t_0}^t u(x,\tau)d\tau=w(x,t)-w(x,t_0)$ and a_{ij}, a_i, a are respectively equal to $a_{ij}^{(1)}$, 0, $a^{(1)}$ for $(x,t)\in Q_T^{(1)}$ and $a_{ij}^{(2)}$, $a_i^{(2)}$, $a^{(2)}$ for $(x,t)\in Q_T^{(2)}$. From this follows

$$\int_\Omega\Big\{\sigma\left(\frac{\partial w}{\partial t}\right)^2+(\nu-2MTn\varepsilon_1)\left(\frac{\partial w}{\partial x_i}\right)^2$$

$$-\left[a(x,t)+2MT(\sigma\sqrt{n}\varepsilon_2+\varepsilon_3)+2T\varepsilon_4\right]w^2\Big\}dx\Big|_{t=t_0}$$

$$\leqslant\int_0^{t_0}\int_\Omega\Big\{\sigma M\sqrt{n}\left(\frac{\partial w}{\partial t}\right)^2+M\left(n+2\sqrt{n}+\frac{n}{2\varepsilon_1}+\frac{\sigma\sqrt{n}}{2\varepsilon_2}\right)\left(\frac{\partial w}{\partial x_i}\right)^2 \tag{7}$$

$$+\left[M\left(1+\sqrt{n}+\frac{1}{2\varepsilon_3}\right)+2\varepsilon_4\right]w^2\Big\}dxdt+\frac{1}{\varepsilon_4}\|Au\|_W^2.$$

Here we used the inequality

$$\|\Phi\|^2_{L_2(Q_{t_0})}\leqslant 2\|w\|^2_{L_2(Q_{t_0})}+2t_0\|w\|^2_{L_2(\Omega)}\Big|_{t=t_0},$$

which follows from the representability of Φ by means of w.

By virtue of the conditions of the lemma, $\operatorname{ess\,sup}_{Q_T^{(1)}}a^{(1)}(x,t)=a_0^{(1)}<0$. Therefore, putting $\epsilon_1=\epsilon_2=\epsilon_3=\epsilon_4=\kappa/4T(Mn+1)$, where $\kappa=\min(\nu,-a_0^{(1)})$, we obtain that

$$\nu-2MTn\varepsilon_1,\ -[a^{(1)}(x,t)+2MT\varepsilon_3+2T\varepsilon_4]\geqslant\frac{\varkappa}{2}.$$

We shall assume that $\operatorname{ess\,sup}_{Q_T^{(2)}}[a^{(2)}(x,t)+2MT(\sqrt{n}\epsilon_2+\epsilon_3)+2T\epsilon_4]=\delta_1<0$, since otherwise this can always be achieved by introducing a new function $u_1(x,t)$ in the problem (1)–(5) instead of the unknown function $u(x,t)$ which is related to it by the equality $u(x,t)=u_1(x,t)e^{-\lambda_0 t}$. In this connection there appears a term $-2\lambda_0\partial u/\partial t$ in the hyperbolic equation, the presence of which does not influence the course of the proof in any essential way.

From (7) follows the inequality

$$\int_\Omega\left[\sigma\left(\frac{\partial w}{\partial t}\right)^2+\left(\frac{\partial w}{\partial x_i}\right)^2+w^2\right]dx\Big|_{t=t_0}$$

$$\leqslant C_2\int_0^{t_0}\int_\Omega\left[\sigma\left(\frac{\partial w}{\partial t}\right)^2+\left(\frac{\partial w}{\partial x_i}\right)^2+w^2\right]dxdt+C_3\|Au\|_W^2, \tag{8}$$

where the constants C_2 and C_3 only depend on ν, $a_0^{(1)}$, M, n and T. The last inequality can be rewritten thus:

$$\frac{dy(t_0)}{dt_0}\leqslant C_2y(t_0)+C_3\|Au\|_W^2,$$

where

$$y(t_0)=\int_0^{t_0}\int_\Omega\left[\sigma\left(\frac{\partial w}{\partial t}\right)^2+\left(\frac{\partial w}{\partial x_i}\right)^2+w^2\right]dxdt$$

and consequently

$$\frac{d}{dt_0}\left[y(t_0)e^{-C_2t_0}\right]\leqslant C_3e^{-C_2t_0}\|Au\|_W^2$$

It follows from this that

$$y(t_0)\leqslant\frac{C_3\left(e^{C_2T}-1\right)}{C_2}\|Au\|_W^2.$$

The last inequality together with (8) gives (6). The lemma is proved.

Theorem 1. *The operator A admits a closure, where $R(\overline{A})=\overline{R(A)}=W$. The operator $\overline{A}$ has a bounded inverse $\overline{A}^{-1}$.*

This theorem is an immediate consequence of Lemmas 1 and 2.

Thus the unique solvability of the operator equation

$$\overline{A}u=(f_1^{(1)};f_3^{(2)}) \tag{9}$$

is proved for any $f_1^{(1)}\in L_2(Q_T^{(1)})$ and $f_3^{(2)}\in L_2(Q_T^{(2)})$. The solution of this equation has finite norm (6).

Let us denote by $\hat{W}_T^1(Q_T)$ ($\hat{W}_0^1(Q_T)$) the Hilbert space of all functions $v(x,t)\in L_2(Q_T)$ which have generalized derivatives of the form $\partial v/\partial x_i$ from $L_2(Q_T)$ and a generalized derivative $\partial v/\partial t$ from $L_2(Q_T^{(2)})$ and which satisfy the conditions

$$v|_{S_T}=0,\quad v|_{t=T}=0\quad (v|_{t=0}=0),\quad x\in\Omega_2,$$

with the scalar product

$$(u,v)_{\hat{W}^1(Q_T)}=\int_{Q_T}\left(\sigma\frac{\partial u}{\partial t}\frac{\partial v}{\partial t}+\frac{\partial u}{\partial x_i}\frac{\partial v}{\partial x_i}+uv\right)dxdt$$

and with the norm $\|\cdot\|_{\hat{W}^1(Q_T)}$.

We shall say that the function $u\in\hat{W}_T^{1,t}(Q_T)$ ($\hat{W}_0^{1,t}(Q_T)$), if $u\in\hat{W}_T^1(Q_T)$ ($\hat{W}_0^1(Q_T)$), $\partial u/\partial t\in\hat{W}_T^1(Q_T)$ ($\hat{W}_0^1(Q_T)$) and it satisfies the condition $u|_{t=T}=0$ ($u|_{t=0}=0$), $x\in\Omega$.

It is not difficult to see that a solution of the operator equation (9) satisfies an integral identity

$$\int_{Q_T}\left[-\sigma\frac{\partial w}{\partial t}\frac{\partial^2\Phi}{\partial t^2}+\frac{\partial w}{\partial x_j}\frac{\partial}{\partial t}\left(a_{ij}\frac{\partial\Phi}{\partial x_i}\right)-\sigma\frac{\partial w}{\partial x_i}\frac{\partial}{\partial t}(a_i\Phi)-w\frac{\partial a\Phi}{\partial t}\right]dxdt \tag{10}$$
$$=\int_{Q_T}f\Phi dxdt,$$

where

$$f(x,t)=\begin{cases}f_1^{(1)}(x,t), & (x,t)\in Q_T^{(1)},\\ f_3^{(2)}(x,t), & (x,t)\in Q_T^{(2)},\end{cases}$$

for any $\Phi\in\hat{W}_T^{1,t}(Q_T)$.

A function u which has a finite norm $\max_{0\leq t\leq T}(\|w_t\|^2_{L_2(\Omega_2)}+\|w\|^2_{W_2^1(\Omega)})^{1/2}$ and which satisfies the integral identity (10) for any $\Phi\in\hat{W}_T^{1,t}(Q_T)$ and the condition $w|_{S_T}=0$ will be called a generalized solution of the problem (1)–(5).

THEOREM 1′. *Let the functions* $a_{ij}^{(k)}$, $a_i^{(2)}$, *and* $a^{(k)}$ *be bounded and measurable and have a bounded first derivative with respect to* t, *and let* $\operatorname{ess\,sup}_{Q_T^{(1)}} a^{(1)}(x,t) = a_0^{(1)} < 0$.

Then the problem (1) – (5) *has a generalized solution for any* $f_1^{(1)}(x,t) \in L_2(Q_T^{(1)})$ *and* $f_3^{(2)}(x,t) \in L_2(Q_T^{(2)})$.

Theorem 1′ is proved by the usual means of smoothing the coefficients of the equations and using Theorem 1 and inequality (6).

The following question arises: is the uniqueness theorem preserved under such a definition of a generalized solution of (1) – (5)? The answer to this question will be given in the following section.

§2. The unique solvability of the boundary value problem (1) – (5)

The following concepts and generalizations will be used in this section.

Let H_0 be some complete Hilbert space with scalar product $(\cdot,\cdot)$ and norm $\|\cdot\|_0$. Let us assume that in H_0 there is an everywhere dense linear set H_+ which is a complete Hilbert space with respect to some other scalar product $(\cdot,\cdot)_+$; $\|\cdot\|_+$ is the norm in H_+. We shall suppose that

$$\|u\|_0 \leqq \|u\|_+ \quad (u \in H_+); \tag{11}$$

H_+ is called a subspace with positive norm.

Elements $f \in H_0$ generate antilinear functionals l on H_+ if one puts $l(u) = l_f(u) = (f,u)_0$. These functionals are continuous by virtue of (11). Their norm is calculated in the following way:

$$\|l_f\| = \sup_{u \in H_+} \frac{|(f,\ u)_0|}{\|u\|_+} \leqslant \|f\|_0 \quad (f \in H_0).$$

The completion of the space H_0 with respect to the norm $\|f\|_- = \|l_f\|$ is called a space with negative norm. It will be convenient for us to define a negative norm in various different ways. The equivalence of these definitions will be established below.

Let us consider a bilinear form of $f \in H_0$ and $u \in H_+$: $B(f,u) = (f,u)_0$. By virtue of (11) this bilinear form is continuous with respect to changes in f and u in the corresponding spaces. Therefore it can be represented either by a scalar product in H_0 or by a scalar product in H_+:

$$(f,u)_0 = B(f,u) = (If,u)_+ \quad (f \in H_0, u \in H_+), \tag{12}$$

where I is a distributive operator acting from H_0 into H_+. It is not difficult to see that it is bounded and that $\|I\| \leqq 1$. Indeed, putting $u = If$ in (12) and taking (11) into account, we obtain

$$\|If\|_+^2 = (f, If)_0 \leqq \|f\|_0 \|If\|_0 \leqq \|f\|_0 \|If\|_+.$$

It follows from this that

$$\|If\|_+ \leqq \|f\|_0.$$

Let us introduce a scalar product for $f,g \in H_0$ by putting

$$(f,g)_- = (If,g)_0 = (If, Ig)_+, \tag{13}$$

and let us construct the completion. The resulting Hilbert space is called a space with negative norm. Let us denote the scalar product in it by $(\cdot,\cdot)$, and the norm by $\|\cdot\|_-$. Since $\|I\| \leqq 1$, we have $\|f\|_- \leqq \|f\|_0$ $(f \in H_0)$.

According to (13), the operator I isometrically transforms a set in H_- (namely H_0) into a set in H_+. Extending this operator by continuity to all of H_-, we obtain an operator I which carries H_- into H_+. The range $R(I)$ of I fills out all of H_+: indeed, $R(I)$ is closed; therefore, if $R(I) \neq H_+$, then there exists a $u \in H_+$ such that $0 = (If, u)_+ = (f, u)_0$ for all $f \in H_0$, i.e. $u = 0$. Thus I is an isometric operator which carries all of H_- onto all of H_+, i.e. $D(I) = H_-$, $R(I) = H_+$ and $(\alpha, \beta)_- = (I\alpha, I\beta)_+$ $(\alpha, \beta \in H_-)$.

Let us show that for $\alpha \in H_-$ and $u \in H_+$ the "scalar product" defined by $(\alpha, u)_0$ is a bilinear form which coincides with the scalar product in H_0 for $\alpha \in H_0$. The Cauchy inequality is valid for it:

(14) $$|(\alpha, u)_0| \leqq \|\alpha\|_- \|u\|_+ \quad (\alpha \in H_-,\ u \in H_+).$$

Indeed, let $f \in H_0$. Then for $u \in H_+$ we have

$$|(f, u)_0| = |(If, u)_+| \leqq \|If\|_+ \|u\|_+ = \|f\|_- \|u\|_+;$$

here, passing to the limit (in H_-) as $f \to \alpha$, we obtain (14).

With the aid of a limiting process, it is easy to obtain from (12) and (13) the equality

(15) $$\begin{gathered}(\alpha,\ \beta)_- = (\alpha,\ I\beta)_0 = (I\alpha,\ \beta)_0 = (I\alpha,\ I\beta)_+, \\ (\alpha,\ u)_0 = (I\alpha,\ u)_+ \ (\alpha,\ \beta \in H_-,\ u \in H_+).\end{gathered}$$

Lemma 3. *Each bounded distributive functional $l(u)$ on H_+ is representable in a unique way by the formula*

(16) $$l(u) = (\alpha, u)_0,$$

where α is some fixed element from H_- and $\|l\| = \|\alpha\|_-$.

Proof. By (14) each $\alpha \in H_-$ generates a functional l_α according to the formula $l_\alpha(u) = (\alpha, u)_0$ $(u \in H_+)$. Conversely, if l is a function on H_+, then the representation $l(u) = (v_l, u)_+$ is valid, where v_l is some element from H_+. It follows from the second equality of (15) that $l(u) = (v_l, u)_+ = (\alpha_l, u)_0$, $\alpha_l = I^{-1}v_l \in H_-$; in addition, if $l = 0$, then it is not difficult to see that also $\alpha_l = 0$. Thus the correspondence $l \leftrightarrow \alpha$ between functionals on H_+ and elements from H_- is one-to-one; in addition, by the first of the equalities of (15), $\|l\| = \|v_l\|_+ = \|\alpha\|_-$. The lemma is proved.

From Lemma 3 follows the equivalence of the two definitions of spaces with negative norm which were cited above.

Let us put $H_0 = L_2(Q_T)$ and $H_+ = \hat{W}_\mu^1(Q_T)$ $(\mu = 0, T)$. The Hilbert space with negative norm which corresponds to it will be denoted by $\hat{W}_\mu^{-1}(Q_T)$.

For the study of the problem (1) – (5) it will be necessary for us to describe the following Banach spaces.

$B_0(Q_T)$ is the Banach space obtained by the closure of the linear manifold of twice continuously differentiable functions satisfying the conditions

(17) $$u|_{t=0} = 0, \quad u_t|_{t=0} = 0, \quad x \in \Omega_2,$$

(18) $$u|_{S_T} = 0, \quad [u]|_{\Gamma_T} = 0, \quad \left[\frac{\partial u}{\partial N}\right]\Big|_{\Gamma_T} = 0,$$

with respect to the norm $\|u\|_{B(Q_T)} = \max_{0 \leqq t \leqq T}(\|u_t(x,t)\|^2_{L_2(\Omega_2)} + \|u(x,t)\|^2_{W_2^1(\Omega)})^{1/2}$.

$B_T(Q_T)$ is the Banach space obtained by the closure of the linear manifold of twice continuously differentiable functions satisfying the conditions

$$u|_{t=T}=0, \quad u_t|_{t=T}=0, \quad x\in\Omega_2, \tag{19}$$

and the conditions (18), with respect to the norm $\|\cdot\|_{B(Q_T)}$.

We shall say that $u\in B_0^t(Q_T)$ $(B_T^t(Q_T))$, if

$$u\in B_0(Q_T)\,(B_T(Q_T)), \quad \frac{\partial u}{\partial t}\in B_0(Q_T)\,(B_T(Q_T))$$

and it satisfies the condition $u|_{t=0}=0$ $(u|_{t=T}=0)$, $x\in\Omega$.

For the proof of "energy inequalities" we shall use averaging operators. The simplest averaging operator is an operator of the form

$$J_\rho u(x,t)=\frac{1}{\rho}\int_0^T\omega\left(\frac{t-t'}{\rho}\right)u(x,t')\,dt',$$

where $\omega(\xi)$ is an even, infinitely differentiable function of one variable; in addition $\omega(\xi)\geqq 0$, $\omega(\xi)\equiv 0$ for $|\xi|\geqq 1$ and $\int_{-1}^1\omega(\xi)d\xi=1$. It possesses the essential deficiency that even in the case when u initially satisfies the condition $u|_{t=0}=0$ or $u|_{t=T}=0$, the averaged function $J_\rho u$ does not, in general, satisfy either of these conditions. Therefore we shall use special averaging operators:

$$J_\rho^- u(x,t)=\frac{1}{\rho}\int_0^T\omega\left(\frac{t-t'-2\rho}{\rho}\right)u(x,t')\,dt'$$

and

$$J_\rho^+ u(x,t)=\frac{1}{\rho}\int_0^T\omega\left(\frac{t-t'+2\rho}{\rho}\right)u(x,t')\,dt'.$$

It is easy to see that J_ρ^+ is conjugate to J_ρ^- and that for any $u\in L_2(Q_T)$ we have $J_\rho^- u|_{t=0}=0$ and $J_\rho^+ u|_{t=T}=0$.

The following assertions are known (cf., for example, [**6**]).

LEMMA 4. *For all functions $u(x,t)$ from $W_2^1(Q_T)$ equal to zero for $t=0$,*

$$\frac{\partial}{\partial t}J_\rho^- u=J_\rho^-\frac{\partial u}{\partial t}$$

and for all $u\in W_2^1(Q_T)$ which are equal to zero for $t=T$,

$$\frac{\partial}{\partial t}J_\rho^+ u=J_\rho^+\frac{\partial u}{\partial t}.$$

LEMMA 5. *Let $u,v\in L_2(Q_T)$, and let $a(x,t)$ have a bounded derivative of first order with respect to t. Then*

$$\left(\frac{\partial}{\partial t}(aJ_\rho u),\, v\right)_{L_2(Q_T)}=-\left(u,\, a\frac{\partial}{\partial t}J_\rho^* v+\eta_\rho\right)_{L_2(Q_T)}, \tag{20}$$
$$\|\eta_\rho\|_{L_2(Q_T)}\to 0 \quad as \quad \rho\to 0,$$

where J_ρ is any of the special averaging operators cited above, and J_ρ^* is the operator conjugate to it.[2)]

[2)] In particular, if $a(x,t)\equiv 1$, then

$$\left(\frac{\partial}{\partial t}J_\rho u,\, v\right)_{L_2(Q_T)}=-\left(u,\,\frac{\partial}{\partial t}J_\rho^* v\right)_{L_2(Q_T)}.$$

In this section we shall assume that the functions $a_{ij}^{(k)}$ and $a_i^{(2)}$ have bounded derivatives with respect to x_i, $i=1,\cdots,n$, and that $a_i^{(2)}|_{\Gamma_T}=0$.

Let $D(A)$ be the set of all elements $u(x,t)$ from $L_2(Q_T)$ which respectively belong to the spaces $W_2^2(Q_T^{(1)})$ and $W_2^2(Q_T^{(2)})$ in the regions $Q_T^{(1)}$ and $Q_T^{(2)}$ and which satisfy the conditions (17) and (18). On this set let us define A in the following way:

$$Au=(L_1^{(1)}u,(x,t)\in Q_T^{(1)};\ L_3^{(2)}u,(x,t)\in Q_T^{(2)}). \tag{21}$$

Let us construct an extension of A defined on $D(A)$. For this let us introduce in $\widetilde{W}_0(Q_T)$ the operator

$$\bar{k}^*u=-\int_0^t u(x,\tau)\,d\tau,^{3)}$$

which is an extension of the operator k^* which is initially defined on the set M, and let us distinguish the class of functions $u(x,t)$ such that $\bar{k}^*u=w\in B_0(Q_T)$.

On such $u(x,t)$ the operator $\bar{A}$ is defined by the identity

$$\begin{aligned}(\bar{A}u,\ \Phi)_0=\int_{Q_T}\Big[&-\sigma\frac{\partial w}{\partial t}\frac{\partial^2\Phi}{\partial t^2}+\frac{\partial w}{\partial x_j}\frac{\partial}{\partial t}\Big(a_{ij}\frac{\partial\Phi}{\partial x_i}\Big)\\ &-\sigma\frac{\partial w}{\partial x_i}\frac{\partial}{\partial t}(a_i\Phi)-w\frac{\partial a\Phi}{\partial t}\Big]dQ_T,\quad \Phi\in\hat{W}_T^{1,t}(Q_T).\end{aligned}\tag{22}$$

Since $\hat{W}_T^{1,t}(Q_T)$ is dense in $\hat{W}_T^1(Q_T)$, it follows that by defining the norm of $\bar{A}u$ as the norm of the functional $(\bar{A}u,\Phi)_0$ on $\hat{W}_T^1(Q_T)$ one can write

$$\|\bar{A}u\|_{\hat{W}_T^{-1}(Q_T)}=\sup_{\Phi\in\hat{W}_T^{1,t}(Q_T)}\frac{|(\bar{A}u,\ \Phi)_0|}{\|\Phi\|_{\hat{W}^1(Q_T)}}.$$

If the functional $(\bar{A}u,\Phi)_0$ is bounded with respect to the norm of $\hat{W}_T^{-1}(Q_T)$, we shall consider it to be extended by continuity to all of $\hat{W}_T^1(Q_T)$. It is not difficult to see that $\bar{A}$ is an extension of A.

Together with A we shall consider the formally adjoint operator A^+ which is defined by the rule

$$A^+\Phi=(L_1^{(1)}\Phi,\ (x,\ t)\in Q_T^{(1)};\ \ (L_3^{(2)})^*\Phi,\ \ (x,\ t)\in Q_T^{(2)}),$$

where $(L_3^{(2)})^*\Phi=\partial^2\Phi/\partial t^2-(\partial/\partial x_i)(a_{ij}^{(2)}\,\partial\Phi/\partial x_j)+(\partial/\partial x_i)(a_i^{(2)}\Phi)-a^{(2)}\Phi$ on the set $D(A^+)$ consisting of all functions $\Phi(x,t)$ from $L_2(Q_T)$ which in the regions $Q_T^{(1)}$ and $Q_T^{(2)}$ respectively belong to the spaces $W_2^2(Q_T^{(1)})$ and $W_2^2(Q_T^{(2)})$ and which satisfy conditions (18) – (19).

Let us denote by $\widetilde{W}_T(Q_T)$ the Hilbert space with scalar product

$$[u,\ v]_T=\int_{Q_T}\Big(\sigma\frac{\partial\bar{k}u}{\partial t}\frac{\partial\bar{k}v}{\partial t}+\frac{\partial\bar{k}u}{\partial x_i}\frac{\partial\bar{k}v}{\partial x_i}+\bar{k}u\bar{k}v\Big)dxdt,$$

where k is the operator defined in the Hilbert space $\widetilde{W}_T(Q_T)$ by the rule $ku=\int_T^t u(x,\tau)d\tau$ and with norm $|\cdot|_T$.

The extension of A^+ defined on $D(A^+)$ is constructed in an analogous

[3)] Here the integral must be understood as the integral of a generalized function $u(x,t)$.

manner to the extension of A. Namely, one distinguishes a class of functions Φ from $\widetilde{W}_T(Q_T)$ such that $\bar{k}\Phi = \psi \in B_T(Q_T)$. The operator A^* is defined on them by means of the identity

$$(23)\quad (u, A^*\Phi)_0 = \int_{Q_T} \Big[\sigma \frac{\partial \psi}{\partial t}\frac{\partial^2 u}{\partial t^2} - \frac{\partial \psi}{\partial x_j}\frac{\partial}{\partial t}\Big(a_{ij}\frac{\partial u}{\partial x_i}\Big) - \sigma a_i \frac{\partial \psi}{\partial t}\frac{\partial u}{\partial x_i} + \psi \frac{\partial}{\partial t}(au)\Big] dQ_T,\ u \in \hat{W}_0^{1,\,t}(Q_T).$$

If the functional $(u, A^*\Phi)_0$ is bounded with respect to the norm of $\hat{W}_0^{-1}(Q_T)$, we shall consider it to be extended by continuity to all of $\hat{W}_0^1(Q_T)$. It is obvious that A^* is an extension of A^+.

The following is valid.

Lemma 6. *If the conditions* a) – d) *are satisfied, then for any function* $u(x,t)$ *for which* $\bar{k}^*u \in B_0(Q_T)$ (*for any* $\Phi(x,t)$ *for which* $\bar{k}\Phi \in B_T(Q_T)$) *the inequalities*

$$(24)\qquad \|\bar{k}^*u\|_{B(Q_T)} \leqslant C_4 \|\bar{A}u\|_{\hat{W}_T^{-1}(Q_T)},$$

$$(25)\qquad \|\bar{k}\Phi\|_{B(Q_T)} \leqslant C_5 \|A^*\Phi\|_{\hat{W}_0^{-1}(Q_T)}$$

are valid.

Proof. Let us first prove (24). Let us remark that if $u \in \hat{W}_0^1(Q_T)$, then $\bar{k}^*u = w \in B_0(Q_T)$. Indeed, for this it suffices to consider a sequence of twice continuously differentiable functions u_n which satisfy (17) – (18) and which converge to u in $\hat{W}_0^1(Q_T)$. Then $\bar{k}^*u$ will be the limit of $\bar{k}^*u_n$ in the metric of $B(Q_T)$. In this case the functional $(\bar{A}u, \Phi)_0$ is given by writing (22) in the form

$$(26)\qquad (\bar{A}u,\ \Phi)_0 = \int_{Q_T} \Big(-\sigma \frac{\partial u}{\partial t}\frac{\partial \Phi}{\partial t} + a_{ij}\frac{\partial u}{\partial x_i}\frac{\partial \Phi}{\partial x_j} - \sigma a_i \frac{\partial u}{\partial x_i}\Phi - au\Phi\Big) dQ_T,$$

which is bounded since

$$(27)\qquad |(\bar{A}u,\ \Phi)_0| \leqslant C_6 \|u\|_{\hat{W}^1(Q_T)} \|\Phi\|_{\hat{W}^1(Q_T)},$$

where C_6 only depends on the M from condition c), and consequently it admits an extension by continuity to all of $\hat{W}_T^1(Q_T)$.

First let $\bar{k}^*u = w \in B_0^t(Q_T)$. Then the functional $(\bar{A}u, \Phi)_0$ is defined on all of $\hat{W}_T^1(Q_T)$, and for it the representation

$$(28)\qquad (\bar{A}u,\ \Phi)_0 = \int_{Q_T} \Big(\sigma \frac{\partial^2 w}{\partial t^2}\frac{\partial \Phi}{\partial t} - a_{ij}\frac{\partial^2 w}{\partial t \partial x_i}\frac{\partial \Phi}{\partial x_j} + \sigma a_i \frac{\partial^2 w}{\partial t \partial x_i}\Phi + a\frac{\partial w}{\partial t}\Phi\Big) dQ_T$$

is valid.

In the role of the function $\Phi(x,t)$ in (28) let us choose

$$(29)\qquad \Phi(x,\ t) = \begin{cases} -\int_{t_0}^{t} u(x,\ \tau)\, d\tau = w(x,\ t) - w(x,\ t_0), & 0 \leqslant t \leqslant t_0, \\ 0, & t_0 \leqslant t \leqslant T. \end{cases}$$

It is admissible since $\Phi \in \hat{W}_T^1(Q_T)$. Substituting it into (28) and integrating by parts, we transform it into the following form:

$$(\bar{A}u,\ \Phi)_0=\frac{1}{2}\int\limits_{\Omega_2}\left(\frac{\partial w}{\partial t}\right)^2 dx\,|_{t=t_0}+\frac{1}{2}\int\limits_{\Omega}\left(a_{ij}\frac{\partial w}{\partial x_i}\frac{\partial w}{\partial x_j}-aw^2\right)dx\,|_{t=t_0}$$

$$+\int\limits_0^t\int\limits_{\Omega}\left[\frac{1}{2}\frac{\partial a_{ij}}{\partial t}\frac{\partial w}{\partial x_i}\frac{\partial w}{\partial x_j}-\sigma\left(\frac{\partial a_i}{\partial t}\frac{\partial w}{\partial x_i}w+a_i\frac{\partial w}{\partial x_i}\frac{\partial w}{\partial t}\right)-\frac{1}{2}\frac{\partial a}{\partial t}w^2\right]dxdt$$

$$+\int\limits_0^{t_0}\int\limits_{\Omega}\left[-\frac{\partial a_{ij}}{\partial t}\frac{\partial w(x,\ t)}{\partial x_i}\frac{\partial w(x,\ t_0)}{\partial x_j}+\sigma\frac{\partial a_i}{\partial t}\frac{\partial w(x,\ t)}{\partial x_i}w(x,\ t_0)+\frac{\partial a}{\partial t}w(x,t)\,w(x,t_0)\right]dxdt.$$

From this follows

$$\int\limits_{\Omega}\left\{\sigma\left(\frac{\partial w}{\partial t}\right)^2+(\nu-2MTn\varepsilon_1-2T\varepsilon_4)\left(\frac{\partial w}{\partial x_i}\right)^2-\left[a(x,\ t)+2MT(\sigma\sqrt{n}\,\varepsilon_2+\varepsilon_3)\right.\right.$$

$$\left.\left.+\,2T\varepsilon_4\right]w^2\right\}dx\,|_{t=t_0}\leqslant\int\limits_0^{t_0}\int\limits_{\Omega}\left\{\sigma(M\sqrt{n}+2\varepsilon_4)\left(\frac{\partial w}{\partial t}\right)^2\right.\tag{30}$$

$$+\left[M\left(n+2\sqrt{n}+\frac{n}{2\varepsilon_1}+\frac{\sigma\sqrt{n}}{2\varepsilon_2}\right)+2\varepsilon_4\right]\left(\frac{\partial w}{\partial x_i}\right)^2$$

$$\left.+\left[M\left(1+\sqrt{n}+\frac{1}{2\varepsilon_3}\right)+2\varepsilon_4\right]w^2\right\}dxdt+\frac{1}{\varepsilon_4}\|\bar{A}u\|^2_{\hat{W}_T^{-1}(Q_T)}.$$

Here we used the inequality

$$\|\Phi\|^2_{\hat{W}^1(Q_{t_0})}\leqslant 2\|w\|^2_{\hat{W}^1(Q_{t_0})}+2t_0\|w\|^2_{W_2^1(\Omega)}\Big|_{t=t_0},$$

which follows from the representability of Φ by means of w (cf. (29)).

By the conditions of the lemma, $\operatorname{ess\,sup}_{Q_T^{(1)}} a^{(1)}(x,t)=a_0^{(1)}<0$. Therefore, putting $\epsilon_1=\epsilon_2=\epsilon_3=\epsilon_4=\kappa/4T(Mn+1)$, where $\kappa=\min(\nu,\,-a_0^{(1)})$, we obtain that

$$\nu-2MTn\varepsilon_1-2T\varepsilon_4,\quad -[a^{(1)}(x,\ t)+2MT\varepsilon_3+2T\varepsilon_4]\geqslant\frac{\varkappa}{2}.$$

Without loss of generality we shall assume that

$$\operatorname*{ess\,sup}_{Q_T^{(2)}}\left[a^{(2)}(x,\ t)+2MT(\sqrt{n}\,\varepsilon_2+\varepsilon_3)+2T\varepsilon_4\right]=\delta_1<0.$$

From (30) follows the inequality

$$\int\limits_{\Omega}\left[\sigma\left(\frac{\partial w}{\partial t}\right)^2+\left(\frac{\partial w}{\partial x_i}\right)^2+w^2\right]dx\,|_{t=t_0}$$

$$\leqslant C_7\int\limits_0^{t_0}\int\limits_{\Omega}\left[\sigma\left(\frac{\partial w}{\partial t}\right)^2+\left(\frac{\partial w}{\partial x_i}\right)^2+w^2\right]dxdt+C_8\|\bar{A}u\|^2_{\hat{W}_T^{-1}(Q_T)},\tag{31}$$

where the constants C_7 and C_8 only depend on ν, $a_0^{(1)}$, M, n and T. Inequality (24) is established from the last inequality in an analogous manner as in Lemma 2.

Now let $w\in B_0(Q_T)$. We use the averaging

$$J_\rho^- w(x,\ t)=\frac{1}{\rho}\int\limits_0^T\omega\left(\frac{t-t'-2\rho}{\rho}\right)w(x,\ t')\,dt'.$$

Then $J_\rho^- w\in B_0^t(Q_T)$, $\|J_\rho^- w-w\|_{B(Q_T)}\to 0$ as $\rho\to 0$ and

(32)
$$\| J_\rho^- w \|_{B(Q_T)} \leqslant C_9 \sup_{\Phi \in \hat{W}_T^{1,t}(Q_T)} \frac{|(\bar{A}k^{*-1} J_\rho^- w,\ \Phi)_0|}{\|\Phi\|_{\hat{W}^1(Q_T)}}.$$

Rewriting the numerator of the right side of (32) and using Lemmas 4 and 5 and a permutation of the averaging with respect to t with the differentiation with respect to x_i, we obtain

(33)
$$\int_0^T\int_\Omega \left(\sigma \frac{\partial^2 J_\rho^- w}{\partial t^2} \frac{\partial \Phi}{\partial t} - a_{ij} \frac{\partial^2 J_\rho^- w}{\partial t \partial x_i} \frac{\partial \Phi}{\partial x_j} + \sigma a_i \frac{\partial^2 J_\rho^- w}{\partial t \partial x_i} \Phi + a \frac{\partial J_\rho^- w}{\partial t} \Phi\right) dxdt$$
$$= \int_0^T\int_\Omega \Big[-\sigma \frac{\partial w}{\partial t} \frac{\partial^2 J_\rho^+ \Phi}{\partial t^2} + \frac{\partial w}{\partial x_i} \frac{\partial}{\partial t}\left(a_{ij} \frac{\partial J_\rho^+ \Phi}{\partial x_j}\right) - \sigma \frac{\partial w}{\partial x_i} \frac{\partial}{\partial t}(a_i J_\rho^+ \Phi)$$
$$- w \frac{\partial}{\partial t}(a J_\rho^+ \Phi) - \eta_{j\rho} \frac{\partial \Phi}{\partial x_j} - \eta_\rho \Phi\Big] dxdt,$$

where

$$\|\eta_{j\rho}\|_{L_2(Q_T)} \to 0, \quad \|\eta_\rho\|_{L_2(Q_T)} \to 0 \quad \text{as} \quad \rho \to 0.$$

It is known (cf., for example, [6]) that for any $\Phi \in L_2(Q_T)$ the inequality

(34)
$$\| J_\rho^+ \Phi \|_{L_2(Q_T)} \leqslant \| \Phi \|_{L_2(Q_T)}$$

is valid.

If $\Phi \in \hat{W}_T^1(Q_T)$, then

(35)
$$\| J_\rho^+ \Phi \|_{\hat{W}^1(Q_T)} \leqslant \| \Phi \|_{\hat{W}^1(Q_T)}.$$

The latter inequality follows from Lemma 4 and a permutation of the averaging with respect to t with the differentiation with respect to x_i.

It follows from (33) that

(36)
$$|(\bar{A}k^{*-1} J_\rho^- w,\ \Phi)_0| \leqslant \|\bar{A}u\|_{\hat{W}_T^{-1}(Q_T)} \| J_\rho^+ \Phi\|_{\hat{W}^1(Q_T)}$$
$$+ \left(\|\eta_{j\rho}\|^2_{L_2(Q_T)} + \|\eta_\rho\|^2_{L_2(Q_T)}\right)^{\frac{1}{2}} \|\Phi\|_{\hat{W}^1(Q_T)}.$$

Here we used the fact that $J_\rho^+ \Phi \in \hat{W}_T^{1,t}(Q_T)$; passing from $J_\rho^+\Phi$ to Φ, we extend the class of admissible functions.

From (32), by virtue of (35) and (36), it follows that

$$\| J_\rho^- w \|_{B(Q_T)} \leqslant C_9 \left[\|\bar{A}u\|_{\hat{W}_T^{-1}(Q_T)} + \left(\|\eta_{j\rho}\|^2_{L_2(Q_T)} + \|\eta_\rho\|^2_{L_2(Q_T)}\right)^{\frac{1}{2}}\right].$$

Since the limit of the left side of the last equation exists as $\rho \to 0$ and is equal to $\|w\|_{B(Q_T)}$, we obtain, by passing to the limit, the validity of (24) for any u for which $\bar{k}^* u \in B_0(Q_T)$.

Inequality (25) is proved in an analogous manner. Namely, as in the proof of (24), it is first proved for functions $\Phi(x,t)$ such that $\bar{k}\Phi = \psi \in B_T^t(Q_T)$ by writing the functional (23) in the form

(37)
$$(u,\ A^*\Phi)_0 = \int_0^T\int_\Omega \left(-\sigma \frac{\partial^2 \psi}{\partial t^2} \frac{\partial u}{\partial t} + a_{ij} \frac{\partial^2 \psi}{\partial t \partial x_j} \frac{\partial u}{\partial x_i} - \sigma a_i \frac{\partial \psi}{\partial t} \frac{\partial u}{\partial x_i} - a \frac{\partial \psi}{\partial t} u\right) dxdt.$$

Here in the role of $u(x,t)$ one must take

$$u(x,\ t)=\begin{cases} 0, & 0\leqslant t\leqslant t_0,\\ \int\limits_{t_0}^{t}\Phi(x,\ \tau)\,d\tau=\psi(x,\ t)-\psi(x,\ t_0), & t_0\leqslant t\leqslant T.\end{cases}$$

By means of averaging operators the validity of this inequality is further proved for any function $\psi\in B_T(Q_T)$. The lemma is completely proved.

For any $f(x,t)\in\hat{W}_T^{-1}(Q_T)$ we shall call *a generalized solution of the problem* (1) – (5) a function $u(x,t)$ for which $\bar{k}^*u\in B_0(Q_T)$ and $(\bar{A}u,\Phi)_0=(f,\Phi)_0$ for any $\Phi\in\hat{W}_T^1(Q_T)$.

Theorem 2. *If the conditions of Lemma* 6 *are satisfied, then for any* $f\in\hat{W}_T^{-1}(Q_T)$ *a generalized solution of the problem* (1) – (5) *exists and is unique.*

Proof. The uniqueness of a generalized solution immediately follows from (24). By the same inequality one can assert that the range of $\bar{A}$ is closed in $\hat{W}_T^{-1}(Q_T)$. Therefore for the proof of the theorem one must only show that the range of $\bar{A}$ is dense in the entire space $\hat{W}_T^{-1}(Q_T)$. It was shown above that if $u\in\hat{W}_0^1(Q_T)$, then $\bar{k}^*u=w\in B_0(Q_T)$. Therefore it suffices to show that the collection of elements $\bar{A}u$, as u runs through the space $\hat{W}_0^1(Q_T)$, is dense in $\hat{W}_T^{-1}(Q_T)$. Otherwise there would exist an element $\Phi\in\hat{W}_T^1(Q_T)$ such that for any function $u(x,t)$ from $\hat{W}_0^1(Q_T)$

$$(\bar{A}u,\Phi)_0=0. \tag{38}$$

We shall show that for $\Phi\in\hat{W}_T^1(Q_T)$, $u\in\hat{W}_0^1(Q_T)$

$$(\bar{A}u,\Phi)_0=(u,A^*\Phi)_0.$$

Indeed, if $\bar{k}\Phi=\psi$, then by virtue of (26)

$$(\bar{A}u,\ \Phi)_0=\int\limits_0^T\int\limits_\Omega\left(-\sigma\frac{\partial u}{\partial t}\frac{\partial\Phi}{\partial t}+a_{ij}\frac{\partial u}{\partial x_i}\frac{\partial\Phi}{\partial x_j}-\sigma a_i\frac{\partial u}{\partial x_i}\Phi-au\Phi\right)dxdt$$

$$=\int\limits_0^T\int\limits_\Omega\left(-\sigma\frac{\partial^2\psi}{\partial t^2}\frac{\partial u}{\partial t}+a_{ij}\frac{\partial^2\psi}{\partial t\partial x_j}\frac{\partial u}{\partial x_i}-\sigma a_i\frac{\partial\psi}{\partial t}\frac{\partial u}{\partial x_i}-a\frac{\partial\psi}{\partial t}u\right)dxdt=(u,\ A^*\Phi)_0.$$

Now, taking (38) and (25) into account, we obtain that $\|\bar{k}\Phi\|_{B(Q_T)}=0$. But then, since

$$\bar{k}\Phi=\int\limits_T^t\Phi(x,\ \tau)\,d\tau,$$

we also have $\Phi(x,t)\equiv 0$ in Q_T. Hence we conclude that the set of elements $\bar{A}u$, as u runs through $\hat{W}_0^1(Q_T)$, is dense in $\hat{W}_T^{-1}(Q_T)$. The theorem is proved.

§3. Concluding remarks

It was proved in the preceding sections that the boundary value problem (1) – (5) has a unique solution for any $f_1^{(1)}(x,t)\in L_2(Q_T^{(1)})$ and $f_3^{(2)}(x,t)\in L_2(Q_T^{(2)})$, if one looks for it in the class of functions $u(x,t)$ for which $\bar{k}^*u\in B_0(Q_T)$. The question naturally arises: do all generalized solutions $u(x,t)$ of (1) – (5) for which $\bar{k}^*u\in B_0(Q_T)$ belong to the Hilbert space $L_2(Q_T)$ if the functions $f_1^{(1)}(x,t)$ and $f_3^{(2)}(x,t)$ are only square integrable in the regions

$Q_T^{(1)}$ and $Q_T^{(2)}$ respectively? The answer to this question has not been obtained except in the case when (1) – (5) "decomposes" into separate boundary value problems. Here we shall consider this case.

Below we shall prove two lemmas which are refinements of known facts.

In the space $W_2^1(\Omega_1)$ let us introduce a new scalar product

$$[u,\ v]=\int_{\Omega_1}\left[a_{ij}^{(1)}(x,\ t)\frac{\partial u}{\partial x_i}\frac{\partial v}{\partial x_j}-a^{(1)}(x,\ t)\,uv\right]dx.$$

It is known (cf., for example, [7]) that the norm $|\cdot|$ corresponding to it, by our assumptions on $a_{ij}^{(1)}$ and $a^{(1)}$, is equivalent to the norm of the space $W_2^1(\Omega_1)$.

Let us denote by $H_{L_1^{(1)}}(\Omega_1,t)$ the Hilbert space consisting of all functions of $W_2^1(\Omega_1)$ which satisfy the integral identity

$$\int_{\Omega_1}\left[a_{ij}^{(1)}(x,\ t)\frac{\partial v}{\partial x_i}\frac{\partial \eta}{\partial x_j}-a^{(1)}(x,\ t)\,v\eta\right]dx=0 \tag{39}$$

for any $\eta\in \overset{\circ}{W}{}_2^1(\Omega_1)$.

Lemma 7. *The Hilbert space $W_2^1(\Omega_1)$, with the new scalar product, is an orthogonal sum of the subspaces $\overset{\circ}{W}{}_2^1(\Omega_1)$ and $H_{L_1^{(1)}}(\Omega_1,t)$, i.e.*

$$W_2^1(\Omega_1)=\overset{\circ}{W}{}_2^1(\Omega_1)\oplus H_{L_1^{(1)}}(\Omega_1,\ t).$$

The integral identity (39) shows that the spaces $\overset{\circ}{W}{}_2^1(\Omega_1)$ and $H_{L_1^{(1)}}(\Omega_1,t)$ are orthogonal in the sense of the new scalar product. The representation of an arbitrary element $u\in W_2^1(\Omega_1)$ in the form

$$u=w+v,\quad w\in\overset{\circ}{W}{}_2^1(\Omega_1),\quad v\in H_{L_1^{(1)}}(\Omega_1,\ t) \tag{40}$$

follows from the fact that the problem $L_1^{(1)}v=0$, $v|_\Gamma=u|_\Gamma$ has a unique generalized solution $v(x,t)$ from $W_2^1(\Omega_1)$ (cf., for example, [7]) which satisfies the integral identity (39), i.e. the function v is an element of $H_{L_1^{(1)}}(\Omega_1,t)$ and $w=u-v\in\overset{\circ}{W}{}_2^1(\Omega_1)$.

In $L_2(\Omega_1)$ let us consider the subspace $U(\Omega_1,t)$ which is the closure in $L_2(\Omega_1)$ of all functions $v(x,t)\in H_{L_1^{(1)}}(\Omega_1,t)$; $G(\Omega_1,t)$ is the orthogonal complement of $U(\Omega_1,t)$.

Lemma 8. *If $f_1^{(1)}(x,t)\in G(\Omega_1,t)$, then a generalized solution from $W_2^1(\Omega_1)$ of the problem*

$$L_1^{(1)}u=f_1^{(1)},\quad u|_\Gamma=0 \tag{41}$$

satisfies the integral identity

$$\int_{\Omega_1}\left(a_{ij}^{(1)}\frac{\partial u}{\partial x_j}\frac{\partial \eta}{\partial x_i}-a^{(1)}u\eta\right)dx=\int_{\Omega_1}f_1^{(1)}\eta\,dx \tag{42}$$

for any function $\eta\in W_2^1(\Omega_1)$.

Proof. By Lemma 7, any function $\eta\in W_2^1(\Omega_1)$ is uniquely representable in the form

$$\eta=\eta_1+\eta_2,\quad \eta_1\in H_{L_1^{(1)}}(\Omega_1,\ t),\quad \eta_2\in\overset{\circ}{W}{}_2^1(\Omega_1).$$

Further, it is known (cf., for example, [7]) that for any $f_1^{(1)}(x,t)\in L_2(\Omega_1)$

the problem (41) has a unique generalized solution from $\overset{\circ}{W}{}_2^1(\Omega_1)$ which satisfies the integral identity

$$\int_{\Omega_1}\left(a_{ij}^{(1)}\frac{\partial u}{\partial x_j}\frac{\partial \eta_2}{\partial x_i}-a^{(1)}u\eta_2\right)dx=\int_{\Omega_1}f\eta_2dx \tag{43}$$

for any $\eta_2\in \overset{\circ}{W}{}_2^1(\Omega_1)$.

Since $\eta_1\in H_{L_1^{(1)}}(\Omega_1,t)$ and $u\in \overset{\circ}{W}{}_2^1(\Omega_1)$, we have

$$\int_{\Omega_1}\left(a_{ij}^{(1)}\frac{\partial u}{\partial x_j}\frac{\partial \eta_1}{\partial x_i}-a^{(1)}u\eta_1\right)dx=0. \tag{44}$$

Combining (43) and (44) and taking into account the fact that

$$\int_{\Omega_1}f\eta_1dx=0,$$

we obtain (42). The lemma is proved.

Let us denote by W_1 (W_2) the Hilbert space consisting of all elements of W which have the form $(f^{(1)};0)$ $((0;f^{(2)}))$. It is not difficult to see that W is the orthogonal sum of the subspaces W_1 and W_2:

$$W=W_1\oplus W_2.$$

It is shown in [4] that for any element $(0;f_3^{(2)})\in W_2$ a generalized solution of (1)–(5) belongs to $\hat{W}{}_0^1(Q_T)$ (even to $B_0(Q_T)$). Therefore for elucidation of the question formulated above it suffices to consider the problem (1)–(5) for all elements of W_1.

We have

Theorem 3. *Let* $(f_1^{(1)};0)\in W_1$, *and for almost all* $t\in[0,T]$ *let* $f_1^{(1)}(x,t)\in G(\Omega_1,t)$. *Then a generalized solution of the problem* (1)–(5) *belongs to* $\hat{W}_0(Q_T)$. *For this solution the inequality*

$$\|u\|_{\hat{W}^1(Q_T)}\leq C_{10}\|f_1^{(1)}\|_{L_2(Q_T^{(1)})} \tag{45}$$

is valid.

Proof. A generalized solution of (1)–(5) for any element $(f_1^{(1)};0)\in W_1$ which satisfies the condition that $f_1^{(1)}(x,t)\in G(\Omega_1,t)$, for almost all $t\in[0,T]$ will be constructed in the following way.

Let us denote by $W_2^{1,0}(Q_T^{(1)})$ the Hilbert space with scalar product

$$(u,v)_{W_2^{1,0}(Q_T^{(1)})}=\int_{Q_T^{(1)}}\left(uv+\frac{\partial u}{\partial x_i}\frac{\partial v}{\partial x_i}\right)dxdt.$$

Let $\psi_k(x)$ $(k=1,2,\cdots)$ be a complete orthonormal system in $W_2^1(\Omega_1)$, and let $\Phi(x,t)$ be an arbitrary element of $W_2^{1,0}(Q_T^{(1)})$.

We expand $\Phi(x,t)$ by its Fourier series with respect to the functions $\psi_k(x)$:

$$\Phi(x,t)=\sum_{k=1}^{\infty}\varphi_k(t)\psi_k(x).$$

By the conditions of the theorem and Lemma 8, for almost all $t\in[0,T]$ a generalized solution from $W_2^1(\Omega_1)$ of the problem (41) satisfies the integral identity

(46)
$$\int_{\Omega_1}\left(a_{ij}^{(1)}\frac{\partial u}{\partial x_j}\frac{\partial \psi_k}{\partial x_i}-a^{(1)}u\psi_k\right)dx=\int_{\Omega_1}f_1^{(1)}\psi_k dx$$

for any $\psi_k(x)$ $(k=1,2,\cdots)$. Multiply (46) by $\varphi_k(t)$ and sum over k from 1 to N. With the aid of an integration with respect to t from 0 to T we reduce them to the form

$$\int_{Q_T^{(1)}}\left(a_{ij}^{(1)}\frac{\partial u}{\partial x_j}\frac{\partial \Phi_N}{\partial x_i}-a^{(1)}u\Phi_N\right)dxdt=\int_{Q_T^{(1)}}f_1^{(1)}\Phi_N dxdt,$$

where $\Phi_N(x,t)=\sum_1^N\varphi_k(t)\psi_k(x)$. Passing to the limit as $N\to\infty$ in the latter identity, we establish the validity of the integral identity

$$\int_{Q_T^{(1)}}\left(a_{ij}^{(1)}\frac{\partial u}{\partial x_j}\frac{\partial \Phi}{\partial x_i}-a^{(1)}u\Phi\right)dxdt=\int_{Q_T^{(1)}}f_1^{(1)}\Phi dxdt$$

for any function $\Phi\in W_2^{1,0}(Q_T)$. Now let us extend the function $u(x,t)$ by zero to all of Q_T. The resulting function satisfies the integral identity

$$\int_{Q_T}\left(-\sigma\frac{\partial u}{\partial t}\frac{\partial \Phi}{\partial t}+a_{ij}\frac{\partial u}{\partial x_i}\frac{\partial \Phi}{\partial x_j}-\sigma a_i\frac{\partial u}{\partial x_i}\Phi-au\Phi\right)dxdt=\int_{Q_T^{(1)}}f_1^{(1)}\Phi dxdt$$

for any $\Phi\in \hat{W}_T^1(Q_T)$ and the conditions $u|_{t=0}=0$, $x\in\Omega_2$ and $u|_{S_T}=0$. Moreover, since for a generalized solution from $W_2^1(\Omega_1)$ of the problem (41) the inequality

$$\|u\|_{W_2^1(\Omega_1)}\leqslant C_{11}\|f_1^{(1)}\|_{L_2(\Omega_1)}$$

is valid for almost all $t\in[0,T]$ (cf., for example, [7]), we obtain (45) by integrating the last inequality with respect to t from 0 to T and taking into account the fact that $u(x,t)$ has been extended by zero to all of Q_T.

It was proved in the preceding section that a generalized solution of (1) – (5) is unique. Consequently it coincides with the solution $u(x,t)$ constructed here. The theorem is proved.

As the example cited below shows, one can expect that for any $(f_1^{(1)};0)\in W_1$ (i.e. without the additional requirement that $f_1^{(1)}(x,t)\in G(\Omega_1,t)$ for almost all $t\in[0,T]$) a generalized solution of (1) – (5) belongs to $\hat{W}_0^1(Q_T)$.

EXAMPLE. Let

(47)
$$-\frac{\partial^2 u}{\partial x^2}=f_1^{(1)}(x,\ t),\quad (x,\ t)\in Q_1^{(1)}=[-1,\ 0]\times[0,\ 1],$$

(48)
$$\frac{\partial^2 u}{\partial t^2}-\frac{\partial^2 u}{\partial x^2}=0,\quad (x,\ t)\in Q_1^{(2)}=[0,\ 1]\times[0,\ 1],$$

(49)
$$u|_{x=-1}=u|_{x=1}=0,\quad [u]|_{x=0}=0,\quad \left[\frac{\partial u}{\partial x}\right]\Big|_{x=0}=0,\quad t\in[0,\ 1],$$

(50)
$$u|_{t=0}=0,\quad u_t|_{t=0}=0,\quad x\in[0,\ 1]$$

be a boundary value problem of the type considered above. Having applied the Laplace transform with respect to t to the problem (47) – (50), we reduce it to the determination of a function $v(x,\lambda)$ satisfying the equations

(51)
$$-\frac{\partial^2 v}{\partial x^2}=\varphi^{(1)}(x,\ \lambda),\quad -1\leqslant x\leqslant 0,$$

(52) $$-\frac{\partial^2 v}{\partial x^2}+\lambda^2 v=0,\quad 0\leqslant x\leqslant 1,$$

and the conditions

(53) $$v|_{x=-1}=v|_{x=1}=0,\quad [v]|_{x=0}=0,\quad \left[\frac{\partial v}{\partial x}\right]\Big|_{x=0}=0,$$

where

$$\varphi^{(1)}(x,\lambda)=\frac{1}{\sqrt{2\pi}}\int_0^\infty F^{(1)}(x,t)\,e^{-\lambda t}dt,$$

and

$$F^{(1)}(x,t)=\begin{cases} f_1^{(1)}(x,t), & 0\leqslant t\leqslant 1,\\ 0 & t>1.\end{cases}$$

A solution of (51) – (53) has the form

$$v(x,\lambda)=\begin{cases} \displaystyle\int_{-1}^{0} G(x,\xi)\,\varphi^{(1)}(\xi,\lambda)\,d\xi-(1+x)\frac{e^{-\lambda}-e^{\lambda}}{\lambda(e^{\lambda}+e^{-\lambda})-(e^{-\lambda}-e^{\lambda})}\\ \displaystyle\qquad\times\int_{-1}^{0}(1+\xi)\,\varphi^{(1)}(\xi,\lambda)\,d\xi, \quad -1\leqslant x\leqslant 0,\\ \displaystyle -\frac{e^{-\lambda(1-x)}-e^{\lambda(1-x)}}{\lambda(e^{\lambda}+e^{-\lambda})-(e^{-\lambda}-e^{\lambda})}\int_{-1}^{0}(1+\xi)\,\varphi^{(1)}(\xi,\lambda)\,d\xi,\quad 0\leqslant x\leqslant 1,\end{cases}$$

where

$$G(x,\xi)=\begin{cases}-(1+x)\,\xi, & x\leqslant\xi,\\ -(1+\xi)\,x, & x\geqslant\xi.\end{cases}$$

If $f_1^{(1)}(x,t)$, together with its derivatives up to the third order in t, is continuous in $\overline{Q}_1^{(1)}$, and if it satisfies the conditions $\partial^k f_1^{(1)}/\partial t^k|_{t=0}=0$, $\partial^k f_1^{(1)}/\partial t^k|_{t=1}=0$, $k=0,1,2$, then $v(x,\lambda)$ depends analytically on λ, is twice continuously differentiable with respect to x, and is subject to the inequalities

$$\left|\frac{\partial^l v}{\partial x^l}\right|\leqslant\frac{C_{12}}{|\lambda|^2}\left\{\sum_{k=1}^{3}\left\|\frac{\partial^k f_1^{(1)}}{\partial t^k}\right\|_{L_2(Q_1^{(1)})}+\left(\int_0^1\left(\frac{\partial^2 f_1^{(1)}}{\partial t^2}\right)^2 dt\right)^{1/2}\right\},$$

$l=0,1,2$, for sufficiently large $\lambda_1=\operatorname{Re}\lambda\geq\lambda_1^0>0$. Therefore the function

$$u(x,t)=\frac{1}{i\sqrt{2\pi}}\int_{\lambda_1-i\infty}^{\lambda_1+i\infty} v(x,\lambda)\,e^{\lambda t}d\lambda$$

has continuous derivatives in $Q_1^{(k)}$ which appear in (47) and (48) for $k=1,2$, and it satisfies equations (47) and (48) and conditions (49) and (50), i.e. it is a solution of (47) – (50).

Let us now remark that for the function $v(x,\lambda)$ one has the inequality

$$\int_{-1}^{1}\left(|v|^2+\left|\frac{\partial v}{\partial x}\right|^2\right)dx+|\lambda|^2\int_0^1|v|^2\,dx\leqslant C_{13}\int_{-1}^{0}|\varphi^{(1)}(x,\lambda)|^2\,dx.$$

Using the Plancherel Theorem on transformations of functions in $L_2(-\infty,\infty)$, we obtain

(54)
$$\int_0^\infty\int_{-1}^1\left[u^2+\left(\frac{\partial u}{\partial x}\right)^2+\sigma\left(\frac{\partial u}{\partial t}\right)^2\right]e^{-2\lambda_1 t}dxdt\leqslant C_{14}\int_0^1\int_{-1}^0(f_1^{(1)})^2\,dxdt.$$

Since the set of functions $f_1^{(1)}(x,t)$ which, together with their derivatives up to and including the third order with respect to t, are continuous in $\overline{Q}_1^{(1)}$ and satisfy the conditions $\partial^k f_1^{(1)}/\partial t^k|_{t=0}=0$, $\partial^k f_1^{(1)}/\partial t^k|_{t=1}=0$, $k=0,1,2$, is dense in $L_2(Q_1^{(1)})$, it follows by (54) that for any $f_1^{(1)}(x,t)\in L_2(Q_1^{(1)})$ there exists a generalized solution of (47) – (50) which has a finite norm

$$\int_0^1\int_{-1}^1\left[u^2+\left(\frac{\partial u}{\partial x}\right)^2+\sigma\left(\frac{\partial u}{\partial t}\right)^2\right]dxdt.$$

There also remains the following open question: is Theorem 2 true in the case when the elliptic operator appearing in (1) – (5) is not formally selfadjoint? Note that in this case existence and uniqueness theorems for the problem (1) – (5) have also not been proved by other methods.

The author is grateful to O. A. Ladyženskaja for discussion of the results of this paper and for valuable advice.

Bibliography

[1] O. A. Ladyženskaja and L. Stupjalis, *Equations of mixed type*, Vestnik Leningrad. Univ. **20** (1965), no. 19, 38-46. (Russian) MR **35** #3275.

[2] L. Stupjalis, *A spectral problem for equations of mixed type*, Vestnik Leningrad. Univ. **22** (1967), no. 13, 86-102. (Russian) MR **36** #4175.

[3] ———, *Boundary value problems for equations of mixed type in the case when* $\lambda = 0$ *is a point of the spectrum for the elliptic part of the equations*, Vestnik Leningrad. Univ. **25** (1970), no. 7, 46-73. (Russian) MR **44** #661.

[4] O. A. Ladyženskaja and L. Stupjalis, *Boundary value problems for mixed equations*, Trudy Mat. Inst. Steklov. **116** (1971), 101-136 = Proc. Steklov Inst. Math. **116** (1971), 102-139.

[5] Ju. M. Berezanskiĭ, *Expansion in eigenfunctions of selfadjoint operators*, "Naukova Dumka", Kiev, 1965; English transl., Transl. Math. Monographs, vol. 17, Amer. Math. Soc., Providence, R. I., 1968. MR **36** #5768; #5769.

[6] K. O. Friedrichs, *Symmetric hyperbolic linear differential equations*, Comm. Pure Appl. Math. **7** (1954), 345-392. MR **16**, 44.

[7] O. A. Ladyženskaja and N. N. Ural′ceva, *Linear and quasilinear equations of elliptic type*, "Nauka", Moscow, 1964; English transl., Academic Press, New York, 1968. MR **35** #1955; **39** #5941.

Translated by
J. D. FABREY